Electrical Power Transmission and Distribution

Electrical Power Transmission and Distribution

Editor: Marko Silver

NY RESEARCH PRESS

New York

Published by NY Research Press
118-35 Queens Blvd., Suite 400,
Forest Hills, NY 11375, USA
www.nyresearchpress.com

Electrical Power Transmission and Distribution
Edited by Marko Silver

International Standard Book Number: 978-1-63238-537-6 (Hardback)

Cataloging-in-Publication Data

Electrical power transmission and distribution / edited by Marko Silver.
p. cm.
Includes bibliographical references and index.
ISBN 978-1-63238-537-6
1. Electric power transmission. 2. Electric power distribution. 3. Electric power systems.
4. Electrical engineering. I. Silver, Marko.
TK3001 .E44 2017
621.319--dc23

Printed in the United States of America.

Contents

Preface

Over the recent decade, advancements and applications have progressed exponentially. This has led to the increased interest in this field and projects are being conducted to enhance knowledge. The main objective of this book is to present some of the critical challenges and provide insights into possible solutions. This book will answer the varied questions that arise in the field and also provide an increased scope for furthering studies.

Electrical power transmission and distribution are an important area of electrical engineering. This book on electrical power transmission and distribution takes into account the layout, design and manufacture of components that form an electrical grid. There has been rapid progress in this field and its applications are finding their way across multiple industries. Contents included in this book aim to facilitate a comprehensive knowledge in the fields of electrical engineering and efficient electricity generation and consumption. This book is a vital tool for all researching or studying electricity transmission as it gives incredible insights into emerging trends and concepts. The readers would gain knowledge that would broaden their perspective about this field.

I hope that this book, with its visionary approach, will be a valuable addition and will promote interest among readers. Each of the authors has provided their extraordinary competence in their specific fields by providing different perspectives as they come from diverse nations and regions. I thank them for their contributions.

Editor

1

Decomposition-coordination model and algorithm for parallel calculation of power system state estimation problem

Mashauri Adam Kusekwa

Electrical Engineering department, Dar es Salaam Institute of Technology (DIT), Dar es Salaam-Tanzania

Email address:
kusekwa_adam@yahoo.com, Kusekwa_adam@dit.ac.dit

Abstract: Power system state estimation is the process of computing a reliable estimate of the system state vector composed of bus voltages' magnitudes and angles from telemetered measurements on the system. This estimate of the state vector provides the description of the system necessary for operation, security monitoring and control. Many methods are described in literature for solving the state estimation problem, the most important of which are the classical weighted least squares and the non-quadratic method. However, both showed drawbacks when it comes to application to large-scale power system networks. In this paper, a new method in the name of decomposition-coordination approach using the weighted least squares is introduced in solving the large-scale power system state estimation problem. The estimation criterion is reformulated; voltage measurement, real and reactive power injections, real and reactive power flows, and real and reactive power flows in tie-line models of a decomposed system are developed. Two level structure of solving the estimation problem is introduced. The first level solves the sub-problem using gradient procedure methods while the second level determines the interconnection variables using predictive method. The positive characteristic of the method is that the coordinator has little work of predicting interconnection variables instead of solving the state estimation problem. The method can be used to solve a multi-area state estimation using parallel or distributed processing architectures.

Keywords: Power Systems, Modelling of Measurement Data, State Estimation, Decomposition-Coordination Method, Algorithm

1. Introduction

The heart of the data processing activities at electrical utility central dispatch centre (CDC) is the power system state estimator using both real-time measurements and historical database. The state estimator detects and identifies errors in the measurements and computes an optimal estimate of the system state vector of bus voltages' magnitudes and angles. This optimal estimate is then used by the security monitoring, operation and control functions [1-5]. The state estimation process is based on a statistical criterion that estimates the true value of the state vector of the system to minimize the selected criterion [6-8].

The most common and familiar criterion used in power industry is the weighted least square method where the objective function is to minimize the sum of the squares of the difference between each measured value and the true estimated value with each squared difference divided or weighted by the variance of meter error[9-10].

State estimation can eliminate the effect of bad data [12] and allow the temporary loss of measurements without significant affecting the quality of the estimated values. It is used to filter redundant data, eliminate incorrect measurements, it allows determination of the power flows in part of the network that are not directly metered and it can produce reliable state estimate

Nowadays, system networks are becoming more and more complicated. In this aspect monitoring and control of these networks feel the necessity of having robust and scalable methods for state estimation that maintain performance of large-scale systems. Recently, there has been increasing interest in improving various types of state estimation algorithms used in the industry [11] to make them applicable in the ever expanding systems. These improved state estimation algorithms have been implemented in various power system central dispatch centres using a centralized estimation algorithm. In this set ups measurements from all sensors are sent to a central estimation unit where the state of the whole system is estimated. Centralized processing has posed challenges such as problems relating to communication between the sensors and the central estimation unit and processing facility (computer) memory limits. These challenge motivated researchers to move toward finding effective, robust and reliable techniques of processing state estimation of large-scale power system networks.

There are two approaches to carry out large-scale power system state estimation. First is to decompose the system into sub-systems and model the neighboring utilities in detail and accurately in one's own state estimator. Second, is to obtain the state estimation output from each sub-system and convert them into global estimation; this set up is known as hierarchical state estimation.

Hierarchical method has been investigated in the past with the aim of reducing computation time, memory requirements and amount of data exchange between sub-systems. Van-Cutsem et al. [12-14] proposed a two-level state estimation algorithm by dividing the system into known overlapping sub-systems which are connected by tie-lines. In the first level each sub-system performs state estimation independently with respect to data and information available in the sub-system. In the second level, the boundary bus states are re-estimated and all voltage angles are coordinated to a global reference.

In [15] a two-level state estimator for multi-area interconnected system is proposed. In the first level of the algorithm, each area runs their own state estimator using measurements from its own area. In the second level of the algorithm, the central coordinator collects the state estimation from each area and coordinates them to get the multi-area state estimation with respect to global reference. In this way the coordinator can use the measurements available from the boundary network such as tie-line power flows, boundary bus injections, boundary bus voltage etc. The coordinator can also use the boundary bus state available from each area state estimators as pseudo measurements to increase the redundancy.

Aguado et al. [16] addressed power system state estimation problem using decomposition-coordination techniques. The whole system network is divided into geographical areas. Then, for each area, an area power system state estimation problem is formulated. The global optimum of the overall system is obtained by iteratively coordinating the solution of area state estimation sub-problems. By using decomposition-coordination techniques, the integrated optimum solution can be achieved by only sharing a reduced amount of information of tie-lines. The techniques can be applied within a utility with a transmission network spanning over different regions, in such a set up every region dispatch centre perform a state estimation algorithm in coordination with neighboring dispatch centre. In case of a large-scale power system networks where computation is a concern, a distributed implementation can be an alternative to save computation time by simultaneously running many power system state estimation algorithm.

Interesting is the work by Aguado et al. [16]. However, the work presented in this paper is different in implementation of decomposition-coordination method. In this paper, decomposition process is implemented using bus admittance matrix of a power system network instead of geographical areas. In this way the sub-matrices obtained after decomposition represent sub-systems; in case of power system state estimation they represent sub-problems. Then measurement model for voltage magnitude, real and reactive power injections, real and reactive power flows and real and reactive power flows in tie-lines are developed. The state estimation problem is solved using a two-level structure under decomposition-coordination principle proposed in [17]. Advantages of the method is that there is no need of re-estimating the boundary bus states; these are included in the real and reactive power injection model, also the method reduces the coordinators work to just calculating of sub-system interconnection variables by using prediction method and sent these variables to the first level to be used as measurements in computing sub-system state estimation.

The paper is organized as follows. Section 2 describes problem formulation of the state estimation problem. Section 3 presents state estimation solution problem formulation and solution under two-level structures. Section 4 presents first and second level algorithms. Section 5 Discusses positive characteristic of the proposed method and its advantages in solving power system state estimation problem. Section 6 concludes the paper.

2. Problem Formulation

The power system state estimator processes real-time redundant telemetered from substations and pseudo measurements to provide a complete, coherent and reliable system database, which can describe the current electrical state of the system network [18-19]. The measurements, which include voltage magnitudes, real and reactive power injections and real and reactive line power flows are measured from the network at a certain moment, thus getting an estimate for respective state vector i.e. vector of voltages' magnitude and angles on different buses [20].

Consider an interconnected system decomposed into N_S sub-system shown in Figure 1. Individual sub-system is connected to each sub-system through the tie-line network (Figure 2). The buses in each sub-system can be categorized as internal buses, internal boundary buses and external boundary buses.

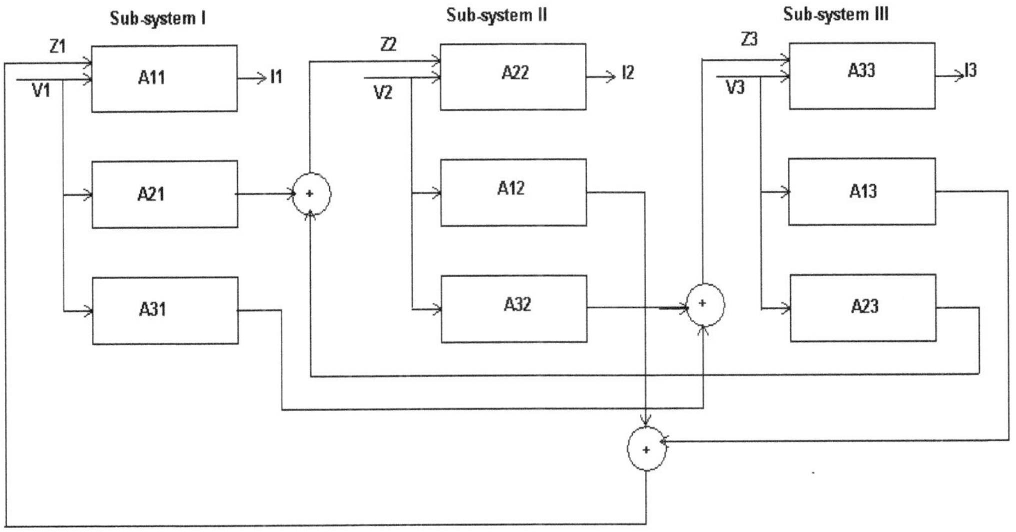

Figure 1. System Decomposed model structure

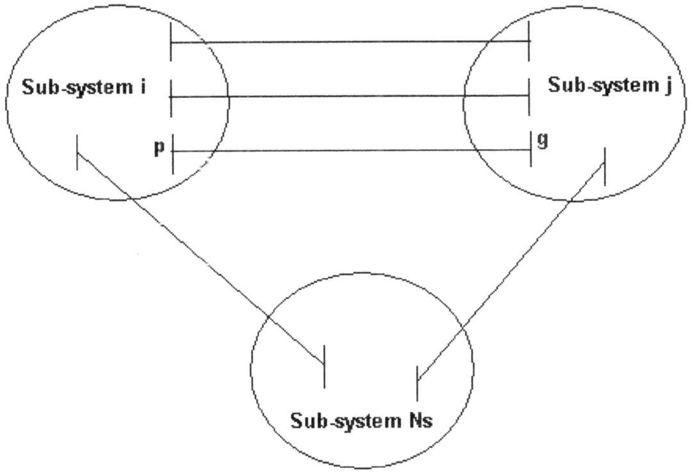

Figure 2. Tie-lines connecting sub-systems

The weighted least squares (WLS) is used in this paper, in this way it is considered that the criterion is a sum of sub-criterion determined for every one of the sub-systems. In the common case, the criterion is written as follows:

$$\min J(V\angle\delta)=\sum_{i=1}^{N_S}\left\{\begin{array}{l}\sum_{k=1}^{M_{k,V}}\frac{\left(V_k^{meas}-V_k^{est}\right)^2}{\sigma_{k,V}^2}+\\+\sum_{k=1}^{M_{k,P}}\frac{\left(P_{k,inj}^{meas}-P_{k,inj}^{est}\right)^2}{\sigma_{k,P}^2}+\\+\sum_{k=1}^{M_{k,Q}}\frac{\left(Q_{k,inj}^{meas}-Q_{k,inj}^{est}\right)^2}{\sigma_{k,Q}^2}+\\+\sum_{k=1}^{M_{k,P}}\frac{\left(P_{k,flow}^{meas}-P_{k,flow}^{est}\right)^2}{\sigma_{k,P}^2}+\\+\sum_{k=1}^{M_{k,Q}}\frac{\left(Q_{k,flow}^{meas}-Q_{k,flow}^{est}\right)^2}{\sigma_{k,Q}^2}\end{array}\right\} \quad (1)$$

Equation (1) can be written in short form as:

$$\min J(V\angle\delta)=\sum_{i=1}^{N_S}J_i(V\angle\delta),\ i=\overline{1,N_S} \quad (2)$$

Where

N_S is the number of sub-systems.

$M_{k,V}, M_{k,P}^{inj}, M_{k,Q}^{inj}, M_{k,P}^{flow}, M_{k,Q}^{flow}$, are the dimension of the corresponding measured variables. The type and number of measurements for different sub-systems can be different. The simplified criterion used in this paper is given by:

$$\min J(V\angle\delta)=\sum_{i=1}^{N_S}\sum_{k=1}^{M_{k,V}}\frac{\left(V_k^{meas}-V_k^{est}\right)^2}{\sigma_{k,V}^2} \quad (3)$$

Where σ is the standard deviation, $V_{k,}^{meas}$: is the measured quantity, $V_{k,}^{est}$: is the estimated quantity (calculated)

2.1. Measurement Model

The global data measurement model is given by

$$z = h(x) + \varepsilon \tag{4}$$

Where ε : is the measurement error.

The model has 4 parts determined by the type of measurements. When the power system is considered to be used for decomposed solution of state estimation problem, these 4 parts of the measurement model can be determined in different ways.

2.1.1. Voltage Magnitude Data Model

Voltage magnitude measurement model is for the whole system and is given by:

$$z_V = |V| \tag{5}$$

Where

$$z_V \in \Re^N$$

N: is the number of model buses.

Decomposition of this model is direct and is determined by the selected number of buses in every sub-system. The *ith* sub-system model can be written as:

$$z_{i,V} = |V_i| \quad i = \overline{1, N_S}, z_{i,V} \in \Re^{n_i} \tag{6}$$

n_i : is the number of buses in the *ith* sub-system

2.1.2. Real and Reactive Power Injection Data Model

The real and reactive power injection data model for the whole power system is decomposed in [21]. The obtained sub-system model is characterized with local for the sub-system state variables and with disturbance input from other sub-system and is given by:

$$P_i = G_i^R V_i + \sum_{\substack{j=1 \\ j \neq i}}^{N_S} G_{ij}^R V_j = G_i^R V_i + y_i^R \tag{7}$$

$$y_i^R = \sum_{\substack{j=1 \\ j \neq i}}^{N_S} G_{ij}^R V_j \quad i = \overline{1, N_S}, j = \overline{1, N_S}, j \neq i \tag{8}$$

$$G_i^R \in \Re^{n_i x n_i}, G_{ij}^R \in \Re^{n_i x n_j}, y_{i,}^R \in \Re^{n_i}$$

$$Q_i = G_i^{im} V_i + \sum_{\substack{j=1 \\ j \neq i}}^{N_S} B_{ij}^{im} V_j = G_i^{im} V_i + y_i^{im} \tag{9}$$

$$y_i^{im} = \sum_{\substack{j=1 \\ j \neq i}}^{N_S} B_{ij}^{im} V_j \tag{10}$$

$$G_i^{im} \in \Re^{n_i x n_i}, B_{ij}^{im} \in \Re^{n_i x n_j}, y_i^{im} \in \Re^{n_i}$$

The type and number of measurements in every sub-system can be different and independent. In the common case the injection data model can be written in matrix form as:

$$\begin{bmatrix} P_i \\ Q_i \end{bmatrix} = \begin{bmatrix} G_i^R \\ G_i^{im} \end{bmatrix} V_i + \begin{bmatrix} y_i^R \\ y_i^{im} \end{bmatrix} = G_{i,inj} V_i + y_{i,inj} \tag{11}$$

$$G_{i,inj} \in \Re^{2n_i x n_i}, y_{i,inj} \in \Re^{2n_i}$$

$$y_{i,inj} = \sum_{\substack{j=1 \\ j \neq i}}^{N_S} \begin{bmatrix} G_{ij}^R \\ B_{ij}^{im} \end{bmatrix} V_j = \sum_{\substack{j=1 \\ j \neq i}}^{N_S} h_{i,inj}(V_i \angle \delta_i, V_j \angle \delta_j) \tag{12}$$

Eqns (10) and (11) can be written in the notation of data model as:

$$z_{i,inj} = h_{i,inj}(V_i \angle \delta_i) + y_{i,inj} + \varepsilon_{i,inj} \tag{13}$$

Where

$$h_{i,inj}(V_i \angle \delta_i) = G_{i,inj} V_i \tag{14}$$

$$y_{i,inj} = \sum_{\substack{j=1 \\ j \neq i}}^{N_S} h_{i,inj}(V_i \angle \delta_i, V_j \angle \delta_j) \tag{15}$$

$$h_{i,inj} \in \Re^{2n_i x n_i}$$

The dimension of the vectors in Eqn (10) and the matrices have the maximal possible value but the number of measurements can be different. For every of the sub-problems a local voltage angle reference bus has to be introduced. In this case the sub-systems are independent.

2.1.3. Real and Reactive Power Flow Data Model

Real and reactive power flows is determined for every two interacting buses separately. This means that this model can be directly decomposed according to the selected dimension of the sub-systems. It is supposed for the power flows between the*pth* and *qth*buses in the ith sub-system (see Figure 2) that real and reactive power flows from bus *p*to bus *q*are:

$$\begin{aligned} P_{i,pq} &= V_{i,p}^2 \left(g_{i,pq}^{sh} + b_{i,pq}\right) - \\ &- V_{i,p} V_{j,q} [g_{i,pq} \cos\delta_{i,pq} - b_{i,pq} \sin\delta_{i,pq}] \end{aligned} \tag{16}$$

$$\begin{aligned} Q_{i,pq} &= -V_{i,pq}^2 \left(b_{i,pq}^{sh} + b_{i,pq}\right) + \\ &+ V_{i,p} V_{j,q} [g_{i,pq} \sin\delta_{i,pq} + b_{i,pq} \cos\delta_{i,pq}] \end{aligned} \tag{17}$$

The models of the flows from bus q to bus p are given as follows:

$$P_{i,qp} = V_{i,q}^2\left(g_{i,qp}^{sh} + g_{i,qp}\right) - \\ -V_{i,q}V_{i,p}[g_{i,qp}\cos\delta_{i,qp} - b_{i,qp}\sin\delta_{i,qp}] \quad (18)$$

$$Q_{i,qp} = -V_{i,q}^2\left(b_{i,qp}^{sh} + b_{i,qp}\right) + \\ +V_{i,q}V_{i,p}[g_{i,qp}\sin\delta_{i,qp} + b_{i,qp}\cos\delta_{i,qp}] \quad (19)$$

Where

$$\delta_{i,pq} = \delta_{i,p} - \delta_{i,q}$$
$$\delta_{i,qp} = \delta_{i,q} - \delta_{i,p}$$

In general the power flow data model can be written as:

$$z_{i,flow} = \left[P_{i,pq},\ Q_{i,pq},\ P_{i,qp},\ Q_{i,qp}\right]^T = \\ = h_{i,flow}(V_i\angle\delta_i) + \varepsilon_{i,flow} \quad i = \overline{1,N_S} \quad (20)$$

Where

$$z_{i,flow} \in \Re^{\left(M_{i,P}^{flow} + M_{i,Q}^{flow} + M_{i,P}^{flow} + M_{i,Q}^{flow}\right)}$$

The type of the measurements and the number of measurements of real and reactive power flows can be different for different sub-systems.

2.1.4. Tie-Line Data Model

The $N_S$sub-systemsshown in Figure 1are connected by tie-lines (electrical transmission lines or transformers). In this paper only electrical transmission lines are considered. The two ends of each tie-line are buses belonging to different sub-systems. The set of these boundary buses define an $(N_S+1)th$ sub-system called interconnection sub-system. The measurement model for the tie-lines between the ith and jth sub-systems, when the number of tie-lines is $N_{tl}$is given in [21] as follows:

$$P_{ij,pg} = G_{ij,pg}V_{i,p}^2 - V_{i,p}V_{j,g}G_{ij,pg} \\ p = \overline{1,N_{tl,ij}}, g = \overline{1,N_{tl,ij}}, i = \overline{1,N_S} \\ j = \overline{1,N_S}, p \neq g, j \neq i \quad (21)$$

$$Q_{ij,pg} = -V_{i,p}^2\left(B_{ij,pg} + b_{ij,pg}^{sh}\right) + V_{i,p}V_{j,g}B_{ij,pg} \quad (22)$$

The vector of measurements for the ith sub-system can be written as

$$z_{tl,ij,pg} = \begin{bmatrix} P_{ij,pg} \\ Q_{ij,pg} \end{bmatrix} = \begin{bmatrix} G_{ij,pg} \\ -B_{ij,pg} - b_{ij,pg}^{sh} \end{bmatrix} V_{i,p}^2 + \\ + V_{i,p}\begin{bmatrix} G_{ij,pg} \\ -B_{ij,pg} \end{bmatrix} V_{j,g} \quad (23)$$

The vector of measurements for the interconnected system is given by:

$$z_{tl} = \left[z_{tl,1}^T,\ z_{tl,2}^T,\ \cdots\ z_{tl,i}^T\ \cdots\ z_{tl,N_S}^T\right] \quad (24)$$

The measurement model is a non-linear and can be written as:

$$z_{tl} = h_{tl}\left(V_i\angle\delta_i, V_j\angle\delta_j\right) \\ i = \overline{1,N_S}, j = \overline{1,N_S}, j \neq i \quad (25)$$

It can be observed that the ijth sub-system tie-line flow measurement model has two parts. The first part depends on the ith sub-system voltages while the second part depends on the jth sub-system. This means that the measurement model can be represented as a sum from a model of the ith sub-system and the model of interconnection with other sub-systems. The first part depends on the voltage $V_{i,p}$ of the ith sub-system and the second depends on $V_{j,g}$ of the jth sub-system. These voltages Participate also in the injection model of the sub-system of the interconnected system. Included in the vector of voltages of the sub-system are also the border buses and voltages. In this way it is not necessary to calculate again the state estimates of the border injections as they are calculated using the injection data model.

Hence, the model of the tie-line data for the ith sub-system can be written in the following way:

$$z_{i,tl} = h_{i,tl}(V_i\angle\delta_i) + y_{i,tl} \quad (26)$$

$$y_{i,tl} = h_{ij,tl}\left(V_j\angle\delta_j\right) \quad (27)$$

The dimension of the measurement vector depends on the number of the tie-lines between the ith sub-system and other sub-system.

Finally the data model(measurement) of the ith sub-system with measurements of the voltage, real and reactive power injections, real and reactive power flows, and real and reactive power flows in the tie-lines can be written as:

$$z_{i,V} = V_i + \varepsilon_{i,V} \\ z_{i,inj} = h_{i,inj}(V_i\angle\delta_i) + y_{i,inj} + \varepsilon_{i,inj} \\ z_{i,flow} = h_{i,flow}(V_i\angle\delta_i) + \varepsilon_{i,flow} \\ z_{i,tl} = h_{i,tl}(V_i\angle\delta_i) + y_{i,tl} + \varepsilon_{i,tl} \quad (28)$$

The model given by (28) is used for formulation and solution of state estimation problem under decomposed environment.

3.State Estimation Problem Solution

3.1. Two Level State Estimation Solution

The solution of a power system state estimation can be obtained by solving the following Lagrangian function

$$L=\sum_{i=1}^{N_S}\left\{\begin{array}{l}\frac{\left(V_i^{meas}-V_i^{est}\right)^2}{\sigma_{i,V}^2}+\\ +\rho_{i,inj}^T[y_{i,inj}-\sum_{\substack{j=1\\ j\neq i}}^{N_S}h_{i,inj}(V_i\angle\delta_i)]+\\ +\lambda_{i,inj}^T[z_{i,inj}-h_{i,inj}(V_i\angle\delta_i)-y_{i,inj}]+\\ +\rho_{i,tl}^T[y_{i,tl}-h_{ij,tl}(V_i\angle\delta_i,V_j\angle\delta_j)]+\\ +\lambda_{i,tl}^T[z_{i,tl}-h_{i,tl}(V_i\angle\delta_i)-y_{i,tl}]\end{array}\right\} \quad (29)$$

where

$\rho_{i,inj}, \rho_{i,tl}, \lambda_{i,inj}, \lambda_{i,tl}$ are vectors of Lagrange multipliers.

The Lagrangian function includes the criterion and the model of the interconnected equations of the sub-system according to power injections and flows. It can be seen from (29) if the interconnections $h_{ij,inj}$ and $h_{ij,tl}$ can be decomposed, then the Lagrangian function can be decomposed and the state estimation problem can be solved in a fully decentralized way.

Such a type of decomposition of the Lagrangian function can be achieved if the problem for state estimation is solved in a two-level structure using the principles of decomposition coordination [17]. The mixed principle of prediction of the aims of the sub-system represented by the Lagrange variables $\rho_{i,inj}$ and $\rho_{i,tl}$, and of prediction of interconnection of the sub-systems $y_{i,inj}$ and $y_{i,tl}$ is applied to the Lagrangian function of (29). This principle is implemented by introducing a coordinator on the second level of the two-level structure of the solution to the problem. The coordinator predicts the values of the Lagrange variables and interconnections as follows:

$$\begin{aligned}\rho_{i,inj}&=\rho_{i,inj}^c\\ \rho_{i,tl}&=\rho_{i,tl}^c\\ y_{i,inj}&=y_{i,inj}^c\\ y_{i,tl}&=y_{i,tl}^c\end{aligned} \quad (30)$$

Where c is the index of the coordinating procedure.

Substitution of the coordinating variables given in (30) into the Lagrangian function (29) allows the interconnection terms to be distributed between the sub-systems in the following way [17]

$$\sum_{i=1}^{N_S}\rho_{i,inj}^T\sum_{\substack{j=1\\ j\neq i}}^{N_S}h_{ij,inj}(V_i\angle\delta_i,V_j\angle\delta_j)=\sum_{i=1}^{N_S}\sum_{\substack{j=1\\ j\neq i}}^{N_S}\rho_{j,in}^T h_{ji,inj}(V_i\angle\delta_i) \quad (31)$$

$$\sum_{i=1}^{N_S}\rho_{i,tl}^T h_{ij,tl}(V_i\angle\delta_i,V_j\angle\delta_j)=\sum_{i=1}^{N_S}\rho_{j,tl}^T h_{ji,tl}(V_i\angle\delta_i) \quad (32)$$

Equations (31) and (32) are possible on the basis that the connection between the primal variables V_j and the dual (Lagrange's) variables ρ_i. As the dual variables have the voltages of other sub-systems can be substituted by the voltages of the ith sub-system and the Lagrange's variables of the *jth* sub-system. In this way the Lagrangian function is a function only of the voltages of the *ith* sub-system and can be fully decomposed. This means that the state estimation can be solved separately. Then, the system solutions in this case depends on the values of the coordinating variables, which means that the optimal solutions will be obtained only when the values of the coordinating variables is computed by an iterative process of coordination based on the necessary conditions for optimality of the Lagrangian function towards the coordinating variables and on the solution of the separate sub-system's problems.

The necessary conditions for optimality of the Lagrangian function towards the coordination variables are given by:

$$\frac{\partial L}{\partial\rho_{i,inj}}=y_{i,inj}^c-\sum_{\substack{j=1\\ j\neq i}}^{N_S}h_{ij,inj}(V_i\angle\delta_i,V_j\angle\delta_j)=0 \quad (33)$$

$$\frac{\partial L}{\partial y_{i,inj}}=\rho_{i,inj}^c-\lambda_{i,inj}=0 \quad (34)$$

$$\frac{\partial L}{\partial\rho_{i,tl}}=y_{i,tl}^c-h_{ij,tl}(V_i\angle\delta_i,V_j\angle\delta_j)=0 \quad (35)$$

$$\frac{\partial L}{\partial y_{i,tl}}=\rho_{i,tl}^c-\lambda_{i,tl}=0 \quad (36)$$

The value of voltages and the value of Lagrange's variables $\lambda_{i,inj}$ and $\lambda_{i,tl}$ can be obtained as solutions of the sub-systems state estimation problems.

Equations (33) to (36) can be solved analytically using the solutions obtained from first level sub-problems in the following way:

$$y_{i,inj}^{c+1}=\sum_{\substack{j=1\\ j\neq i}}^{N_S}h_{ij,inj}(V_i^c\angle\delta_i^c,V_j^c\angle\delta_j^c) \quad (37)$$

$$\rho_{i,inj}^{c+1}=\lambda_{i,inj}^c \quad (38)$$

$$y_{i,tl}^{c+1}=h_{ij,tl}(V_i^c\angle\delta_i^c,V_j^c\angle\delta_j^c) \quad (39)$$

$$\rho_{i,tl}^{c+1}=\lambda_{i,tl}^c \quad (40)$$

The optimal solution of the initial state estimation problem is obtained if the necessary conditions for optimality according to the coordinating variables are fulfilled. This can be checked by calculating of errors

$$\varepsilon_1 = y_{i,inj}^{c+1} - y_{i,inj}^{c} \quad (41)$$

$$\varepsilon_2 = \rho_{i,inj}^{c+1} - \rho_{i,inj}^{c} \quad (42)$$

$$\varepsilon_3 = y_{i,tl}^{c+1} - y_{i,tl}^{c} \quad (43)$$

$$\varepsilon_4 = \rho_{i,tl}^{c+1} - \rho_{i,tl}^{c} \quad (44)$$

If $\varepsilon_1 \le \varphi_1, \varepsilon_2 \le \varphi_2, \varepsilon_3 \le \varphi_3, \varepsilon_4 \le \varphi_4$

Where $\varphi_1 > 0, \varphi_2 > 0, \varphi_3 > 0, \varphi_4 > 0$ are very small pre-defined number, the optimal solution of the coordinating problem and the sub-problems are obtained.

3.2. Formulation of the State Estimation Sub-Problem of the First Level

The state estimation problem for every isolated sub-system is formulated using decomposed Lagrangian function for sub-problem criterion given by:

$$\begin{aligned} L_i = {} & \frac{\left(V_i^{meas} - V_i^{est}\right)^2}{\sigma_{i,V}^2} + \\ & + \rho_{i,inj}^c y_{i,inj}^c - \sum_{\substack{j=1 \\ j \ne i}}^{N_S} \rho_{ji,inj}^c h_{ji,inj}\left(V_i \angle \delta_i\right) - \\ & - \lambda_{i,inj}^T [-y_{i,inj}^c + z_{i,inj} - h_{i,inj}\left(V_i \angle \delta_i\right)] + \\ & + \rho_{i,tl}^c y_{i,tl}^c - \rho_j^c h_{ji}\left(V_i \angle \delta_i\right) + \\ & + \lambda_{i,tl}^T [z_{i,tl} - h_{i,tl}\left(V_i \angle \delta_i\right) - y_{i,tl}^c] + \\ & + \lambda_{i,V}^T [z_{i,v} - V_i] + \lambda_{i,flow}^T [z_{i,flow} - h_{i,flow}\left(V_i \angle \delta_i\right)] \end{aligned} \quad (45)$$

3.2.1. Solution of the First Level Problems

A Lagrangian function is formed for every sub-problem as follows:

$$\begin{aligned} L_i^i = {} & \frac{\left(V_i^{meas} - V_i^{est}\right)^2}{\sigma_{i,V}^2} + \\ & + \rho_{i,inj}^c y_{i,inj}^c - \sum_{\substack{j=1 \\ j \ne i}}^{N_S} \rho_{ji,inj}^c h_{ji,inj}\left(V_i \angle \delta_i\right) - \\ & - \lambda_{i,inj}^T [-y_{i,inj}^c + z_{i,inj} - h_{i,inj}\left(V_i \angle \delta_i\right)] + \\ & + \rho_{i,tl}^c y_{i,tl}^c - \rho_j^c h_{ji}\left(V_i \angle \delta_i\right) + \\ & + \lambda_{i,tl}^T [z_{i,tl} - h_{i,tl}\left(V_i \angle \delta_i\right) - y_{i,tl}^c] + \\ & + \lambda_{i,V}^T [z_{i,v} - V_i] + \lambda_{i,flow}^T [z_{i,flow} - h_{i,flow}\left(V_i \angle \delta_i\right)] \end{aligned} \quad (46)$$

The optimal solution is based on the necessary conditions for optimality as follows:

$$\begin{aligned} \frac{\partial L_i^i}{\partial V_i} = {} & 2\frac{\left(V_i^{meas} - V_i\right)}{\sigma_{i,V}^2} - \\ & - \sum_{\substack{j=1 \\ j \ne i}}^{N_S} \left(\frac{\partial h_{ji,inj}\left(V_i \angle \delta_i\right)}{\partial V_i}\right)^T \rho_{i,inj}^c - \\ & - \left(\frac{\partial h_{i,inj}\left(V_i \angle \delta_i\right)}{\partial V_i}\right)^T \lambda_{i,inj} - \\ & - \left(\frac{h_{ji,tl}\left(V_i \angle \delta_i\right)}{\partial V_i}\right)^T \rho_{j,tl}^c - \\ & - \left(\frac{h_{i,tl}\left(V_i \angle \delta_i\right)}{\partial V_i}\right)^T \lambda_{i,tl} - \lambda_{i,V} - \\ & - \left(\frac{h_{i,flow}\left(V_i \angle \delta_i\right)}{\partial V_i}\right)^T \lambda_{i,flow} = 0 \end{aligned} \quad (47)$$

$$\begin{aligned} \frac{\partial L_i^i}{\partial \delta_i} = {} & -\sum_{\substack{j=1 \\ j \ne i}}^{N_S} \left(\frac{\partial h_{ji,inj}\left(V_i \angle \delta_i\right)}{\partial \delta_i}\right)^T \rho_{j,inj}^c - \\ & - \left(\frac{\partial h_{i,inj}\left(V_i \angle \delta_i\right)}{\partial \delta_i}\right)^T \lambda_{i,inj} - \\ & - \left(\frac{\partial h_{ji,tl}\left(V_i \angle \delta_i\right)}{\partial \delta_i}\right)^T \rho_{j,tl}^c - \\ & - \left(\frac{h_{i,tl}\left(V_i \angle \delta_i\right)}{\partial \delta_i}\right)^T \lambda_{i,tl} - \\ & - \left(\frac{\partial h_{i,flow}\left(V_i \angle \delta_i\right)}{\partial \delta_i}\right)^T \lambda_{i,flow} = 0 \end{aligned} \quad (48)$$

$$\frac{\partial L_i^i}{\partial \lambda_{i,inj}} = z_{i,inj} - y_{i,inj}^c - h_{i,inj}\left(V_i \angle \delta_i\right) = 0 \quad (49)$$

$$\frac{\partial L_i^i}{\partial \lambda_{i,tl}} = z_{i,tl} - h_{i,tl}\left(V_i \angle \delta_i\right) - y_{i,tl}^c = 0 \quad (50)$$

$$\frac{\partial L_i^i}{\partial \lambda_{i,V}} = z_{i,V} - V_i = 0 \quad (51)$$

$$\frac{\partial L_i^i}{\partial \lambda_{i,flow}} = z_{i,flow} - h_{i,flow}\left(V_i \angle \delta_i\right) = 0 \quad (52)$$

The solution of set of equations (47) to (52) gives the necessary conditions for optimality determines the optimal solution of the ith sub-problem. Eqns (47) t0 (52) are non-linear with many variables; they cannot be solved by analytical method. Gradient procedures are used to calculate the values of primal variables (V_i, δ_i) and the dual variables ($\lambda_{i,inj}, \lambda_{i,tl}, \lambda_{i,V}, \lambda_{i,flow}$) as follows:

$$\begin{aligned} V_i^{t+1} &= V_i^t - \alpha_{i,V}\varepsilon_{i,V} \\ \delta_i^{t+1} &= \delta_i^t - \alpha_{i,\delta}\varepsilon_{i,\delta} \\ \lambda_{i,inj}^{t+1} &= \lambda_{i,inj}^t + \alpha_{i,inj}\varepsilon_{i,inj} \\ \lambda_{i,tl}^{t+1} &= \lambda_{i,tl}^t + \alpha_{i,tl}\varepsilon_{i,tl} \\ \lambda_{i,V}^{t+1} &= \lambda_{i,V}^t + \alpha_{i,V}\varepsilon_{i,\lambda} \\ \lambda_{i,flow}^{t+1} &= \lambda_{i,flow}^t + \alpha_{i,flow}\varepsilon_{i,flow} \end{aligned} \quad (53)$$

where

$\varepsilon_{i,V}, \varepsilon_{i,\delta}, \varepsilon_{i,inj}, \varepsilon_{i,tl}, \varepsilon_{i,\lambda}, \varepsilon_{i,flow}$ are errors and α_i is step-length.

The calculations given by eqn (53) are performed under the given by the second level values. The gradient procedure continues until convergence on maximum number of iterations on the first level is attained. Norm [21] of the errors of every iteration is calculated from:

$$\begin{aligned} \varepsilon_1^i &= \left\|\varepsilon_{i,V}\right\| \\ \varepsilon_2^i &= \left\|\varepsilon_{i,\delta_i}\right\| \\ \varepsilon_3^i &= \left\|\varepsilon_{i,inj}\right\| \\ \varepsilon_4^i &= \left\|\varepsilon_{i,tl}\right\| \\ \varepsilon_5^i &= \left\|\varepsilon_{i,flow}\right\| \end{aligned} \quad (54)$$

Norms of the errors are compared with small pre-defined positive numbers given by the following constants $\phi_1, \phi_2, \phi_3, \phi_4, \phi_5$. If $\varepsilon_1^i \le \phi_1, \varepsilon_2^i \le \phi_2, \varepsilon_3^i \le \phi_3, \varepsilon_4^i \phi_4, \varepsilon_5^i \le \phi_5$ is fulfilled, then the optimal solution is attained and the computation process is stopped. When the calculations of the first level are completed, the values of $V_i, \delta_i, \lambda_{i,V} \lambda_{i,tl}$ are sent to the second level and the new values of the coordinating variables are calculated and so on. The computation set up and communication between first level and second level is presented in Figure 3.

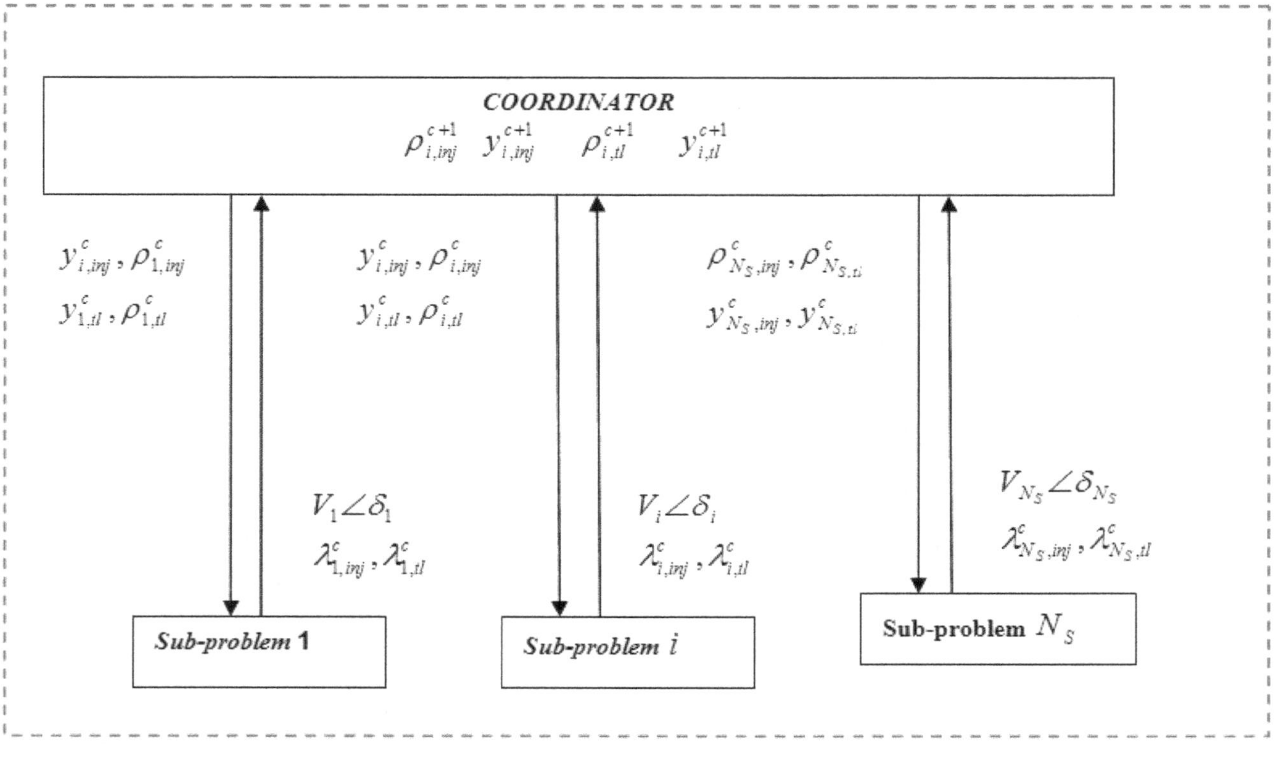

Figure 3. *Communication between Two level structures for solution of state estimation problem*

4. Algorithms

The computation procedure is implemented using the following first and second level algorithms.

4.1. First Level Algorithm

At first level, the optimal operating condition of each sub-system is determined by solving independent N power system state estimation. At this level the interconnection between the sub-systems have not to be considered. Interconnection values are provides by the coordinator. Hence, gradient methods are used to compute the primal and dual variables of the isolated sub-system. The following algorithm is used to calculate the solution of the first level sub-problem. Before starting the algorithm, first the number of iterations is defined.

I. Initialize t= 0

II. Set initial values for $V_i, \delta_i, \lambda_{i,inj}, \lambda_{i,flow.} \lambda_{i,tl}, \lambda_{i,V}$. Initial value of V_i is obtained from load flow program and δ_i is set equal to zero.

III. Obtain the values of the coordinating variables

IV. $\rho_{i,inj}, \rho_{i,tl}, y_{i,inj}, y_{i,tl}$ from the second level

V. Improve values of set variables using Eqn (53)

VI. Check for conditions for convergence using Eqn (54)

If conditions of convergence satisfy (54) stop the procedure and sent the values of $V_i, \delta_i, \lambda_{i,inj}, \lambda_{i,tl}$ to the second level. If not go to step II. First level algorithm is schematically given in Figure 4.

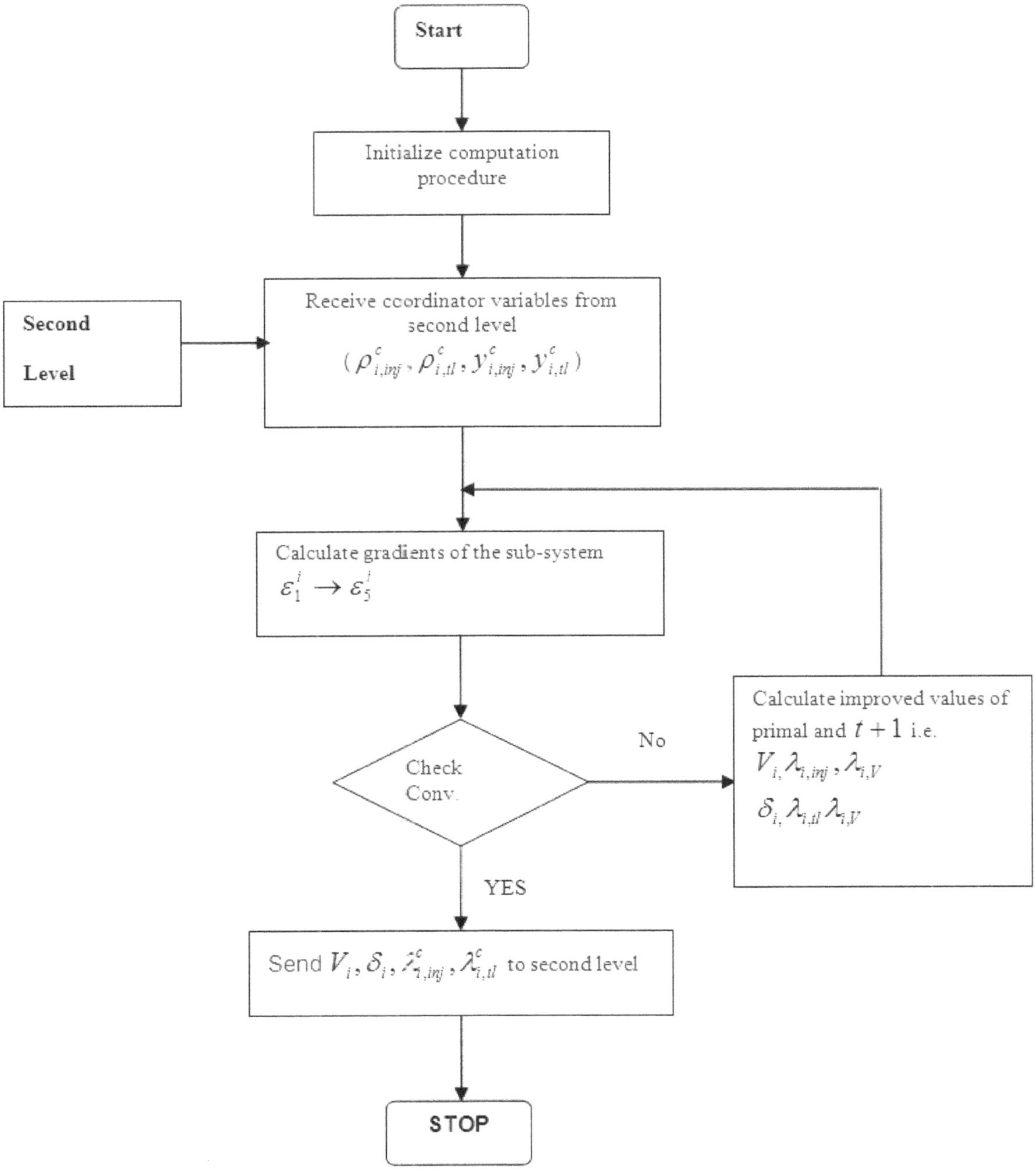

Figure 4. *First level algorithm flowchart*

4.2. Second Level Algorithm

The coordinator does not know or need the detailed operating information of each sub-system. The coordinator executes the following function:

I. Predict the values of Lagrange's variables and interconnection variables and sent predicted variables to the first level

II. Wait all calculations on the first level to be completed and receive the calculated values from first level

III. Compare $\rho_{i,inj}, \rho_{i,tl}$ if the error between them is bigger than pre-defined tolerance, calculate the improved values of the coordinating variables

IV. If the error is smaller than the pre-defined tolerance, stop the procedure.

It can be observed that the coordinator has little work to do. The coordinator does not calculate the state vector but has to predict the Lagrange's variables and variables related to subsystem interconnections. The positive characteristic of this method is that of reducing computation work at coordinator level. The second level algorithm is schematically given in Figure 5.

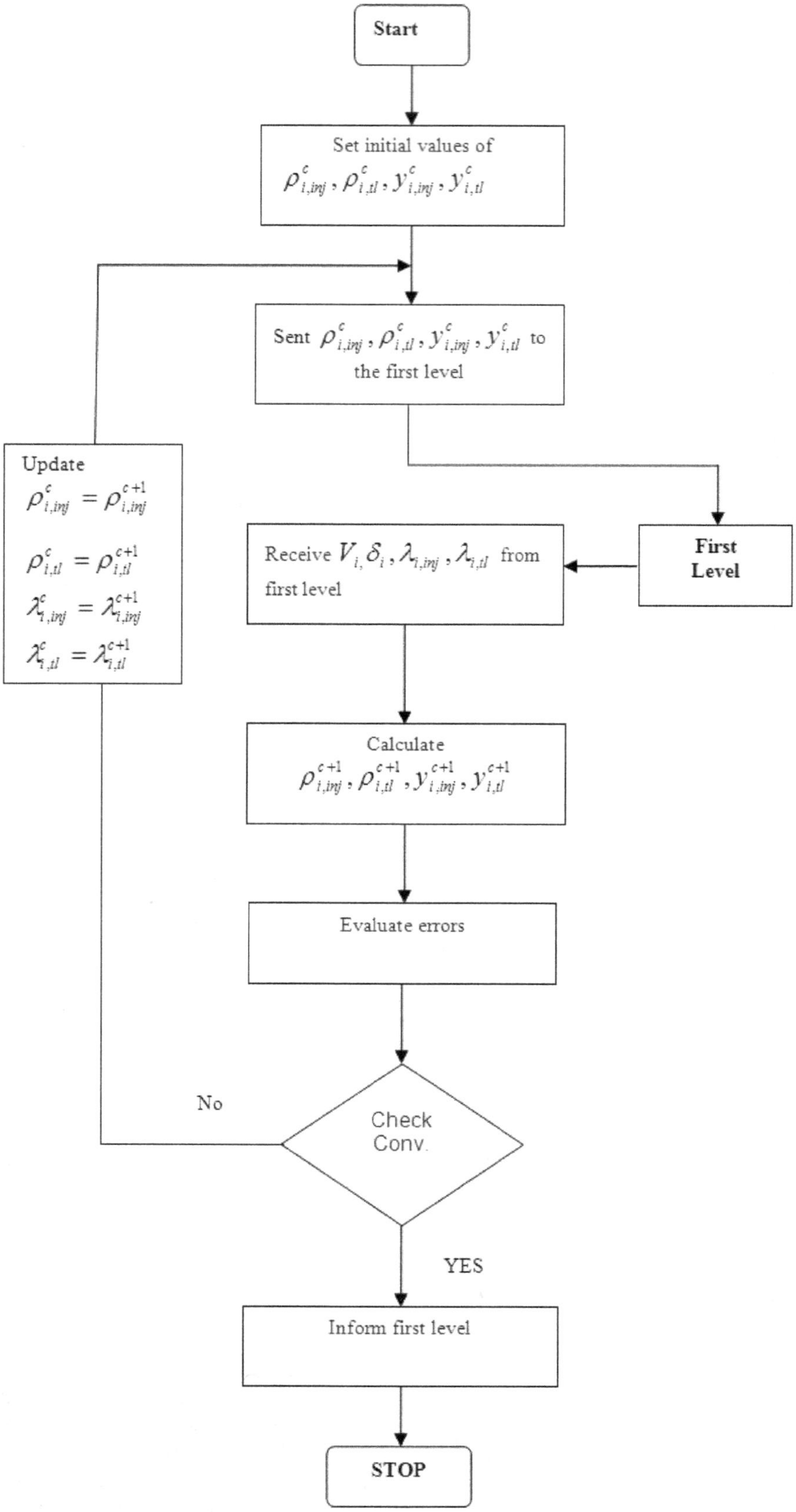

Figure 5. *Second level algorithm flowchart*

5. Discussion

The developed decomposition-coordination model and algorithm in this paper presents positive characteristic in solving the power system state estimation problem. First, the solution of power system state estimation is reduced to the solution of N+1 independent sub-problem. Secondary, the implementation of the algorithm is carried out using parallel and distributed architectures. In this way, the computational task corresponding to the decomposition work can be carried out by a unique parallel computer located in the power system central dispatch centre, using a cluster or by using a suitable distributed computing scheme with several processors located at lower levels in the control hierarchical such as in the regional control centres.

First and second algorithms are applied in solving the estimation problem provided that: the system is decomposed into sub-systems or areas, at least one generating bus is available in a sub-system and the sub-system is observable i.e. measurements from the sub-system are enough to perform the state estimation.

Decomposition-coordination method and algorithm is aimed at facilitating parallel or distributed processing, decentralizing measurement of a large-scale power system networks, reducing amount of measurements sent to central dispatch centre and reducing computation time and complexity on solution of state estimation.

6. Conclusion

In this paper the formulation of the problem for decomposed solution of power system state estimation is presented. Measurement models or voltage magnitude, real and reactive power injections, real and reactive line power flows, and real and reactive power flows in tie-lie lines connecting sub-systems are developed. State estimation problem solution using two-level structure is proposed. First and second level algorithms for implementing the two-level computation are presented. The developed decomposition-coordination method, models and algorithm presents positive characteristic in solving power system state estimation of large-scale power system network and can be implemented using parallel or distributed computing architecture. Further work is still going on to establish the accuracy of the proposed approach. The proposed method will be tested using IEEE 14, IEEE 30, and IEEE 57 buses.

References

[1] J. Gu, K.A. Clements, G. Krumpholz, P.Davis, " The solution of ill-conditioned Power System State Estimation Problem via the method of Peters and Wilkinson" Power Industry Computer Applications Conference Proceedings, Huston, May 1983, pp.239-246

[2] A.Monticelli, "State Estimation in Electric Power Systems: A generalized Approach, Kluweri Academic Publishers, Boston, 1999.

[3] A.Bose, and K.A. Clements, "Real-Time modelling of Power Networks", IEEE Proceeding, Special issue on Computers in Power System Operations, Vol., 75, No. 12, Dec 1987, pp. 1607-1622

[4] B. Stott, O. Alsac and A. Monticelli, "Security Analysis and Optimization", IEEE Proceeding, Vol. 75, No.12, Dec 1987, pp. 1623-1644

[5] Dy Liacco, "System security: The computer role." IEEE Spectrum, Vol.16, No. 6, June 1978, pp. 48-53

[6] A.A. Hossam-Eldin, E.N. Abdallah and M.S.El-Nozahy, "A modified Genetic Based Technique for solving the Power System State Estimation Problem," World Academy of Science, Engineering and Technology, Vol.55, 2009, pp. 311-320.

[7] J.B. Calvalho and F.M. Barbosa, "A parallel Algorithm to Power System State Estimation ," IEEE, 1988, pp. 1213-1217

[8] O. Alsac, N. Vempati, B. Stott, and A. Monticelli, "Generalized State Estimation," IEEE Transactions on Power Systems, Vol. 13, No.3, August 1998, pp. 1069-1075

[9] F.C. Schweppe, J. Wildes, and D. Rom, "PowerSystemStaticState Estimation Parts: I, II and III," Power Industry Computer Conference, PICA, Denver, Colorado, June 1969.

[10] K.A. Clements and B.F. Wollenberg, "An algorithm for Observability Determination in Power System State Estimation," Paper A75 447-3, IEEE/PES Summer Meeting, San Francisco, CA, July 1975.

[11] P. Zarco and A.G. Exposito, "Power System Parameter Estimation: A survey," IEEE Trans. on Power Systems, Vol. 15, No.1, Feb. 2000, pp. 216-222

[12] Th. VanCutsem, J.L. Howard, M. Ribben-Pavella and Y.M. El-Fattah, "Hierarchical State Estimation," International Journal Of Electric Power and Energy Systems, Vol. 2, April 1980, pp. 70-78

[13] Th. VanCutsem, J.L. Howard, and M. Ribben-Pavella, "A Two-level Static State Estimation for Electric Power Systems," IEEE Trans. on PAS, Vol. PAS, August 1981, pp. 3722-3732

[14] M.Y. Patel and A.A. Girgis, "Two level State Estimation for Multi-area Power System," IEEE Power Engineering Society General Meeting, June 2007

[15] S. Lakshminarasimhan, and A.A. Girgis, "Hierarchical State Estimation Applied to Wide-Area Power Systems," IEEE Power Engineering Society General Meeting, June 2000, pp. 1-6.

[16] J.A. Aguado, C.P. Molina and V.H. Quintana, "Decentralized PowerSystemState Estimation: A Decomposition-Coordination Approach,2 IEEE Porto Power Tech. Conference, 10th -13th September, 2001, Porto, Portugal, Vol. 3, pp. 6-11

[17] M. Singh and M. Titli, "System: decomposition, Optimization and Control, Kluweri Publishers, Boston, 1978

[18] M.B. Couto Filho, A.M. Leite da Silva and D.M. Falcao, "Bibliography on Power System State Estimation (1968-1989)", IEEE Trans. on PWR, Vol. 5, No.3, August 1990, pp.950-961

[19] R.D. Masiello and F.C. Schweppe, "A tracking Static State Estimator," IEEE, Vol. PAS-90, March/April 1971, pp.1025-1033

[20] M.A. Kusekwa, "Real-Time State Estimation of a Distributed Electrical Power System Under Conditions of Deregulation," PhD Thesis in Electrical Engineering, Cape Peninsula University of Technology, Cape Town, May 2010.

[21] http://www.mathworld.wolfram.com/Norm.html

2

Transient Analysis and Modelling of Sixphase Asynchronous Machine

Akpama Eko James[1], Linus Anih[2], Ogbonnaya Okoro[3]

[1]Dept. of Elect/Elect/ Engineering Cross River University of Technology, Calabar/Nigeria
[2]Dept. of Electrical Engineering Nniversity of Nigeria, Nsukka, Enugu/Nigeria
[3]Dept. of Elect/Elect/ Engineering, Micheal Okpara University Agriculture, Umudike/Nigeria

Email address:
ekoakpama2004@yaahoo.com (E. J. Akpama). luanih@yahoo.com (L. U. Anih). oiokoro@yahoo.co.uk (O. I. Okoro)

Abstract: Multiphase Induction machine offers numerous advantages compared to the conventional three phase induction machine. Among the different types, the six-phase induction machine is most common, due to the flexibility in converting a three phase to six phase. By splitting the phase belt of a three phase induction machine, a six-phase induction machine is realized. The areas of application of multiphase (>3) machines are enormous, mostly where reliability is paramount. This paper presents the analysis, modeling and simulation of a 5Hp, 50HZ, 4 pole, 24 slots asymmetrical six-phase induction machine for submarine application. With the help of MATLAB software, the models developed are simulated and results in the form of computer traces of the dynamic performance of the machine at start-up are presented and discussed. This investigation shows that the replacement of three phase with a six phase machine is both technically and economically advantageous.

Keywords: Modeling, Six-Phase, Submarines, Computer Simulation, MATLAB, Transients Studies

1. Introduction

Three phase induction machines are asynchronous speed machines. They are operated as motors and generators, comparatively less expensive to equivalent synchronous or dc machines. It is considered as the workhorse of the industry. Induction machines has almost replaced the DC machine in the industry due to the simplicity of design, ruggedness, low-cost, low maintenance cost and direct connection to AC power source compared to the DC motors[1, 2, 3]. In fact 85% of the total power consumption in the industrial sector is from induction motors. Due to the above, new strategies, new methods of analysis and design are being sought that will improve the performance of this machines and increase efficiency. Investigation has shown that multiphase machines are possible and advantageous [4]. Among the groups of multiphase machines, the six-phase has received more attention due the simplicity in converting a three phase machine to six-phase machine [5]. This is achieved by splitting the phase belt of three phase machine, that is, two sets of three phase stator winding of the original three phase, with set I spanning 30^0 electrical from set II, having a common magnetic structure[6-7].

The use of a common magnetic structure shared by two sets of stator winding dated since 1930 [4, 8], where in an attempts to increase the power capability of a large synchronous generator, the stator winding has to be doubled. From that time, research activities of the dual stator induction machine (DSIM) increased. These type of machines are normally constructed by "splitting" the stator winding into two identical windings [9-12]. Phase orders of multiples of 3, i.e. 6,9, etc, are very possible. Presently, there is evidence of current research on higher phase AC machines of phase order 5, 6, 9 even 15 [13-15]. Multiphase induction machine finds its applicability in the area of high degree of reliability demand as in more electric aircrafts, electric ship propulsion, electric vehicles, (EH) and hybrid electric vehicles (HEV).

Particularly for ship propulsion, there is an ongoing research in the application of six phase induction motor by this author, where a six phase motor is to replace two three phase motors.

Analysis of multiphase induction machine is carried out in [16, 17]. Lipo in [8] presented a six-phase induction machine model in d-q transformation, the analysis here uses a split-phase configuration where the 60^0 phase belt was split into two portions each spanning 30^0. The simulated results for three phase and six-phase were compared. B. Kundrotas et al [12] modelled a six phase induction motor in the dynamic mode. In [18], a six-phase induction motor was used to reduce the noise level in an electric traction system. Improved reliability is guaranteed in submarine with the application of two sets of star connected stator winding spatially shifted by 30^0 electrical with isolated neutral point as reported in [19]. A transition design based on the Markov chain was used to analyse the availability index of multiphase induction machine, [20, 21]. The submarine is rapidly emerging in the last couple of years as the main potential application area for multi-phase motor drives. The idea of multiphase suggests a replacement of two electric machines with a single machine of DSIM. A lot of research is ongoing in multiphase induction machines. The concept of multiphase is still in the infant stage, though much research is ongoing. It is the aim of this paper to investigate the transient performance of a six-phase split wound induction machine for submarines.

2. Modelling of Six-Phase Induction Machine

Multiphase electric machine (MPEM) which is more advantageous and has improved performance compared to the three phase counterpart come as both synchronous and asynchronous machines. Figure 1, shows the V_{abcxyz} voltages of a six-phase induction machine with their phase displacement. In order to model the six-phase induction machine, the dq transformation is applied, and the rotor reference frame is adopted [16].

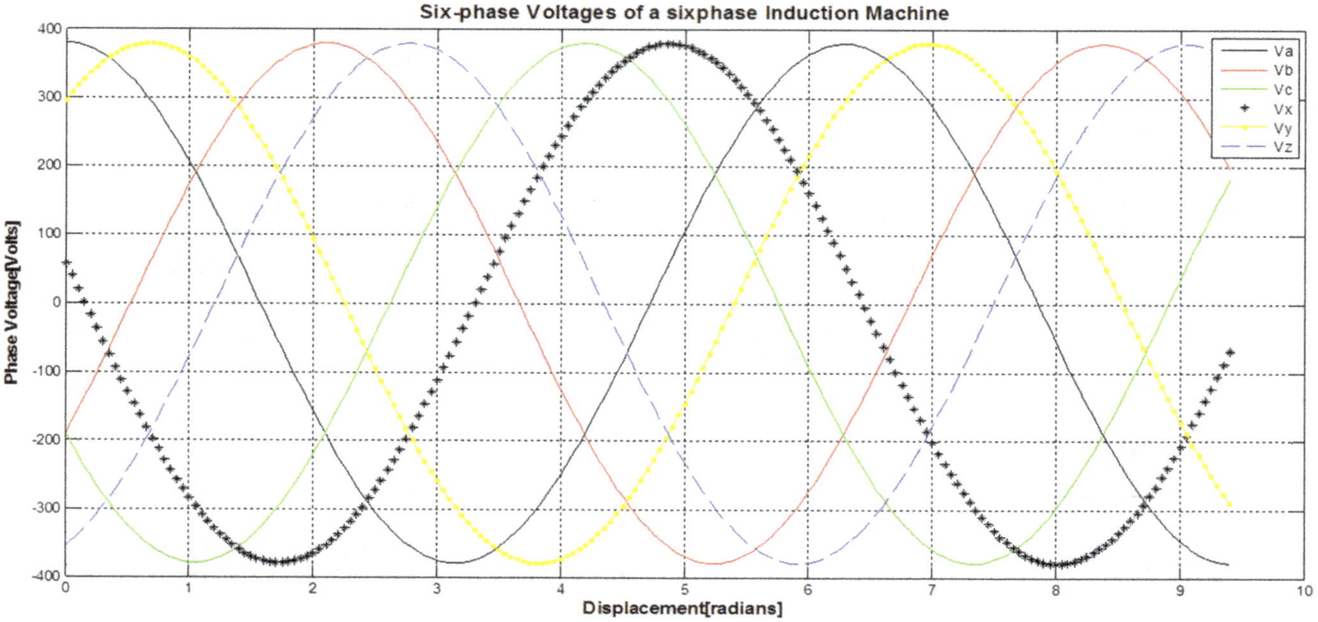

Fig. 1. Six-phase source voltages of six-phase induction machine.

2.1. Electrical Model

The voltage equation of a split phase induction machine is written as, where $\alpha=30^0$ elect

$$V_{as}=V\cos\omega_e t \quad (1)$$

$$V_{bs}=V\cos(\omega_e t-2\pi/3) \quad (2)$$

$$V_{cs}=V\cos(\omega_e t+2\pi/3) \quad (3)$$

$$V_{xs}=V\cos(\omega_e t-\alpha) \quad (4)$$

$$V_{ys}=V\cos(\omega_e t-2\pi/3-\alpha) \quad (5)$$

$$V_{zs}=V\cos(\omega_e t+2\pi/3-\alpha) \quad (6)$$

Using the appropriate transformation, the phase voltage of set I, abc, is transformed to its equivalent d-q axis as below;

$$V_{qs}=2/3(V_{as}-V_{bs}/2-V_{cs}/2) \quad (7)$$

$$V_{ds}=2/3(\sqrt{3}/2[-V_{bs}-V_{cs}]) \quad (8)$$

For a balance system, since

$$V_{as}=-V_{bs}-V_{cs} \quad (9)$$

Then equation (7) becomes

$$V_{qs}=V_{as}=V\cos\omega_e t \quad (10)$$

Simplifying equation (8)

$$V_{ds}=1/\sqrt{3}(V_{cs}-V_{bs}) \quad (11)$$

$$V_{ds}=1/\sqrt{3}(V\cos(\omega_e t+2\pi/3)-V\cos(\omega_e t-2\pi/3) \quad (12)$$

Applying Euler's identity to equation (9), the result becomes

$$V_{ds}=-V\sin\omega_e t \quad (13)$$

The same analysis is carried out on the second set II. The dq voltage equations of a six-phase induction machine are readily written as in [24]:

$$V_{q1}= r_1 i_{q1}+ \omega_k \lambda_{d1}+p\lambda_{q1} \tag{14}$$

$$V_{d1}= r_1 i_{d1}- \omega_k \lambda_{q1}+p\lambda_{d1} \tag{15}$$

$$V_{q2}= r_2 i_{q2}+ \omega_k \lambda_{d2}+p\lambda_{q2} \tag{16}$$

$$V_{d2}= r_2 i_{d2}- \omega_k \lambda_{q2}+p\lambda_{d2} \tag{17}$$

$$V_{qr}=0= r_r i_{qr}+ (\omega_k-\omega_r)\lambda_{dr}+p\lambda_{qr} \tag{18}$$

$$V_{dr}=0= r_r i_{dr}- (\omega_k-\omega_r)\lambda_{qr}+p\lambda_{dr} \tag{19}$$

The flux linkage equations are given below

$$\lambda_{q1}=L_{l1}i_{q1}+L_{lm}(i_{q1}+i_{q2})+L_{dq}i_{d2}+L_{mq}(i_{q1}+i_{q2}+i_{cr}) \tag{20}$$

$$\lambda_{d1}=L_{l1}i_{d1}+L_{lm}(i_{d1}+i_{d2})+L_{dq}i_{q2}+L_{md}(i_{d1}+i_{d2}+i_{cr}) \tag{21}$$

$$\lambda_{q2}=L_{l2}i_{q2}+L_{lm}(i_{q1}+i_{q2})+L_{dq}i_{d1}+L_{mq}(i_{q1}+i_{d2}+i_{qr}) \tag{22}$$

$$\lambda_{d2}=L_{l2}i_{d2}+L_{lm}(i_{d1}+i_{d2})+L_{dq}i_{q1}+L_{md}(i_{d1}+i_{d2}+i_{dr}) \tag{23}$$

$$\lambda_{qr}=L_{lr}i_{qr}+L_{mq}(i_{q1}+i_{q2}+i_{qr}) \tag{24}$$

$$\lambda_{dr}=L_{lr}i_{dr}+L_{md}(i_{d1}+i_{d2}+i_{dr}) \tag{25}$$

Equations 14-25, suggested the equivalent circuit of Fig. 2
Let

$$Ldq = 0 \tag{26}$$

$$L_m=L_{mq}=L_{md} \tag{27}$$

$$L_1=L_{l1}+L_{lm}+L_m \tag{28}$$

$$L_2 = L_{l2}+L_{lm}+L_m \tag{29}$$

$$L_3 = L_{lm}+L_m \tag{30}$$

$$L_r = L_{lr}+L_m \tag{31}$$

Using state variable method, equation (32) is put in state variable form [17];

$$\begin{bmatrix} V_{q1} \\ V_{d1} \\ V_{q2} \\ V_{d2} \\ V_{qr} \\ V_{dr} \end{bmatrix} = \begin{bmatrix} r_1\ \omega_k L_1\ 0\ \omega_k L_2\ 0\ \omega_k L_m \\ -\omega_k L_1\ r_1\ -\omega_k L_2\ -\ 0_2\ -w_k L_m\ 0 \\ 0\ \omega_k L_3\ r_2\ \omega_k L_2\ 0\ \omega_k L_m \\ -\omega_k L_3\ 0\ -\omega_k L_2\ r_2\ -\omega_k L_m\ 0 \\ 0\ \alpha L_m\ 0\ \alpha L_m\ r_r\ \alpha L_r \\ -\alpha L_m\ 0\ -\alpha L_m\ 0\ -\alpha L_r\ r_r \end{bmatrix} \begin{bmatrix} i_{q1} \\ i_{d1} \\ i_{q2} \\ i_{d2} \\ i_{qr} \\ i_{dr} \end{bmatrix} + \begin{bmatrix} Pi_{q1} \\ Pi_{d1} \\ Pi_{q2} \\ Pi_{d2} \\ Pi_{qr} \\ Pi_{dr} \end{bmatrix}$$

$$\begin{bmatrix} L_1\ 0\ L_2\ 0\ L_m\ 0 \\ 0\ L_1\ 0\ L_2\ 0\ L_m \\ L_3\ 0\ L_2\ 0\ L_m\ 0 \\ 0\ L_3\ 0\ L_2\ 0\ L_m \\ L_m\ 0\ L_m\ 0\ L_r\ 0 \\ 0\ L_m\ 0\ L_m\ 0\ L_r \end{bmatrix} \begin{bmatrix} Pi_{q1} \\ Pi_{d1} \\ Pi_{q2} \\ Pi_{d2} \\ Pi_{qr} \\ Pi_{dr} \end{bmatrix} \tag{32}$$

$$\text{idot} = [L][V] - [L]^{-1}\ [G]\ [1] \tag{33}$$

where,

$$[V] = \left[V_{q1}\, V_{d1}\, V_{q2}\, V_{d2}\, V_{cr}\, V_{dr}\right] \tag{34}$$

$$[I] = \left[i_{q1}\, i_{d1}\, i_{q2}\, i_{d2}\, i_{qr}\, i_{dr}\right] \tag{35}$$

$$[G] = \begin{bmatrix} r_1 & \omega_k L_1 & 0 & \omega_k L_2 & 0 & \omega_k L_m \\ -\omega_k L_1 & r_1 & \omega_k L_2 & 0 & -\omega_k L_m & 0 \\ 0 & \omega_k L_3 & r_2 & \omega_k L_2 & 0 & \omega_k L_m \\ -\omega_k L_3 & 0 & -\omega_k L_2 & r_2 & -\omega_k L_m & 0 \\ 0 & \alpha L_m & 0 & \alpha L_m & r_r & \alpha L_m \\ \alpha L_m & 0 & -\alpha L_m & 0 & -\alpha L_m & r_r \end{bmatrix} \tag{36}$$

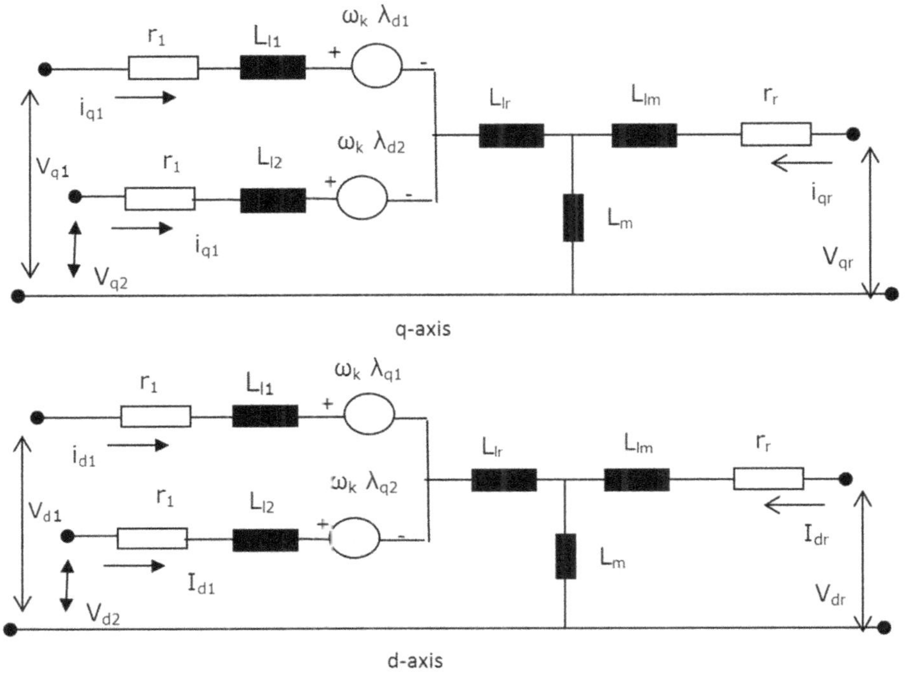

Fig. 2. *DQ equivalent circuit of a six-phase Induction Machine.*

Where

$$\alpha = \omega_r - \omega_k \quad (37)$$

$$[L] = \begin{bmatrix} L_1\ 0\ L_2\ 0\ L_m\ 0 \\ 0\ L_1\ 0\ L_2\ 0\ L_m \\ L_3\ 0\ L_2\ 0\ L_m\ 0 \\ 0\ L_3\ 0\ L_2\ 0\ L_m \\ L_m\ 0\ L_m\ 0\ L_r\ 0 \\ 0\ L_m\ 0\ L_m\ 0\ L_r \end{bmatrix} \quad (38)$$

2.2. Mechanical Model

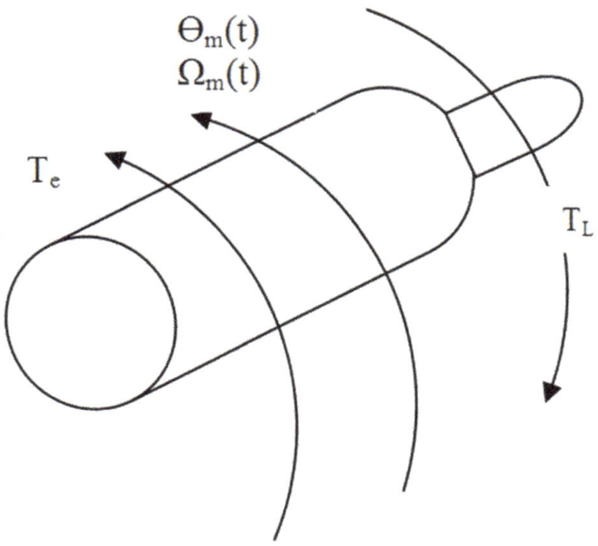

Fig. 3. *Induction Motor Mechanical Model.*

The mechanical model of the six-phase induction machine is the equation of motion of the machine and driven load as in fig. 3, the figure suggest equation (39),

$$J_m\, p^2\, \theta_m = T_e - F\omega_r - T_L \quad (39)$$

The mechanical data of the experimental machine indicates that the combined rotor and load viscous friction 'F' is appropriately zero, so that, equation (38) becomes.

$$J_m p^2 \theta_m = T_e - T_L \quad (40)$$

Breaking equation (40) into two first-order differential equation gives

$$J_m p(\omega_m) = (T_e - T_L) \quad (41)$$

Because

$$p\theta_m = \omega_m \quad (42)$$

We know that

$$\omega_r = \omega_m\, p \quad (43)$$

And

$$\theta_r = \theta_m p \quad (44)$$

Where $P = \frac{d}{dt}$, and ω_m, θ_m, θ_r, ω_r, J_m and T_L represent angular velocity of the rotor, rotor angular position, electrical rotor angular position, electrical angular velocity, combined rotor and load inertia coefficient, and applied load torque respectively. Matlab m-files are developed to simulate the transient performance of a six-phase, 4 pole, 50Hz squirrel cage induction machine.

3. Experiment

In order to get data for simulation, a 5.5 Hp 3Ø, 24 slot, 4 pole induction motor was reconfigured into a split phase (6-phase) motor with two set of 3-phase displaced 30^0 elect from each other. The new motor maintains all specifications of the old, except that, the motor is now a split winding motor (6-phase). The construction motor tested on No-load and On-load. Retardation test was also carried out. The experimental results and the computed results is shown in table 1.

Fig. 4. *Stator winding of the sample motor.*

Table 1. *Simulation Parameters.*

Parameter	Value
Power Rating	5HP
Phase Number	6
No of Poles	4
Efficiency	85%
Power Factor	0.8 lag
Mechanical Speed	1400rev/min
Frequency (f)	50Hz
Stator Resistance (R_s)	0.28Ω
Rotor Resistance (R_r)	2.14 Ω
Stator winding reactance (X_s)	1.24989 Ω
Rotor winding reactance (X_r)	1.24989 Ω
Magnetizing reactance (X_m)	35.718 Ω
Rotor inertia(J_m)	0.25Kgm^2
Load inertia (T_L)	0.52Kgm^2

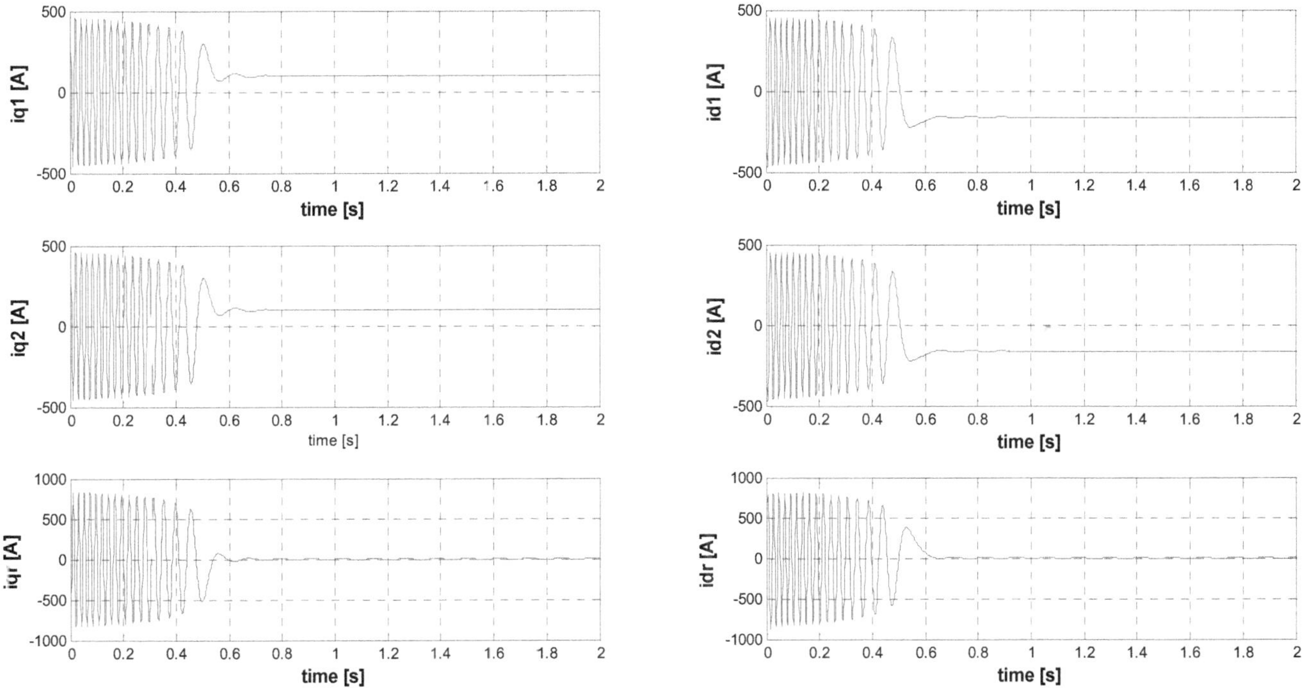

Fig. 5. *A graph of dq currents against time.*

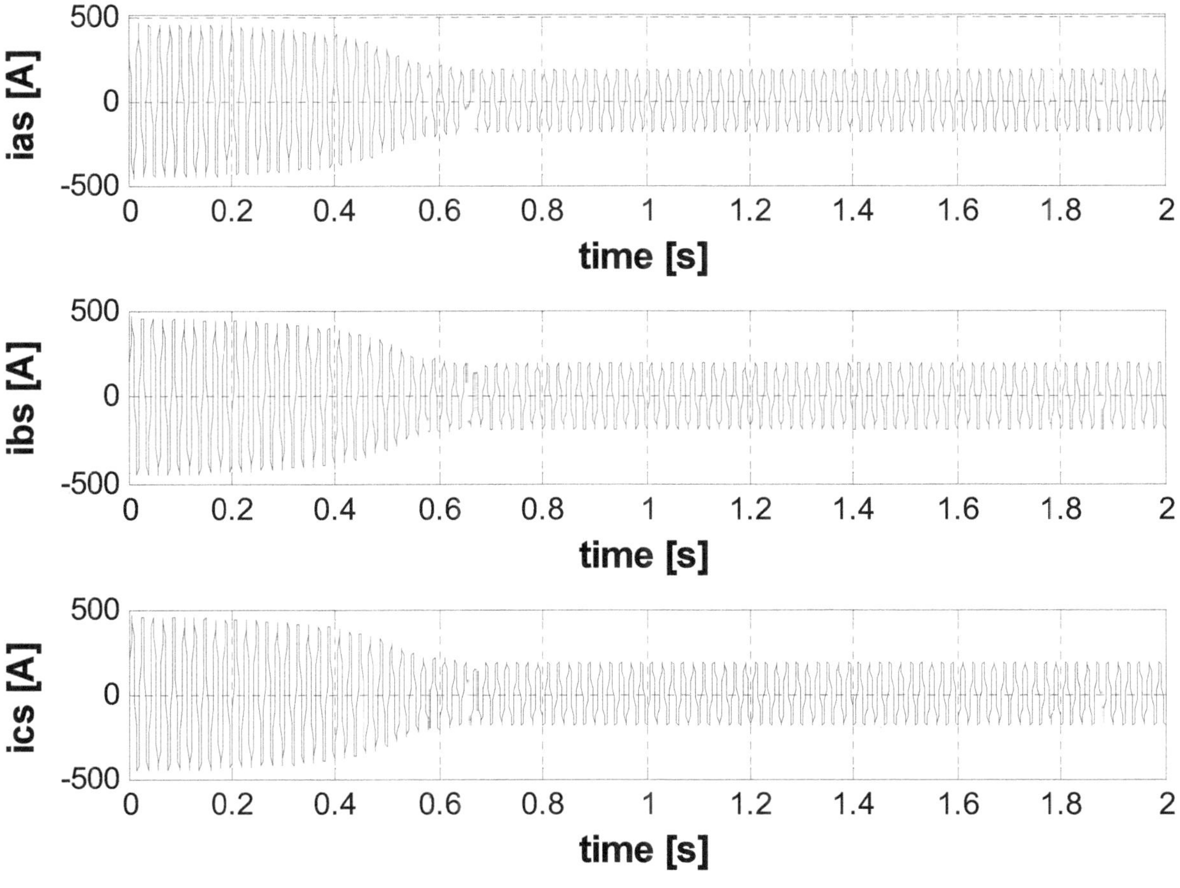

Fig. 6. *A graph of xyz phase currents against time.*

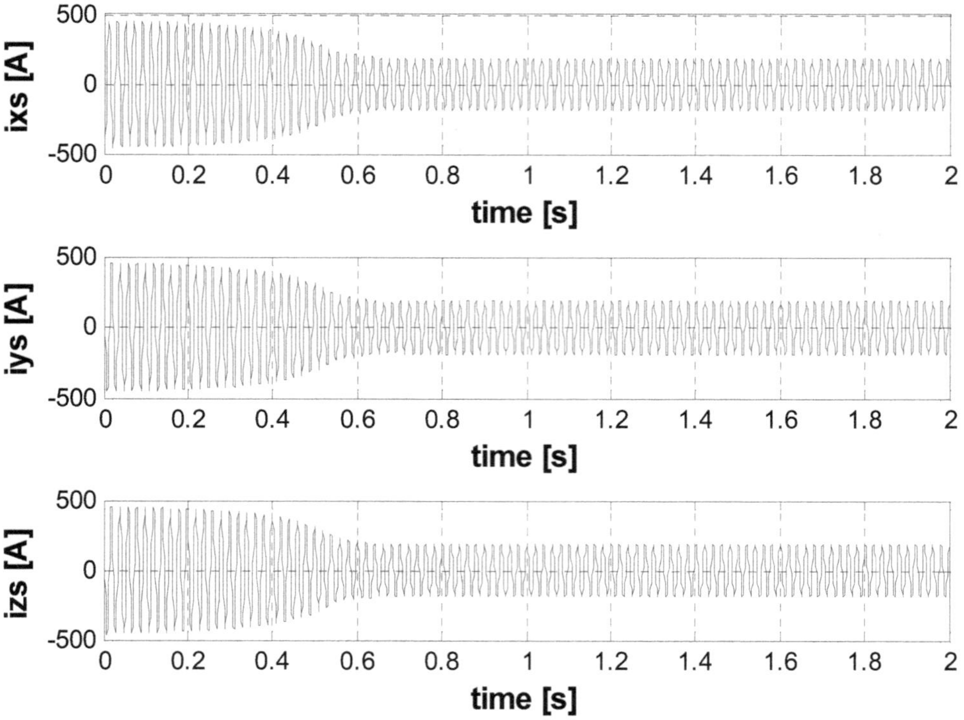

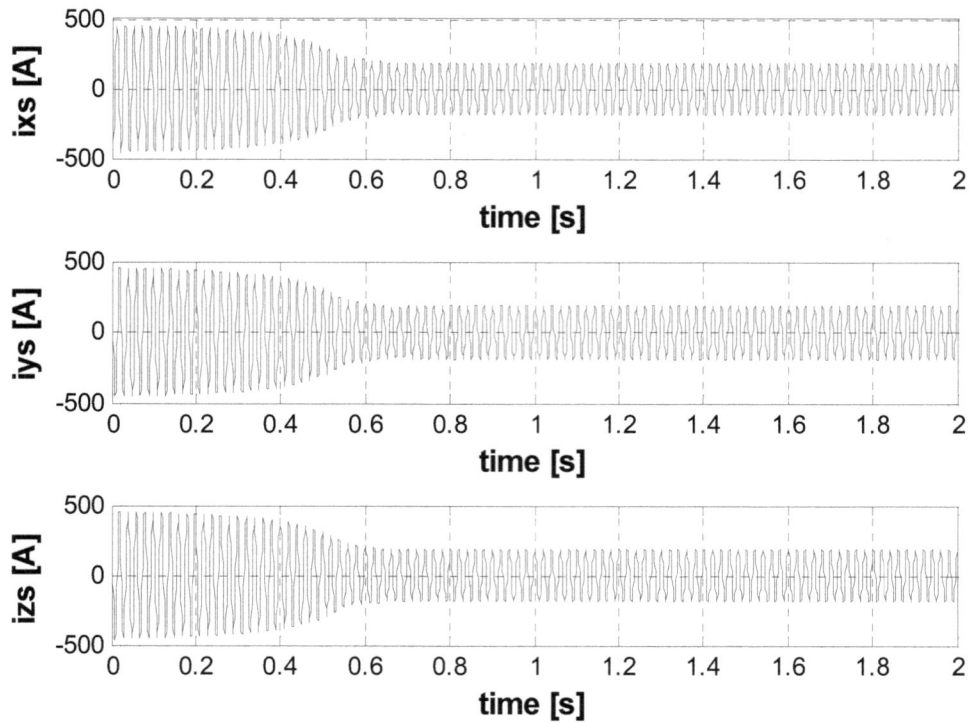

Fig. 7. A graph of xyz phase currents against time.

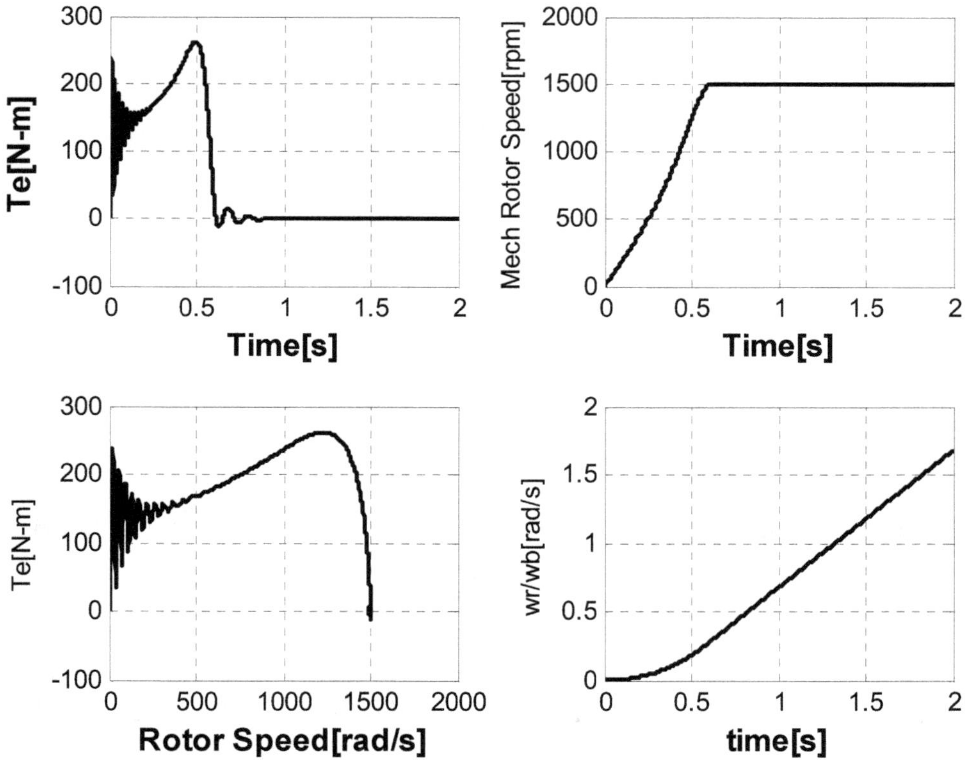

Fig. 8. Mechanical Model plots.

4. Simulation Results

The simulation results in the form of computer traces are presented of the performance of a six-phase induction machine is presented in figs 5-8. The simulation for phase currents and qd currents, mechanical rotor speed and Electromagnetic Torque are simulated and presented using the data in table1.

5. Discusion of Results

Electromagnetic Torque in fig. 8, stabilizes at 0.7s. The mechanical rotor speed reaches synchronous speed at 0.6s, this agrees favorably with the theoretical concept. The phase currents also stabilize at 0.6s. Looking at the simulations, the redesigned motor can work as a three phase machine or/and as six phase. In submarines, the idea here is to use a six phase motor to replace two three phase motors, and money is saved redesigning the machine with higher reliability. The loss of a phase does not stop the motor from running. In submarines mostly Naval War ships, reliability is a criterion in designing a submarines.

6. Conclusion

The simulation result shows that, the split six phase is a dual stator induction machine with sets of three phase currents; I_{abc} and I_{xyz}. This is expected because DSIM is like paralleling two three phase induction motors. So, instead of using two three phase induction motors for propulsion (submarines), a single DSIM can replace the two and the cost is reduced and reliability increased which is the actual aim of this work.

References

[1] P.S. Bimbhra, "Generalized theory of electrical machines", Khanna publishers India fourth ed pp. 179 – 546. 1987

[2] P.C. Krause, "Analysis of electric machinery", MC. Graw Hill book Inc USA 19S86.

[3] P. C. Krause, F. Nozari, T. L. Skvarenina and D.W. Olive. "The theory of neglecting stator transients", IEEE Trans Power Apparatus and systems vol. 98 pp 141-148, Jan/Feb 1979.

[4] P. L. Alger, et al., "Double windings for turbine alternators," AIEE Transactions, vol. 49, January, 1930, pp.226-244.

[5] Hadiouche D., Razik H., Rezzoug A., "Modelling of a double-star induction motor with an arbitrary shift angle between its three phase winding". 9th international conference on EPE, PEMC 2000 Kosice, Slovak Republic.

[6] K. Gopakumar, et al., "Split-phase induction motor operation from pwm voltage source inverter," IEEE Transactions on Industry Applications, vol. 29, no. 5, Sep./Oct. 1993, pp. 927-932.

[7] Fuchs E. F. and Rosenberg L. T., "Analysis of an Alternator with Two Displaced Stator Windings," IEEE Trans. Power Apparatus and Systems, 1974, vol. 93, pp. 1776-1786.

[8] T.A. Lipo, "A d-q model for six phase induction machines," Proc. Int. Conf. on Electrical Machines ICEM, Athens, Greece, pp. 860-867, 1980.

[9] R.H. Nelson, P.C. Krause, Induction machine analysis for arbitrary displacement between multiple winding sets, IEEE Trans. Power Apparatus Syst. PAS-93 (1974) 841–848.

[10] L.J. Hunt, "A new type of induction motor," J. Inst Elect. Eng., vol.39, pp. 648–677, 1907.

[11] Singh G. K., "Multi-Phase Induction Machine Drive Research – A Survey," Electric Power Systems Research, 2002, vol. 61, pp. 139-147.

[12] B. Kundrotas, S. Lisauskas, R. Rinkevičienė. Daugiafazio asinchroninio variklio modelis // Elektronika ir elektrotechnika. – Kaunas: Technologija, 2011. – Nr. 5(111). – P. 111–114.

[13] Bugenis S. J., Vanagas J., Gečys S. Optimal phase number of induction motor with the integrated frequency converter // Electronics and Electrical Engineering. – Kaunas: Technologija, 2008. – No. 8(88). – P.67–70.

[14] E.E. Ward, H. H¨arer, "Preliminary investigation of an inverter-fed 5-phase induction motor", Proc. IEE 116 (6) (1969) 980–984.

[15] M.A. Abbas, R. Christen, T.M. Jahns, Six-phase voltage source inverter driven induction motor, IEEE Trans. Ind. Appl. IA-20 (5) (1984) 1251–1259.

[16] O.I. Okoro and T.C. Nwodo; "Simulation Tools for electrical machine modeling", Teaching and Research, International conference on power system operation and planning – v1 (ICPSOP), pp. 120 – 124. 2005.

[17] O.I. Okoro and T.C. Nwodo; "Simulation Tools for electrical machine modeling", Teaching and Research, International conference on power system operation and planning – v1 (ICPSOP), pp. 120 – 124. 2005.

[18] E.J. Akpama and O.I.Okoro, "Simulating Asynchronous Machine with Saturation effect", Proceedings of ESPTAEE 2008 National Conference, University of Nigeria Nsukka, pp130-135, June 2008.

[19] A.N. Golubev and S.V. Ignatenko, "Influence of number of stator-winding phases on the noise characteristics of an asynchronous motor," Russian Electrical Engineering, vol. 71, no. 6, pp. 41-46, 2000.

[20] C. Hodge, S. Williamson and S. Smith, "Direct drive marine propulsion motors," Proc. Int. Conf. on Electrical Machines ICEM, Bruges, Belgium, CD-ROM paper 087, 2002.

[21] M. Molaei, H. Oraee, and M. Fotuhi-Firuzabad, "Markov Model of Drive-Motor Systems for Reliability Calculation," in Proc. IEEE Int. Symp. on Ind. Electron., 2006, pp. 2286-2291.

[22] Jahns T. M., "Improved Reliability in Solid-State AC Drives by Means of Multiple Independent Phase-Drive Units," IEEE Trans. Ind. Application, May/June 1980, vol. IA-16, pp.321-331.

3

An Experimental Approach to Determine the Effect of Different Orientation of Dimples on Flat Plates

Amjad Khan[1], Mohammed Zakir Bellary[1], Mohammad Ziaullah[1], Abdul Razak Kaladgi[2, *]

[1]Department of Electronics and communication Engineering, P.A College of Engineering, Mangalore, Karnataka, India
[2]Department of Mechanical Engineering, P.A College of Engineering, Mangalore, Karnataka, India

Email address
amjad_ece@pace.edu.in (A. Khan), mdzakir87@gmail.com (M. Z. Bellary), mdziya1990@gmail.com (M. Ziaullah), abdulkaladgi@gmail.com (A. R. Kaladgi)

Abstract: Dimples play a very important role in the heat transfer enhancement of electronic cooling systems. In the current paper, the flow and heat transfer characteristics of spherical dimples of non uniform diameter were investigated. The experiment was carried out under laminar forced convection conditions using air as a working fluid. The overall Nusselt number and heat transfer coefficient at different dimple structures were obtained for various inlet air flow rates. From the obtained results, it was observed that the heat transfer coefficient and Nusselt number were high for the plate in which the diameter of dimples increases centrally in the direction of flow as compared to the other cases.

Keywords: Forced Convection, Electronic Cooling, Dimples, Passive Techniques

1. Introduction

The development of integrated electronic devices with increased level of miniaturization, higher performance and output has increased the cooling requirement of chips considerably. As the chip temperature increases, the stability and efficiency Issues will increase and the problem of heat dissipation will become a bottleneck for the development of chips in the electronic industry [1].Passive heat transfer enhancement techniques are used in electronic cooling devices. In these techniques, passive augmented heat transfer devices such as rib-tabulators, concavities (dimples), extended surfaces or fins, and protrusions are used. Among these, the dimples (concavities) can be considered important because they not only enhance/augment heat transfer rate but also produce minimum pressure drop penalties which is important for pumping power requirements [1]. The dimple usually produces vortex pairs, causes flow separation, creates reattachment zones and hence enhances the heat transfer rate. And as they do not protrude into the flow so they contribute less to the foam drag, to produce minimum pressure drop penalties [2]. Another advantage is that in dimple manufacture the removal of material takes place and reduces the cost and weight of the equipments.

Kuethe [3] can be considered as the first person to use dimples on flat surfaces. He observed that the dimples produces rapid or turbulent mixing in the flow, acting as vortex generator & increases the heat transfer rate. Afanasyev et al [4] conducted an experimental where he used spherical dimples on flat plates. He observed an increase of 30-40% in the heat transfer rate along with minimum pressure drop. In another study, Chyu et al [5] used tear drop type dimples along with hemispherical dimples to study the heat transfer distribution in the channel and observed a considerable increment in the heat transfer rate for the surfaces having dimples. Mahmood et al [6] conducted an experiment to investigate the effect of dimples on heat transfer using the flow visualization techniques and concluded that the periodic nature of shedding off of vortices is the main cause of enhancement of heat transfer (much more pronounced at the downstream rims of the dimples).Mahmood et al [7] studied the effect of Reynolds number, aspect ratio, temperature ratio & flow structure in a channel having dimples at the bottom. And observed through the flow visualization techniques that the secondary vortices that are shed off from the dimples become stronger as the non-dimensional channel height to dimple diameter (H/D) ratio decreases and increases the local Nusselt number in these regions. Xie et al [8] carried out a numerical investigation to study the effect of heat transfer enhancing devices (pin-fins protrusions & dimples) on turbine blade tip wall. And concluded that though the dimples have a simple geometry but they are best suited for cooling of blade tip especially at low Reynolds numbers

From the literature, it is very much clear that dimples (vortex generators) have high potential to enhance the heat transfer along with the production of lower pressure drop penalties. The other advantages include low weight, cost and fouling rates [9]; however, most of the researchers conducted numerical or experimental work on spherical dimples of uniform diameter [5, 10]. Also most of the research is confined to flow in the channel or Internal flow, with a very few studies on external flow [10].So the main aim of this project is to experimentally study the effect of spherical dimples (non-uniform diameter) on aluminum plates under external laminar forced flow conditions .

2. Experimental Setup

The fabricated experimental setup used in this study is as shown in the figure below

Figure 1. *Experimental setup.*

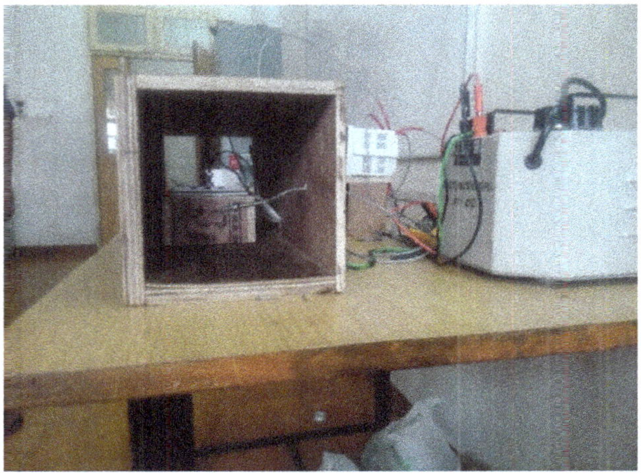

Figure 2. *Experimental setup(side view).*

The main components of the experimental setup used are aluminum test plates of dimensions 100x100x2 mm, a calibrated orifice flow meter, Strip plate heater (100 watts capacity), Dimmer stat, Digital temperature indicator, voltmeter, and ammeter with J type thermocouple and a centrifugal blower.

Table 1. *Components and Specifications.*

Components	Specification
Test plate	10x10x2 cm aluminium plates
Blower	110W, 0.4BHP, 280rpm
Heater	100W, 4"x4"
Dimmer stat	6A,230V
Digital Temperature Indicator	6 channel,12000C, 230V
Orifice plate	12mm dia.
Manometer	"U-tube" glass manometer
Casing	A wooden casing of size of 8"x8" and 2feet long.
Thermocouple	K-Type, 3000C, 1m long.
Digital Multi-meter	Voltmeter, Ammeter

3. Results and Discussion

Experiments were conducted on aluminum test plates with spherical dimples of non uniform diameters made on flat plates. The dimples were arranged in a staggered fashion with different arrangements like

Case a. Centrally increasing the diameter of dimples in the direction of flow & maintaining the left & right column with constant diameter dimples.

Case b.Gradual Increase in the diameter of dimples in the direction flow.

Case c. Gradual Decrease in the diameter of dimples in the direction flow (reverse case).

The data obtained were used to find important heat transfer parameters like Nusselt number, heat transfer coefficient, and heat transfer rate. And the experimental findings have been plotted in the form of graphs, mainly

- Nusselt number(Nu) vs Reynolds number(Re)
- Heat transfer coefficient(h) vs Reynolds number(Re)
- Heat transfer rate Q vs Reynolds number(Re)

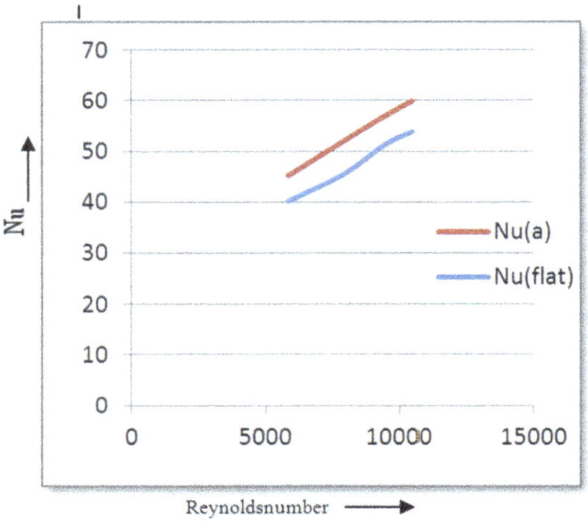

Figure 3. *Variation of Nusselt number with Reynolds number(case.a).*

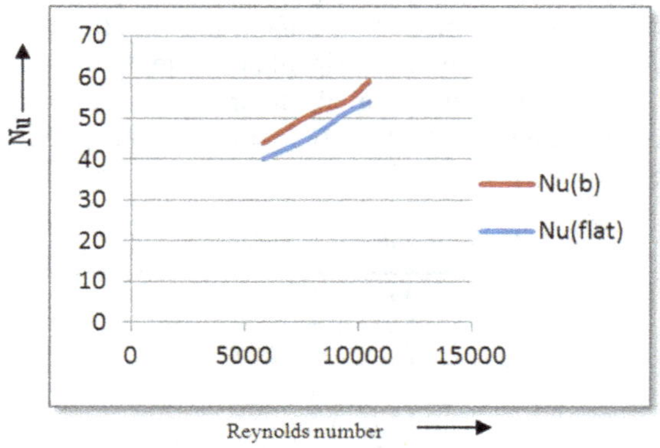

Figure 4. *Variation of Nusselt number with Reynolds number(case.b.).*

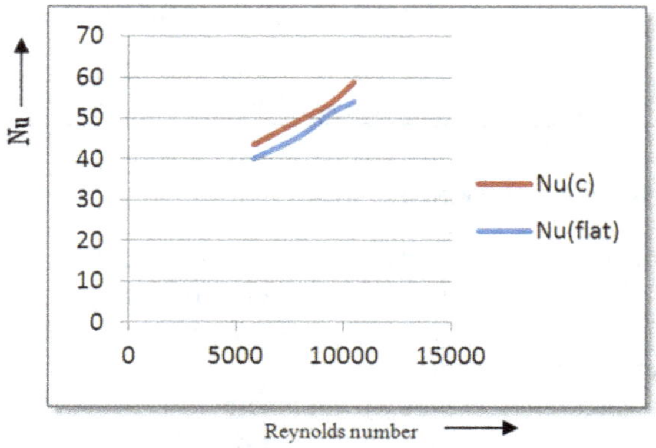

Figure 5. *Variation of Nusselt number with Reynolds number(case.c.).*

Figure 3, 4, 5 shows variation of Nusselt number 'Nu' with Reynolds number 'Re' for the cases considered. It can be seen that as expected the Nusselt number increases as Reynolds number increases. This is due to direct flow impingement on the downstream boundary of the plate and strengthened flow mixing by vortices at the downstream of the plate [1, 11]. The formation of vortex pairs, periodically shedding off from the dimples, a large up wash regions with some fluids coming out from the central regions of the dimples are the other reasons of enhancement of Nusselt number & is more pronounced near the downstream rims of the dimples [6].It can also be seen that the variation in the Nusselt number is gradual with Reynolds number as expected [12, 13, 16].

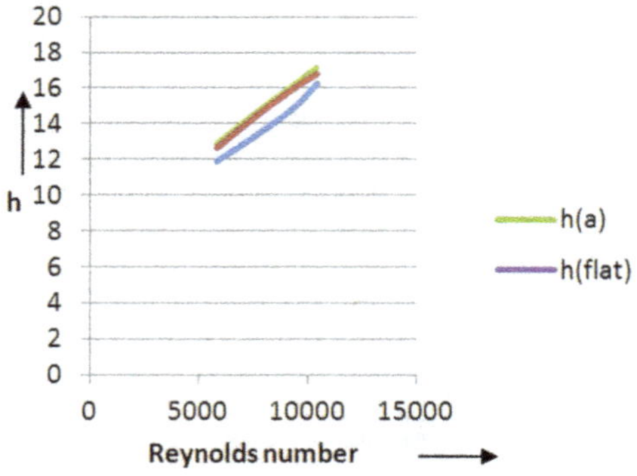

Figure 6. *Variation of Heat transfer coefficient with Reynolds number(case.a.).*

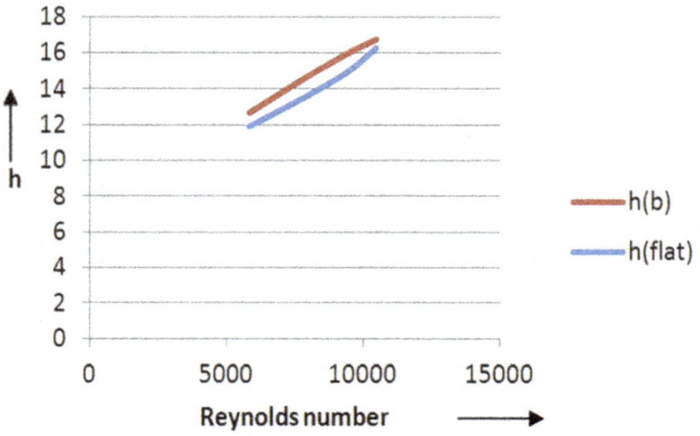

Figure 7. *Variation of Heat transfer coefficient with Reynolds number(case.b.).*

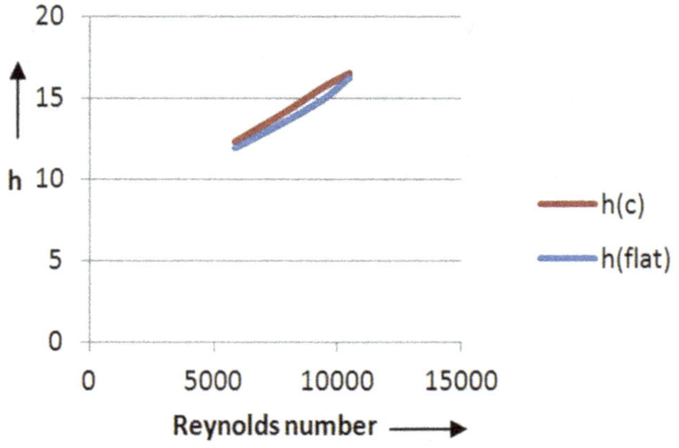

Figure 8. *Variation of Heat transfer coefficient with Reynolds number(case.c.).*

Figure 6, 7, 8 shows the variation of heat transfer coefficient 'h' with Reynolds number 'Re' for the various cases considered. It is obvious that 'h' increases with 'Re'as expected because the development of the thermal boundary layer is delayed or disrupted & hence enhances the local heat transfer in the reattachment region and wake region and increases the heat transfer coefficient [1].

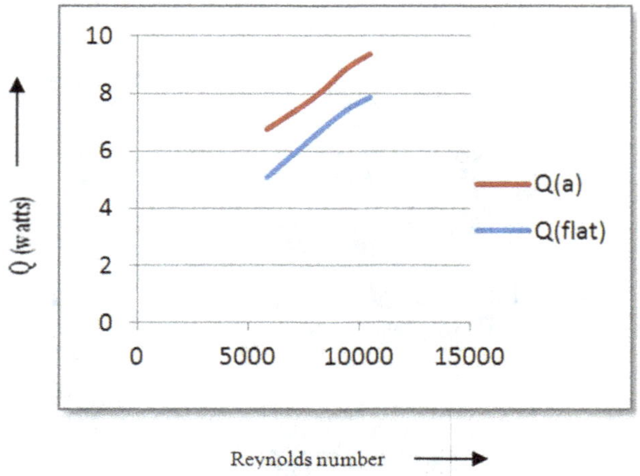

Figure 9. *Variation of Heat transfer rate with Reynolds number(case.a.).*

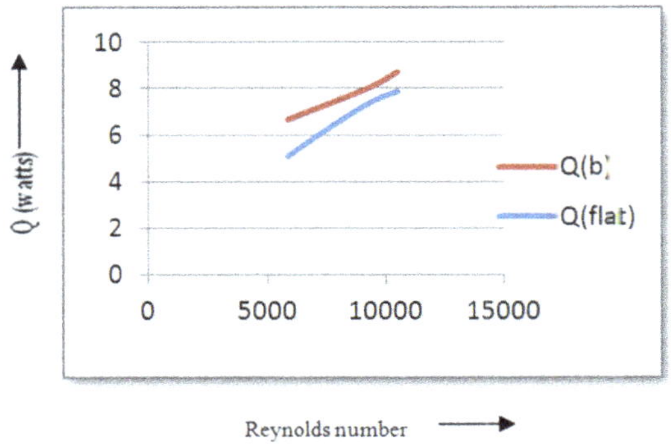

Figure 10. *Variation of Heat transfer rate with Reynolds number(case.b.).*

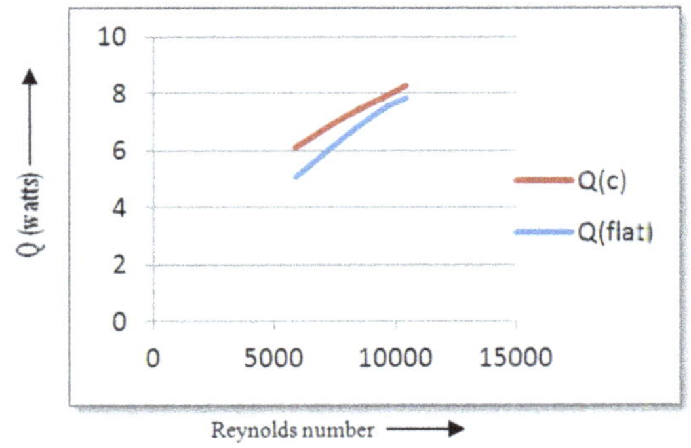

Figure 11. *Variation of Heat transfer rate with Reynolds number(case.c.).*

Figure 9, 10, 11 shows variation of Heat transfer rate 'Q' with Reynolds number 'Re' for the various cases considered. It can be seen that again 'Q' increases as 'Re' increases in all the three cases. Because the near-wall turbulent mixing intensity downstream the dimple increases due to the vortex flow shedding from the dimple [6, 15] and hence increases the heat transfer rate. It is also seen that 'Q' is very much higher for case 'a' (dimples diameter decreasing centrally) because of increase in the level of turbulence downstream the dimples. So it can be concluded that case 'a' helps in better enhancing the heat transfer compared to other cases.

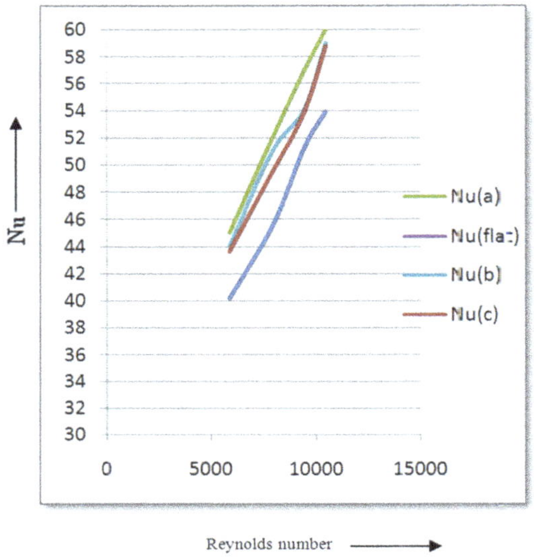

Figure 12. *Variation of Nusselt number with Reynolds number.*

Figure 12 shows the comparison of Nusselt number 'Nu' with Reynolds number 'Re' for the all the three cases considered. It can be seen that 'Nu' increases as 'Re' increases in all the three cases. It can also be seen that the variation of Nusselt number 'Nu' for the last two cases is very less especially for high Reynolds number flows. Also the 'Nu' is low for the last two cases as compared to the first case may be due the fact that the dimple diameter is not increased or decreased centrally where the pronounce effect of heat transfer will occur. Also this configuration produces strongest turbulent mixing near the wall and hence lowest end wall temperatures [14].

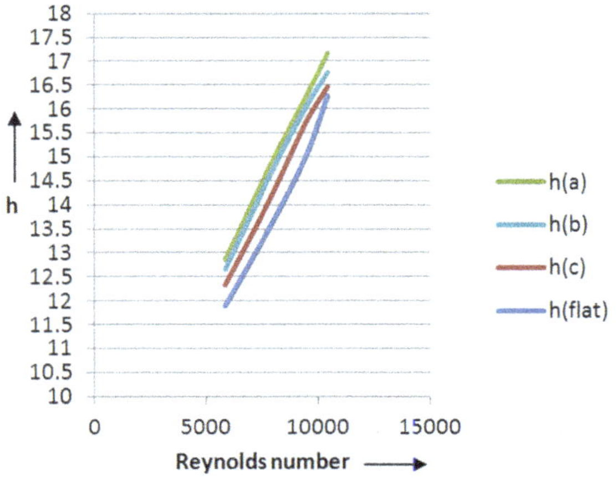

Figure 13. *Variation of Heat transfer coefficient with Reynolds number.*

Figure 13 shows the comparison of the variation of heat transfer coefficient 'h' with Reynolds number 'Re' for the various cases considered. It is obvious that 'h' increases with 'Re' as expected and it is also observed that heat transfer coefficient is high for the case'a' (case of dimple diameter increasing centrally) due to higher turbulent mixing occurring at the central region where the fluid flow rate is highest.

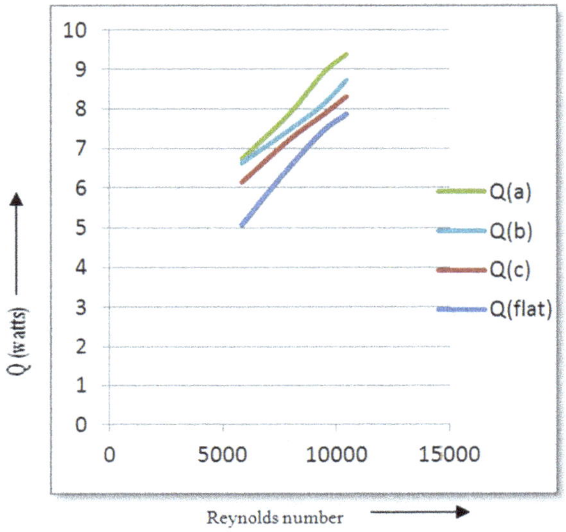

Figure 14. *Variation of Heat transfer rate with Reynolds number.*

Figure 14 shows the variation of Heat transfer rate 'Q' with Reynolds number 'Re' for the various cases considered. It can be seen that again 'Q' increases as 'Re' increases in all the three cases as expected because the vortex flow shedding from the dimples significantly increases the turbulence level in the flow near the downstream wall, and hence increases the convective heat transfer, especially near the rear rim of the dimple. And therefore the end wall temperature of dimple surfaces decreases [14].

4. Conclusion

In this experimental work an investigation of the effect of air flow over a flat plate with different diameter dimples is carried out. The main conclusions of the work were:

- Nusselt number increases with Reynolds number for all the three cases of dimple arrangement considered due to direct flow impingement on the downstream boundary and strengthened flow mixing by the vortices at the downstream.
- Case 'a' dimple arrangement has highest Nusselt number because it produces strongest turbulent mixing near the wall and hence produces lowest end wall temperatures. Case'b'&'c' dimple arrangement gives nearly the same value of Nusselt number at high Reynolds numbers.
- Heat transfer coefficient increases with Reynolds number for all the three cases of dimples arrangement considered due to the disruption of the thermal boundary layer development & hence enhance the local heat transfer in the reattachment and wake regions.
- Case 'a' dimple arrangement gives slightly higher value of heat transfer coefficient as compared to case 'b' & 'c' dimples.
- Case 'a' dimple arrangement has better heat transfer enhancing capacity as compared to other cases because the level of turbulent mixing is highest in case 'a'. However, the Augmentation depends on the configuration [10].

References

[1] Zhang, D., Zheng, L., Xie, G., and Xie, Y.,An Experimental Study on Heat Transfer enhancement of Non-Newtonian Fluid in a Rectangular Channel with Dimples/Protrusions, Transactions of the ASME, Vol. 136, pp.021005-10, 2014.

[2] Beves, C.C., Barber, T.J., and Leonardi, E., An Investigation of Flow over Two-Dimensional Circular Cavity. In 15th Australasian Fluid Mechanics Conference, the University of Sydney, Australia, pp.13-17, 2004.

[3] Kuethe A. M., Boundary Layer Control of Flow Separation and Heat Exchange. US Patent No. 1191, 1970.

[4] Afanasyev, V. N., Chudnovsky, Y. P., Leontiev, A. I., and Roganov, P. S., Turbulent flow friction and heat transfer characteristics for spherical cavities on a flat plate. Experimental Thermal Fluid Science, Vol. 7, Issue 1, pp. 1–8, 1993.

[5] Chyu, M.K., Yu, Y., Ding, H., Downs, J.P., and Soechting, F.O., Concavity enhanced heat transfer in an internal cooling passage. In Orlando international Gas Turbine & Aero engine Congress & Exhibition, Proceedings of the 1997(ASME paper 97-GT-437), 1997.

[6] Mahmood, G. I., Hill, M. L., Nelson, D. L., Ligrani, P. M., Moon, H. K and Glezer, B., Local heat transfer and flow structure on and above a dimpled surface in a channel. J Turbo mach, Vol.123, Issue 1, pp: 115–23, 2001.

[7] Mahmood, G. I., and Ligrani, P. M., Heat Transfer in a Dimpled Channel: Combined Influences of Aspect Ratio, Temperature Ratio, Reynolds Number, and Flow Structure. Int. J. Heat Mass Transfer, Vol.45, pp.2011–2020, 2002.

[8] Xie, G. N., Sunden, B., and Zhang, W. H., Comparisons of Pins/Dimples Protrusions Cooling Concepts for an Internal Blade Tip-Wall at High Reynolds Numbers. ASME J. Heat Transfer, Vol.133, Issue 6, pp. 0619021-0619029, 2011.

[9] Gadhave, G., and Kumar.P. Enhancement of forced Convection Heat Transfer over Dimple Surface-Review. International Multidisciplinary e - Journal .Vol-1, Issue-2, pp.51-57, 2012

[10] Katkhaw, N., Vorayos, N., Kiatsiriroat, T., Khunatorn, Y., Bunturat, D., and Nuntaphan., A. Heat transfer behavior of flat plate having 450 ellipsoidal dimpled surfaces. Case Studies in Thermal Engineering, vol.2, pp. 67–74, 2014

[11] Patel, I.H., and Borse, S.H. Experimental investigation of heat transfer enhancement over the dimpled surface. International Journal of Engineering Science and Technology, Vol.4, Issue 6, pp.3666–3672, 2012.

[12] Faheem Akhtar, Abdul Razak R Kaladgi and Mohammed Samee, Heat transfer augmentation using dimples in forced convection -An experimental approach.Int. J. Mech. Eng. & Rob. Res. Vol 4, Issue 1, pp 150-153, 2015.

[13] Faheem Akhtar, Abdul Razak R Kaladgi and Mohammed Samee, Heat transfer enhancement using dimple surfaces under natural convection—An experimental study, Int. J. Mech. Eng. & Rob. Res. Vol 4, Issue 1, pp 173-175, 2015.

[14] Yu Rao ,Yamin Xu,Chaoyi Wan,A Numerical Study of the Flow and Heat Transfer in the Pin Fin-Dimple Channels With Various Dimple Depths, Journal of Heat Transfer, Transactions of the ASME, Vol. 134, pp-071902-1-9, 2012.

[15] P.M. Ligrani, J.L. Harrison, G.I. Mahmood, M.L. Hill, Flow structure due to dimple depression on a channel surface, Phys. Fluids 13 (2001) 3442–3451.

[16] Hasibur Rahman Sardar and Abdul Razak Kaladgi Forced Convection Heat Transfer Analysis through Dimpled Surfaces with Different Arrangements, American Journal of Energy Engineering, Vol 3, Issue 3, pp. 37-45, 2015.

4

Smart Generator Monitoring System in Industry Using Microcontroller

S. Boopathi[1], M. Jagadeeshraja[2], L. Manivannan[2], M. Dhanasu[3]

[1]Embedded System Technologies, Knowledge Institute of Technology, Tamilnadu, India
[2]Department of Electrical and Electronics Engineering, Knowledge Institute of Technology, Tamilnadu, India
[3]Steel Authority of India, Tamilnadu, India

Email address:
bootengineer@gmail.com (S. Boopathi)

Abstract: The electrical power systems are highly non-linear, extremely huge and complex networks. On the other hand, all the developed and countries have not sufficient supply of power. My Project focuses the detection of power failure and takes reflex action to solve the problem with help of GSM communication. The power failure will be detect by relay, and it communicates to Microcontroller to alerts the authorized person. In addition to that, parameters of Generator like Fuel level, Oil level, Temperature, battery status, etc., are monitored and communicated to authorized person. The acquired parameters are processed and recorded in the system memory. If there is any abnormality in their process, according to some predefined instruction and policies that are stored on the embedded system EEPROM then GSM alerts to concerned person immediately.

Keywords: Microcontroller, GSM Modem (SIM 300), Fuel Level (PH606), Oil Level (R Series), Temperature Sensors (LM35)

1. Introduction

The use of Generators has become a very common in almost every passive infrastructure companies, Industries, hospitals, Townships etc. while using these Generators a number of challenges are faced by the user such as maintaining the Quality of grid power, asset protections, generator maintenance, capturing real time data, Remotely monitoring of the generator, fuel theft monitoring, Data collection Analysis issues, Human dependency etc.

The Generator Monitoring System (GMS) is designed specifically for emergency power generators to monitor engine operations and detect pre-alarms or failures. This insures you of increased generator availability and a rapid response to service problems. The GMS monitors the power generators placed at the remote areas and increases its Efficiency by monitoring the various parameters of generator, Reporting critical Problems minimizes downtime and maximizes availability by sending generator failure messages instantly to you for diagnosis and emergency service dispatch if required.

It works on GSM technology, GMS can monitor various parameters such as external power supply, the battery voltage, fuel level, etc.

This system provides ideal solution to the problems caused in situations when a wired connection between a remote appliance/device and the control unit might not be feasible. The project is aimed to analyzing and testing the use of mobile phones to remotely monitor an appliance control system through GSM based wireless communication.

2. Literature Survey

Amit sachen *et al* have discussed the user can send commands in the form of SMS messages to read the remote electrical parameters. This system also can automatically send the real time electrical parameters periodically (based on time settings) in the form of SMS. This system can be designed to send SMS alerts whenever the Circuit Breaker trips or whenever the Voltage or Current exceeds the predefined limits. This project makes use of an onboard computer which is commonly termed as microcontroller [1].

Mallikarjun *et al* proposed this system is a specially designed computer system that is completely encapsulated by the device it controls. The embedded system has specific requirements and performs pre-defined tasks. The diesel generator is used when electricity is not readily available, or when power failures occur due to natural disasters such as typhoons or floods, or during other unexpected crises. Generally, the diesel generator operates in analog. The analog type controller cannot be processed precisely due to the distortions and noises coming from the data. In order to increase data accuracy, the controller needs to be digitalized [2]

Vimalraj *et al* have described a distribution transformers have a long service life if they are operated under good and rated conditions. However, their life is significantly reduced if they are overloaded, resulting in unexpected failures and loss of supply to a large number of customers thus effecting system reliability. This system provides flexible control of load parameters accurately and also provides effective means for rectification of faults if any abnormality occurs in power lines using SMS through GSM network [3].

Andriy Palamar *et al* proposed the system the Cellular phone containing SIM (Subscriber's Identifying Module) card has a specific number through which communication takes place. The mode of communication is wireless and mechanism works on the GSM (Global System for Mobile communication) technology. Here, the communication is made bi- directional where the user transmits and also receives instructions to and from the system in the form of SMS [4].

Kwang Seon Ahn *et al* have discussed the Using remote management; you can check operating hours, oil pressure, battery status, coolant temperatures, generated power output, fuel level, GPS position and more. A notification also could be generated whenever a critical level has been reached, such as when a generator has been running more than expected, or when the running hours exceed the service interval [5].

Henrik Arleving proposed system by using a cloud-based remote management solution with a communication gateway can help reduce costs, avoid fuel theft and improve power generator control. It can be difficult to focus on the right actions, simply because there isn't enough information on fuel levels, oil pressure or battery status for each generator. With a cloud-based remote management solution, we can have immediate access to generator parameters via a regular web browser being able to analyse each generator remotely enables you to better understand their health and more efficiently schedule field service visits and Fuel theft can be a significant problem[6].

Chetan Patil *et al* have discussed the design of BTS safety and fault management system the measures are taken to rectify these problems. The method makes use of GSM modem which gives the instant message about the each activity happening in the site. The temperature sensors will sense the temperature of the room and if it rises above the threshold value the GSM module will send the message to the master mobile which is already set in the system [7].

Y Jaganmohan Reddy *et al* is discussed the model of combination of Photo Voltaic (PV) cell System, Wind turbine system, Fuel cell (FC), and Battery systems for power generation, and to improve power quality they proposing Motor-Generator model instead of using static converters, and an energy management and control unit using Programmable Logic Controller (PLC). The power transformer is regarded as the heart of any electrical transmission and distribution system [8].

A.P.Agalgaonkar *et al* have discussed the measurement and control of temperature, humidity and the other parameters at different places. The Data Acquisition is defined as the process of taking a real-world signal as input, such as a voltage or current any electrical input, into the computer, for processing, analysis, storage or other data manipulation or conditioning[9].

3. Block Diagram

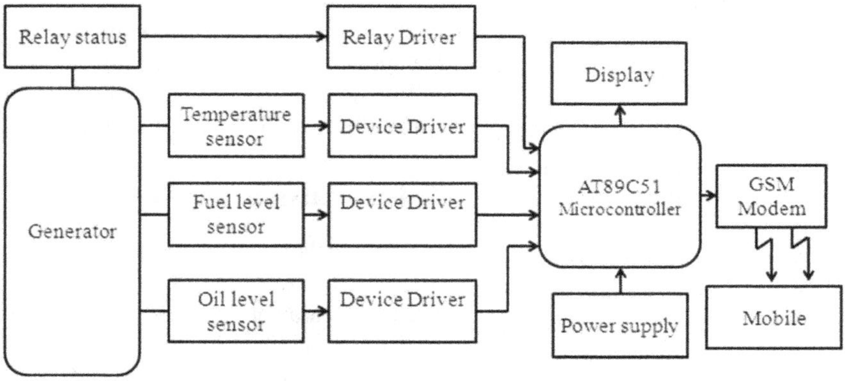

Fig. 1. *Block Diagram of Proposed System.*

4. Description of Proposed Method

The system has two parts, namely; hardware and software. The hardware architecture consists of a stand-alone embedded system that is based on Microcontroller a GSM handset with GSM Modem and a driver circuit. The GSM modem provides the communication by means of SMS messages. The SMS message consists of commands to be executed. The SMS message is sent to the GSM modem via the GSM public networks as a text message with a definite predefined format. Once the GSM modem receives negative signal from the EB supply, it sends the SMS to the user consisting of non-availability of power supply, fuel level, temperature of the coolant, etc.

The user can decide whether to switch the generator on/off and issue the command. Based on the message, the commands sent will be extracted and executed by the Microcontroller. In this case, if the EB power supply resumes, again the user is made to know the status of on-site..

4.1. Microcontroller

The AT89S51 is a low-power, high-performance CMOS 8-bit microcontroller with 8K bytes of in-system programmable Flash memory. The device is manufactured using Atmel's high-density nonvolatile memory technology and is compatible with the industry standard 80C51 instruction set and pin out. The on-chip Flash allows the program memory to be reprogrammed in-system or by a conventional nonvolatile memory programmer.

4.2. Sensor Used

Tab 1. Description of Sensor types

Sensor Types	Range	Description
1.Oil Level(R SERIES)	Detecting Range:10-2000mm	The oil mainly used in generator for purpose of using cooling of generator. When temperature of generator goes high, oil level in generator tank decreases due to heating effect.
2.Temperature(LM35D)	Detecting Range: -55° to +150℃	The temperature sensor which is used to monitoring the generator temperature and when the generator temperature exceeds predefined limits it is known through temperature sensor.
3 .Fuel level(PH606)	Detecting Range : 10 -2000mm	The fuel level sensor using to monitoring the fuel level of generator. The generator to maintain the level of fuel and an abnormal decrease in content could indicate fuel is being stolen.

4.3. Relay

Relay is an electrically operated switch. Many relays use an electromagnet to mechanically operate a switch, but other operating principles are also used, such as solid-state relays. Relays are used where it is necessary to control a circuit by a low-power signal (with complete electrical isolation between control and controlled circuits), or where several circuits must be controlled by one signal. The first relays were used in long distance telegraph circuits as amplifiers; they repeated the signal coming in from one circuit.

4.4. GSM Modem (SIM 300)

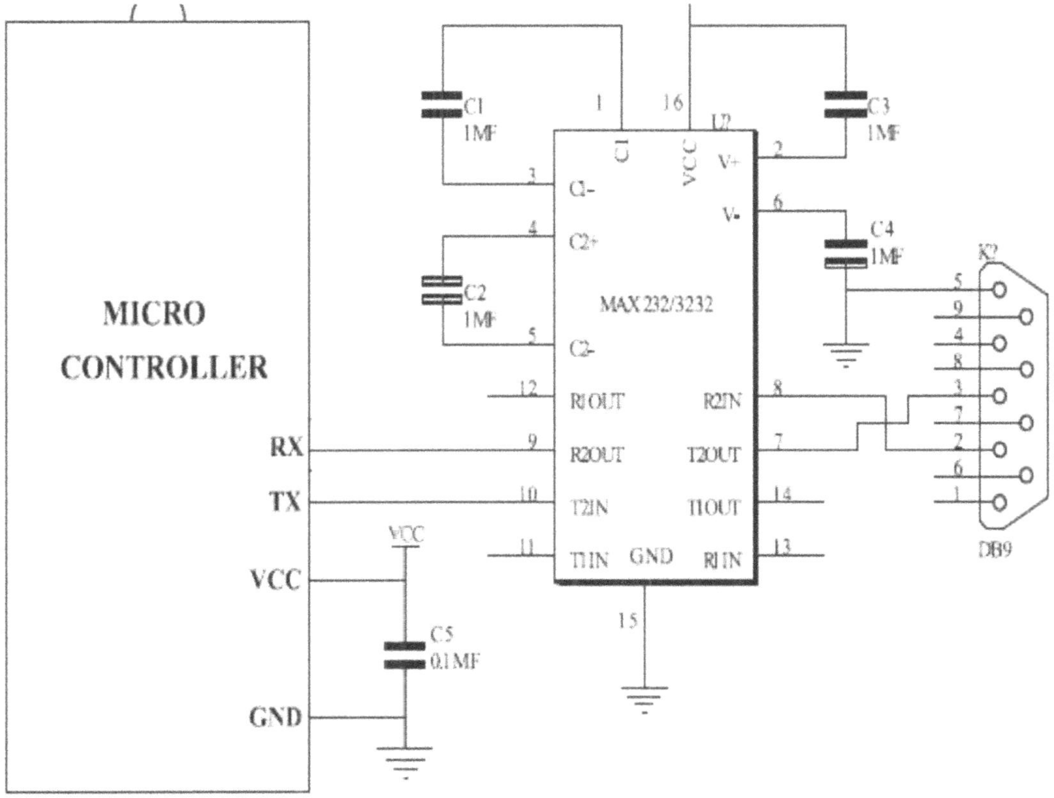

***Fig 2.** Connect MCU&MAX232 communicate with GSM Modem.*

This GSM Modem can accept any GSM network operator SIM card and act just like a mobile phone with its own unique phone number. Advantage of using this modem will be that you can use its RS232 port to communicate and develop embedded applications. Applications like SMS Control, data transfer, remote control and logging can be developed easily. The modem can either be connected to PC serial port directly or to any microcontroller. It can be used to send and receive SMS or make/receive voice calls.

It can also be used in GPRS mode to connect to internet and do many applications for data logging and control.

The SIM300 is a complete Tri-band GSM solution in a compact plug-in module. Featuring an industry-standard interface, the SIM300 delivers GSM/GPRS900/1800/1900Mhz performance for voice, SMS, data and Fax in a small form factor and with low power consumption.

4.5. Flow Chart

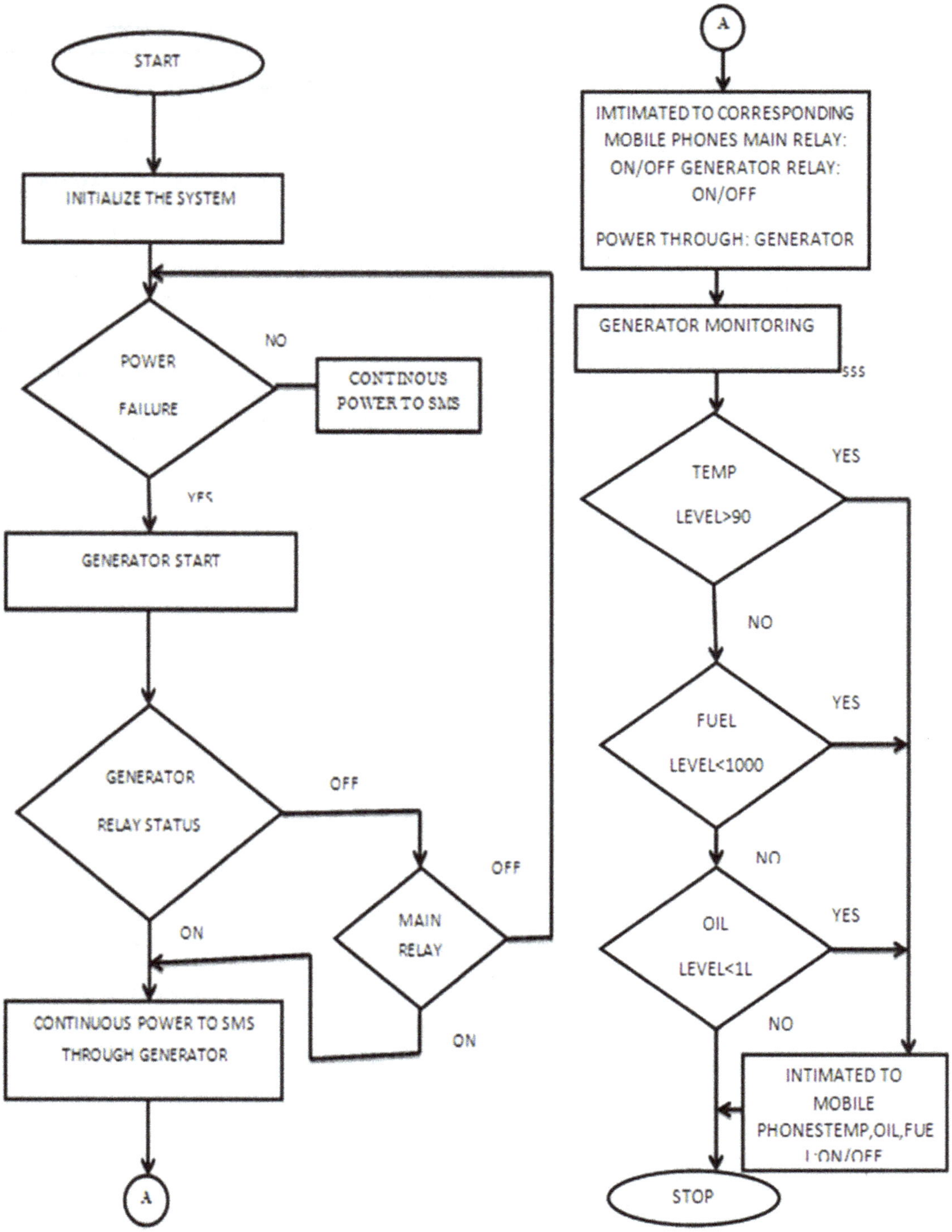

Fig. 3. Flow Chart.

4.6. Algorithm

Step 1: Start the program

Step 2: To initialize the system

Step 3: Get Hardware Software for relevant application.

Step 4: To monitoring the generator status and EB power and if any abnormal conditions occur it is automatically intimated to authorized person.

Step 5: If new SMS received on mobile and go to step3 else, go to step1

Step 6: Receive SMS

Step 7: Check SMS pattern

Step 8: Control the device based on status and operator can making a decision.

Step 9: Notify end user

Step 10: Go to step1

5. Output

To monitor temperature range and fuel, oil level and circuit breaker status of generator.

Under normal condition

Temperature Range: 100 degree Celsius

Fuel Level: 1000 Liter

Oil Level: 4 Liter

Taking the data of previous fault condition and intimated automatically when they exist their limit. Regarding taking threshold value, we have to take account the normal fuel, temperature, oil of generator and associated errors.

Fig. 4. *Electronic Control Unit arrangement.*

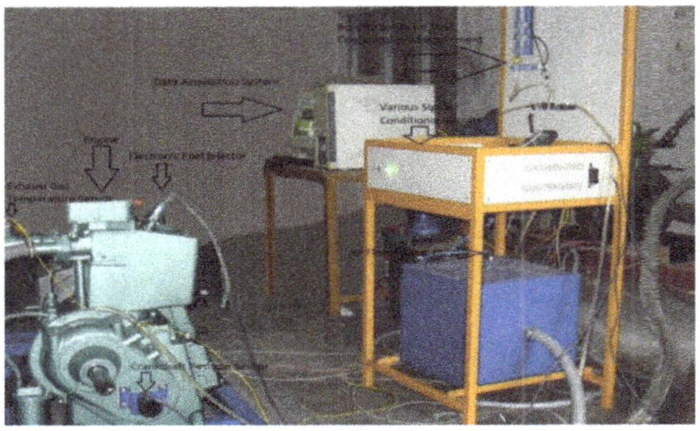

Fig. 5. *Various sensor connections.*

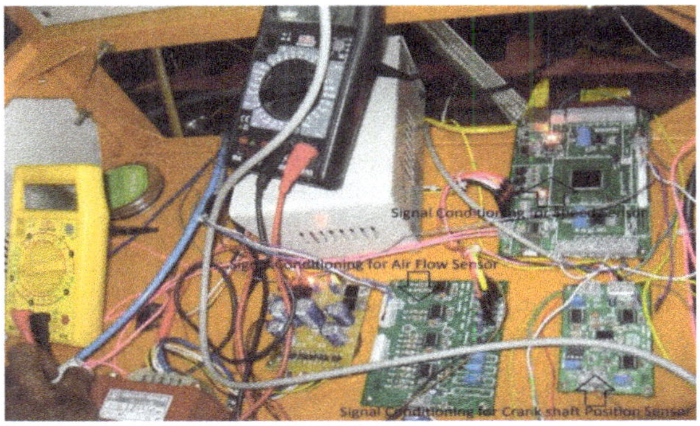

Fig. 6. *Experimental Setup.*

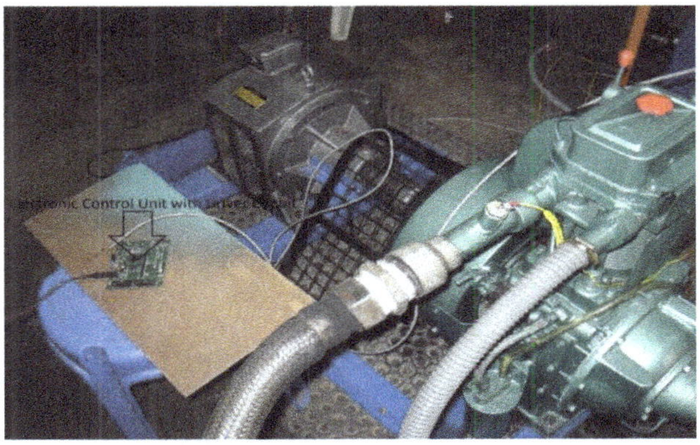

Fig. 7. *Various Signal Conditioning Boards.*

6. Conclusion

The hunch delineated of this project is immense in the ever changing technological world. It allows a greater degree of freedom to an individual to sway via GSM. In particular the suggested system will be a powerful, flexible and secure tool that will offer this service at any time, and from anywhere with the constraints of the technologies being applied. The embedded controllers are capable of sensing and controlling the various parameter of generator in normal and abnormal condition .this proposed system provides the immediate solution for catastrophic failure of generator using GSM communication. The embedded controller offers a wide scope of application in the field of remote digital controllers in the diesel generator industry.

References

[1] Amit Sachan, "Microcontroller based Based Substation Monitoring and Control System with Gsm Modem" IOSR Journal of Electrical and Electronics Engineering, ISSN: 2278-1676 Volume 1, Issue 6 (July-Aug. 2012).

[2] Mallikarjun Sarsamba "The Load Monitoring and Protection on Electricity Power lines using GSM Network" International Journal of Advanced Research in Computer Science and Software Engineering, Volume 3, Issue 9, September 2013 ISSN: 2277 128X.

[3] S.Vimalraj, Gausalya.R.B, "GSM Based Controlled Switching Circuit between Supply Mains and Captive Power Plant" International Journal of Computational Engineering Research, Vol, 03, Issue, 4.April 2013.

[4] Andriy Palamar "Control System for a Diesel Generator and UPS Based Microgrid", Scientific Journal of Riga Technical University Power and Electrical Engineering, Volume 27, 2010.

[5] Kwang Seon Ahn "Digital Controller of a Diesel Generator using an Embedded System" International Journal of Information Processing Systems, Vol.2, No.3, December 2006.

[6] Henrik arleving "ways to cut power generator maintanence"the journal, December 2013.

[7] Chetan Patil, Channabasappa Baligar, "Base Transceiver Station (BTS) Safety and Fault Management", International Journal of Innovative Technology and Exploring Engineering (IJITEE) ISSN: 2278-3075, Volume-3, Issue-7, December 2013.

[8] Y Jaganmohan Reddy, Y V Pavan Kumar, K Padma Raju, Anilkumar Ramsesh, "PLC Based Energy Management and Control Design for an Alternative Energy Power System with Improved Power Quality", International Journal of Engineering Research and Applications (IJERA) ISSN: 2248-9622 Vol. 3, Issue 3, May-Jun 2013.

[9] Alper T. Alan "A Field Study of Human-Agent Interaction for Electricity Tariff Switching", Agents, Interaction and Complexity Group, University of Southampton, Southampton, UK.

[10] J. Pierce and E. Paulos. Beyond energy monitors: interaction, energy, and emerging energy systems. In Proc. CHI'12. ACM, 2012.

5

Reduction of Total Harmonic Distortion for A Three Phase Fault in a Distribution Network by Using PID, Fuzzy & Hybrid PID-Fuzzy Controller Based DVR

Danish Chaudhary[1], Aziz Ahmed[1], Anwar Shahzad Siddiqui[2]

[1]Dept. of Electrical & Electronics Engg, Alfalah University, Dhauj, Faridabad, India
[2]Dept. of Electrical Engineering, Jamia Milia Islamia, New Delhi, India

Email address:
Danishchaudhary89@gmail.com (D. Chaudhary), azizjmi98@gmail.com (A. Ahmed), anshsi@yahoo.co.in (A. S. Siddiqui)

Abstract: The use of electric energy is, in developed countries around the world, a natural part of life. It is used everywhere for living, work and travelling, at any residence, commercial building, industry and so on. The number of electrical devices connected to the power system, during the century, has increased enormously, with the main increase having been during the last 20 – 25 years. The total power demand has also increased but not at the same rate as the number of devices connected. This is due to more power efficient equipment being used, both for new devices and older replaced devices. There are differences for different voltage levels and of course a large variation among different countries. Due to complexity of power system combined with other factors such as increasing susceptibility of equipment. With electricity demand growing, low power quality is on the rise & becoming notoriously difficult to remedy [1]. Distribution system needs to be protected against voltage sags, dips & swells that adversely affect the reliability & quality of power supply at the utility end. The Dynamic voltage restorer (DVR), which has been utilized in optimized way so as to improve performance, has been put under new technique of sag detection. The applications of Fuzzy logic controller have taken new dimension in various fields. In this paper, the essentials of control scheme with immediate voltage generation to regulate the unbalance voltage phase in three phase system and a tested method to improve the reliability within the distribution system is presented. The 13kV distribution system is having a three phase fault which controlled by non-linear techniques and their performance levels are compared. The capability of DVR is demonstrated using MATLAB/SIMULINK simulation models. This paper emphasizes the importance of DVR application for better power quality, by comparing the mitigated voltage and current THD values among PID, Fuzzy, PID-Fuzzy have been compared on account of the amount of compensation being injected into the system under voltage sag condition for non-linear loads.

Keywords: PID, Fuzzy, PID-Fuzzy, DVR, Total Harmonic Distortion, PWM, Voltage Sags/Swells

1. Introduction

The high quality sinusoidal waveform is produced at power stations. The widespread applications of power electronic based non-linear devices and faults cause deviation from pure sinusoidal waveform. These situations facing electricity customers and suppliers have increased the popularity and development of power quality devices. Users need constant sine wave shape, constant frequency and symmetrical voltage with a constant root mean square (rms) value to continue the production. To satisfy these demands, the disturbances must be eliminated from the system. The typical power quality disturbances are voltage sags, voltage swells, interruptions, phase shifts, harmonics and transients. Power electronic based devices provide protection for industry and commercial customers from power quality problems basically sags, swells & harmonics. These devices are known as custom power devices and they can increase the availability of sensitive loads in the system and supply reliable power. Custom power devices are typically building on the distribution system to provide higher power quality and most economical solution. One of those is DVR used to mitigate voltage sag and well. DVR is a series connected device. It is best custom power device for mitigation impacts of upstream voltage disturbances on sensitive loads [1].

The economy invested in the distribution system is large enough to take into account the concept of equipment protection against various disturbances that affects the reliability of not only the distribution system but the entire power system incorporating generation & transmission too. The wide acceptance of sophisticated electronic devices at the utility end deteriorates the quality of supply & utility is suffering from its bad effects on large scale. The results of various faults with three phase fault being the most severe among all, starting of induction motor which is most often used due to its rugged construction, switching off large loads and energizing of capacitor banks. This paper attempts to explain the various control strategies providing a reliable solution to the faulted system with the help of DVR (Dynamic Voltage Restorer).

Various control techniques are available to obtain a controlled output voltage, to be injected into the system. They are known as Non-linear techniques like PI controller, PID controller, Fuzzy based controller, by using Artificial Neural Network (ANN) and Hybrid controllers of all above explained controllers. A PID controller with a linear structure offers satisfactory performance over a wide range of operation [5]. The problem encountered by the controller is the setting of PID parameters i.e. the gains (K_P, K_I, K_D). In the influence of varying parameters and operating conditions, the fixed gains of linear controller don't adapt accordingly to give good dynamic response. To overcome the problems faced by a non-linear technique is an effective solution [10]. The recommended system uses the PID, Fuzzy and Hybrid PID-Fuzzy [10] controllers to investigate the performance level of various controllers in a regard to increase the capability of the existing system by creating immunity from disturbances. Simulation results of voltage sag condition for a non-linear load are presented.

2. Total Harmonic Distortion

The degree of the voltage distortion varies with the impedance of the electrical power distribution system and the number and type of non-linear loads connected. In order to compare these two distortion levels, it is necessary to quantitatively describe the distortion. Harmonic analysis is used to provide this description. The level of voltage distortion that is acceptable depends on the sensitivity of the equipment installed in the building. In harmonic analysis, any repetitive wave form can be described mathematically as a series of pure sine waves. These sine waves consist of a fundamental frequency and multiples of that frequency, called harmonics.

Total harmonic distortion (THD) is an important figure of merit used to quantify the level of harmonics in voltage or current waveforms [8]. Total harmonic distortion (THD) is often used as a percentage, this single number is calculated by adding the square of each relative harmonic value and taking the square root [9].

$$\text{THD} = \sqrt{\sum_{n=2}^{\infty}\left(\frac{H_n}{H_1}\right)^2} * 100\%$$

The most detailed method describes the amplitude of each individual harmonic component, either in absolute units (such as volts) or as a percentage of the fundamental component. With this, it is possible to determine the source of harmonic distortion. For example, in a balanced electrical system, the only harmonics that can be generated by a symmetrical three phase load are those that are not multiples of 2 or 3 (the 5th, 7th, 11th, and similar harmonics). If a third harmonic is present in the system, it is likely the result of single phase loads or phase imbalances.

Voltage and current harmonics have undesirable effects on power system operation and power system components. In some instances, interaction between the harmonics and the power system parameters (R–L–C) can cause harmonics to multiply with severe consequences. Voltage harmonics are mostly caused by current harmonics. The voltage provided by the voltage source will be distorted by current harmonics due to source impedance. If the source impedance of the voltage source is small, current harmonics will cause only small voltage harmonics [9]. Harmonics provides a mathematical analysis of distortions to a current or voltage waveform. Based on Fourier series, harmonics can describe any periodic wave as summation of simple sinusoidal waves which are integer multiples of the fundamental frequency.

Power system transients are fast, short-duration events that produce distortions such as notching, ringing, and impulse. The mechanisms by which transient energy is propagated in power lines, transferred to other electrical circuits, and eventually dissipated are different from the factors that affect power frequency disturbances [23]. In electrical engineering, oscillation is an effect caused by a transient response of a circuit or system. It is a momentary event preceding the steady state (electronics) during a sudden change of an event.

3. Dynamic Voltage Restorer

3.1. Introduction

Among the voltage disturbances, voltage sag is most severe that adversely affects the performance of the system. The one such efficient & reliable solution is the DVR.

DVR is a static series compensator that injects voltage in series to the distribution system, regulating the load side voltage. It is connected between the supply and the sensitive load to compensate the line voltage harmonics, reduction of transients in addition to compensation of voltage sags & swells.

3.2. Principle of DVR Operation

The main aim of DVR is to regulate the voltage at the load terminals irrespective of sag, distortion or unbalance in the supply voltage. The basic operating principle is to inject a voltage of required magnitude & frequency to restore the load voltage under voltage sag or distortion. Generally; it employs solid state power electronic switches such as GTO, IGBT or IGCT in the VSI, which can be operated in various pulse width modulation techniques such SPWM(sinusoidal pulse width modulation), MSPWM(multiple sinusoidal pulse width modulation). They inject a set of three phase AC voltage in series & synchronism with the distribution system.

4. Configuration of DVR

The vital components of DVR are the power circuit which injects the desired voltage & control circuit that controls the load voltage of the system within prescribed limits. Its schematic diagram explains the various components as the constituents of DVR as shown in Figure1.

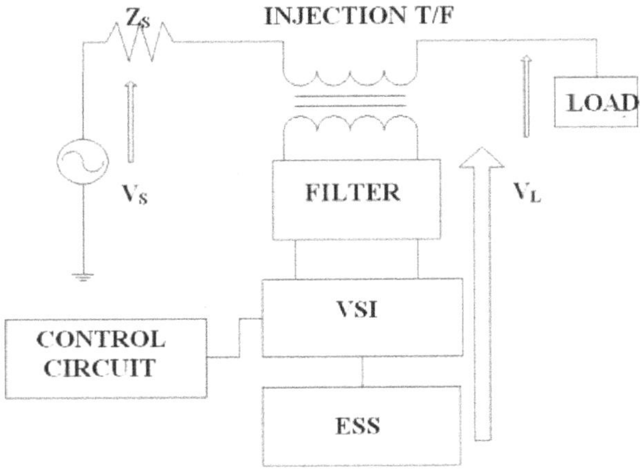

Figure 1. *Block Diagram of DVR.*

4.1. Voltage Source Inverter

It forms the building block of compensating device. It performs the power conversion process from DC to AC. VSI consists of fully controlled semiconductor power switches to form a single phase or three phase topologies. For medium power inverters, IGBT"s are used and GTO"s or IGCT"s due to compact size & fast response for high power inverters are employed. The single phase VSI topology encompasses a low-range power applications and medium to high power applications are covered by the three phase topology [10]. Single phase VSI consists of four semiconductor switches (in 2 legs) to generate the ac output waveform. Three phase VSI is a six step bridge inverter that uses a minimum of six thyristors, where a step means a change in the firing from one thyristor to the next thyristor in proper sequence. For one cycle of 360 degree, each step is of 60 degree for a six step inverter.

4.2. Series Injection Transformer

It provides electrical isolation & voltage boost to the system. In a 3-phase system, either 3 single phase units of isolating transformer or 3-phase isolating transformer can be employed for the purpose of voltage injection. While selecting the injection transformer, the determination of expected maximum output voltage is prime significance, both economically & technically. Prior to the level of the distribution system being compensated by DVR & largest sag to be compensated by VSI at the minimum DC-link voltage decides the turn ratio of the series injection transformer. The effects of higher order harmonics on the transformer are related to the positioning of filtering system, i.e. inverter filtering side system & line side filtering.

4.3. Filter

These are electronic circuits comprising of combination of passive elements; resistors, inductors & capacitors. They perform signal processing functions to remove the unwanted frequency signals to enhance the desired signal output. LC type of filters corrects the harmonic output from VSI to provide compensation in the required phase of the 3 phase system boosted by DVR.

4.4. Energy Storage Unit

The purpose of storage systems is to protect sensitive equipments from shutdown caused by voltage sags or interruptions. They provide necessary energy to the VSI via a dc link for the generation of injected voltages. There are different types of storage systems such as superconducting magnetic energy storage system (SMES), DC batteries, flywheel energy storage system, battery energy storage system (BESS) etc. Capacity of the storage system directly determines the duration of the sag which can be mitigating by the DVR. Among the above mentioned storage systems, Batteries are more common & can be highly effective if high voltage configuration is used. There are different types of battery energy storage technologies such as lead-acid battery, flooded type battery, valve regulated type battery (VLRA), Sodium Sulphur battery (NaS) etc. [6].

4.5. Control Circuit

Several techniques & control philosophy of the DVR have been implemented for power quality improvement in the distribution system. The DVR is equipped with a control system to mitigate voltage sags/swells. The control of the DVR is very important as it involves the detection of voltage sags (start, end & depth of voltage sag) by appropriate detection algorithm [7]. The control strategy can depend on the type of load connected. Its main purpose is to maintain constant voltage magnitude at the point where the sensitive load is connected under system disturbances. Three basic control strategies of DVR can be stated as:

4.6. Pre-Sag Compensation Method

In this method, both magnitude & phase angle are to be compensated. The supply voltage is continuously tracked & load voltage is compensated to the pre-sag condition by injecting voltage equal to the difference of voltage under pre-sag & sag condition as in Fig 2. Though, it gives a nearly undisturbed load voltage but suffers a drawback of exhausting the rating of the DVR.

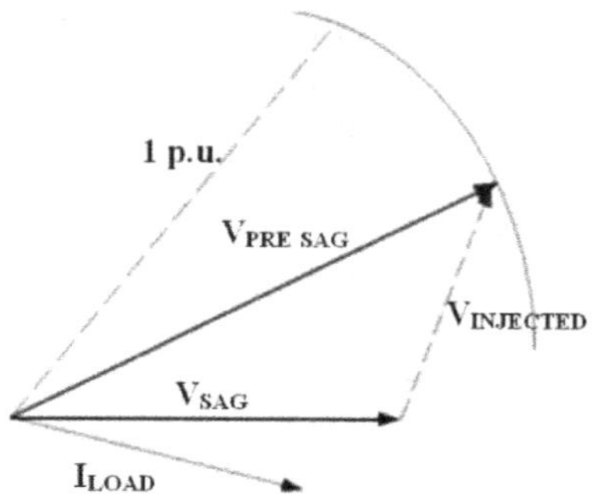

Figure 2. Pre-Sag Compensation Method.

4.7. In Phase Compensation Method

In this method, when the source voltage drops due to sagging condition, the VSI injects a voltage called missing voltage based on the drop of voltage magnitude as in fig 3. The generated Voltage of the DVR is always in phase with the measured supply voltage regardless of the load current and the pre-sag voltage.

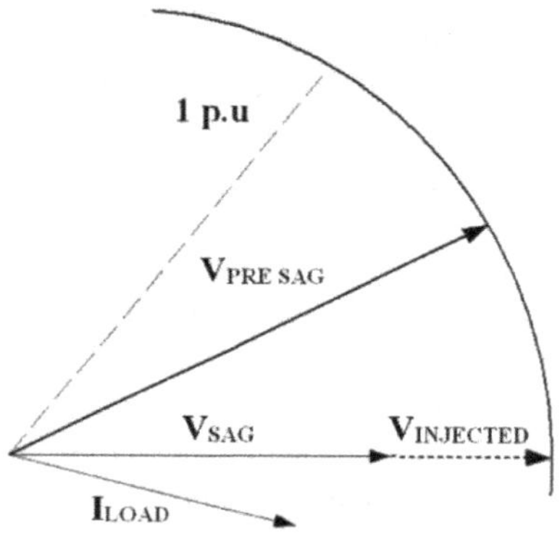

Figure 3. In-Phase Compensation Method.

4.8. Reactive Power Compensation

This is also known as the minimum energy injection, which depends on maximizing the active power supplied by the network (keeping the apparent power constant and decreasing the network reactive power) by minimizing the active power supplied by the compensator (increasing the reactive power supplied by the compensator). In this injection method the injected voltage is in quadrature with load current.

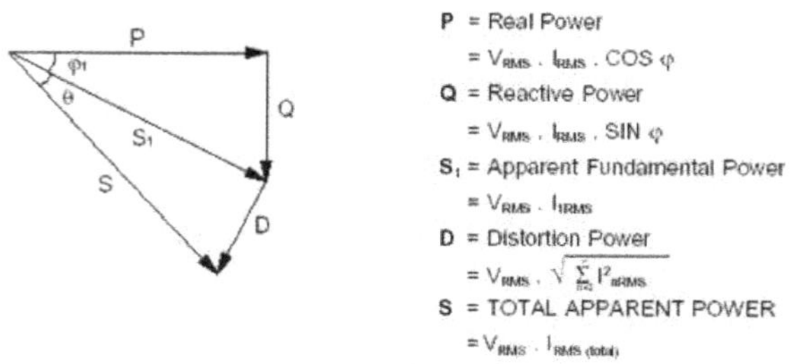

Figure 4. Reactive Power Compensation.

5. Control Philosophies of DVR

5.1. Introduction

Voltage sags are one of the most severe power quality problems & DVR is an effective solution to mitigate it. The purpose of control scheme is to control the system output by generating an appropriate control signal prior to the unbalanced condition prevailing in the system. It generates the signals to enable the VSI (voltage source inverter) by providing proper firing sequence to the circuit. In this work, different control strategies for dynamic voltage restorer are investigated with emphasis on voltage sag compensation. Three promising control methods to compensate voltage sags are tested & compared with simulation of DVR on 13kV system. The comparison of the performance of three control strategies is made on basis of voltage waveforms & its frequency spectrum analysis. Their performance level is presented in the decreasing order of their compensation capability & better performance in mitigating voltage sag over a broader range for different faults. Three control philosophies have been used namely, PID, Fuzzy & PID-Fuzzy. These are discussed as below:

5.2. Proprotional-Integral-Derivative (PID) Controller Based DVR

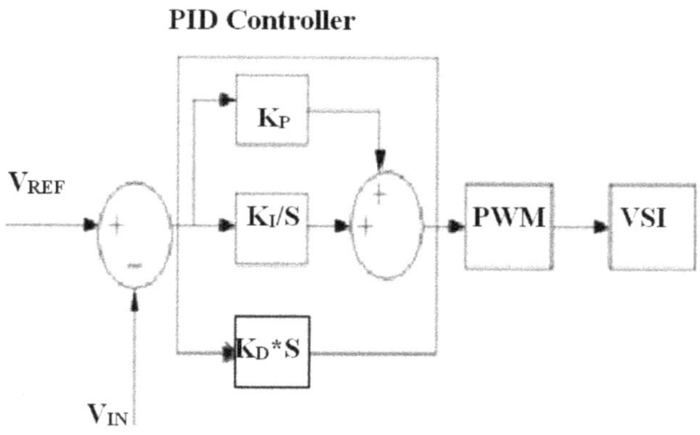

Figure 5. Control Strategy of PID Controller.

PID is a feedback controller that uses the weighted sum of error & its integral value to perform the control operation. The proportional response can be adjusted by multiplying the error by constant K_P, called proportional gain. The contribution from integral term is proportional to both the magnitude of error and duration of error. The error is first multiplied by the integral gain, K_I and then was integrated to give an accumulated offset that have been corrected previously [8]. Derivative term is also proportional to both error magnitude and duration of error. The error is first multiplied by differential gain K_D and then was differentiated to give an accumulate offset that have been corrected. This derivative gain increases the speed of error correction while integral gain removes the error completely. The input to the PID controller is difference between the reference value & error value of voltage. As per the comparison of reference value & error value of voltage, linear PID adjusts its proportional, integral and differential gains K_P, K_I and K_D in order to reduce the steady state error to zero for a step input as shown in Figure5. It is widely used due to simple control structure but suffers a disadvantage of fixed gains i.e. it cannot adapt itself to the varying parameters & conditions of the system.

5.3. Fuzzy Controller Based DVR

The drawback suffered by PID controller is overcome by Fuzzy. In comparison to the linear PID controller, this is a non-linear controller that can provide satisfactory performance under the influence of changing system parameters & operating conditions [8] [9]. The function fuzzy controller is very useful as relieves the system from exact & cumbersome mathematical modeling & calculations. The performance of fuzzy controller is well established for improvements in both transient & steady state [10]. The fuzzy controller comprises of four main functional modules namely; Knowledge base, Fuzzification, Inference mechanism & Defuzzification as in figure 6.

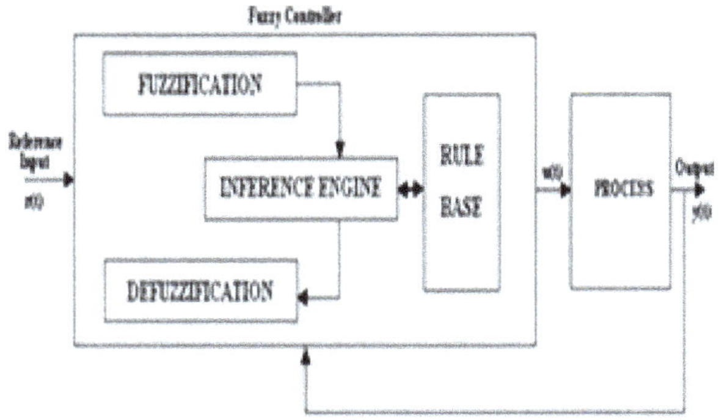

Figure 6. *Schematic Diagram of Fuzzy Logic.*

5.4. Knowledge Base

It consists of data base & rule base that maps all the input & output with certain degree of uncertainty in process parameters & external disturbances to obtain good dynamic response. Data base scales the input-output variables in the form of membership functions that defines it in a range appropriate to provide information to the fuzzy rule-based system & output variables or control actions to the system under observation. Fuzzy rule-based system utilizes a collection of fuzzy conditional statements derived from a knowledge base to approximate and construct the control surface.

5.5. Fuzzification

It is the process of defining a crisp data or digital data operating on discrete values of either 0 or 1 in terms of logical variables that take on continuous values between 0 and 1 i.e. fuzzy set. Fuzzy set maps the input-output variables into membership functions & truth values as in figure 7 & figure 8.

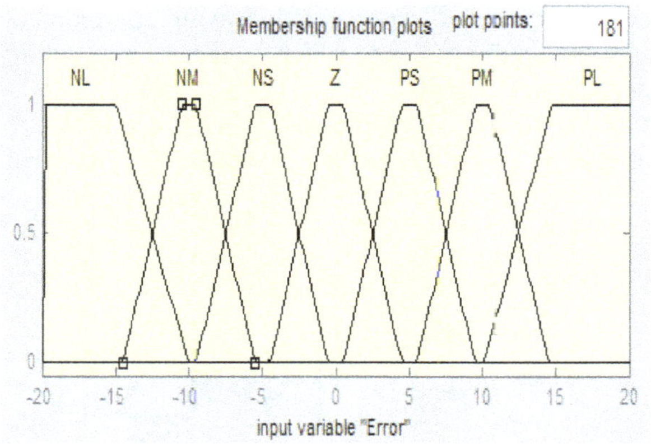

Figure 7. *Input Membership Function of "Error".*

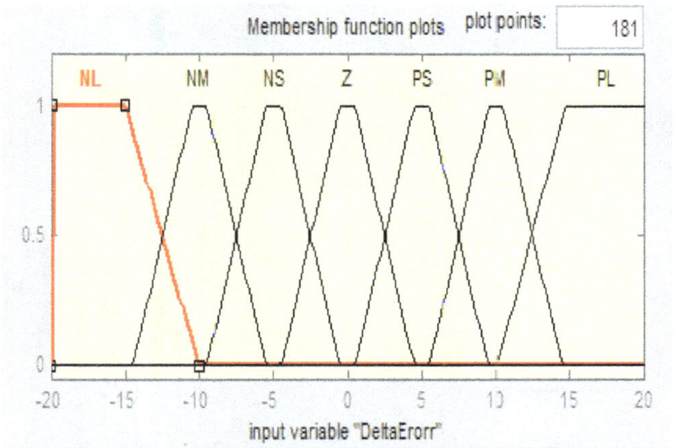

Figure 8. *Input Membership Function of "Change in Error".*

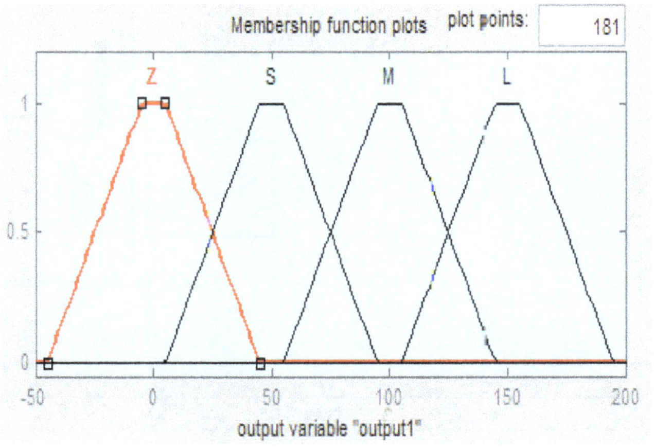

Figure 9. *Output Membership Function.*

5.6. Inference Mechanism

It is referred to as approximate reasoning that uses knowledge to conduct deductive inference of IF-THEN rules. This mechanism encodes knowledge about a system in statements form of linguistic IF-THEN propositions with antecedents & consequents.

- Defuzzification

It is a conversion process of fuzzy quantity to a precise quantity and is reverse process of fuzzification. A logical union of two or more membership functions in the universe of discourse requires a crisp decision with approximate solution for the output of fuzzy which is uncertain in nature to be a single scalar quantity

The FLC controller of the tested system exploits the Mamdani type of inference method. It defuzzifies the crisp input-output variables into fuzzy trapezoidal membership function and reverse process of Defuzzification is based upon the Centroid method. The controller core is the fuzzy control rules as shown in table I. which are mainly obtained from intuitive feeling and experience [11].

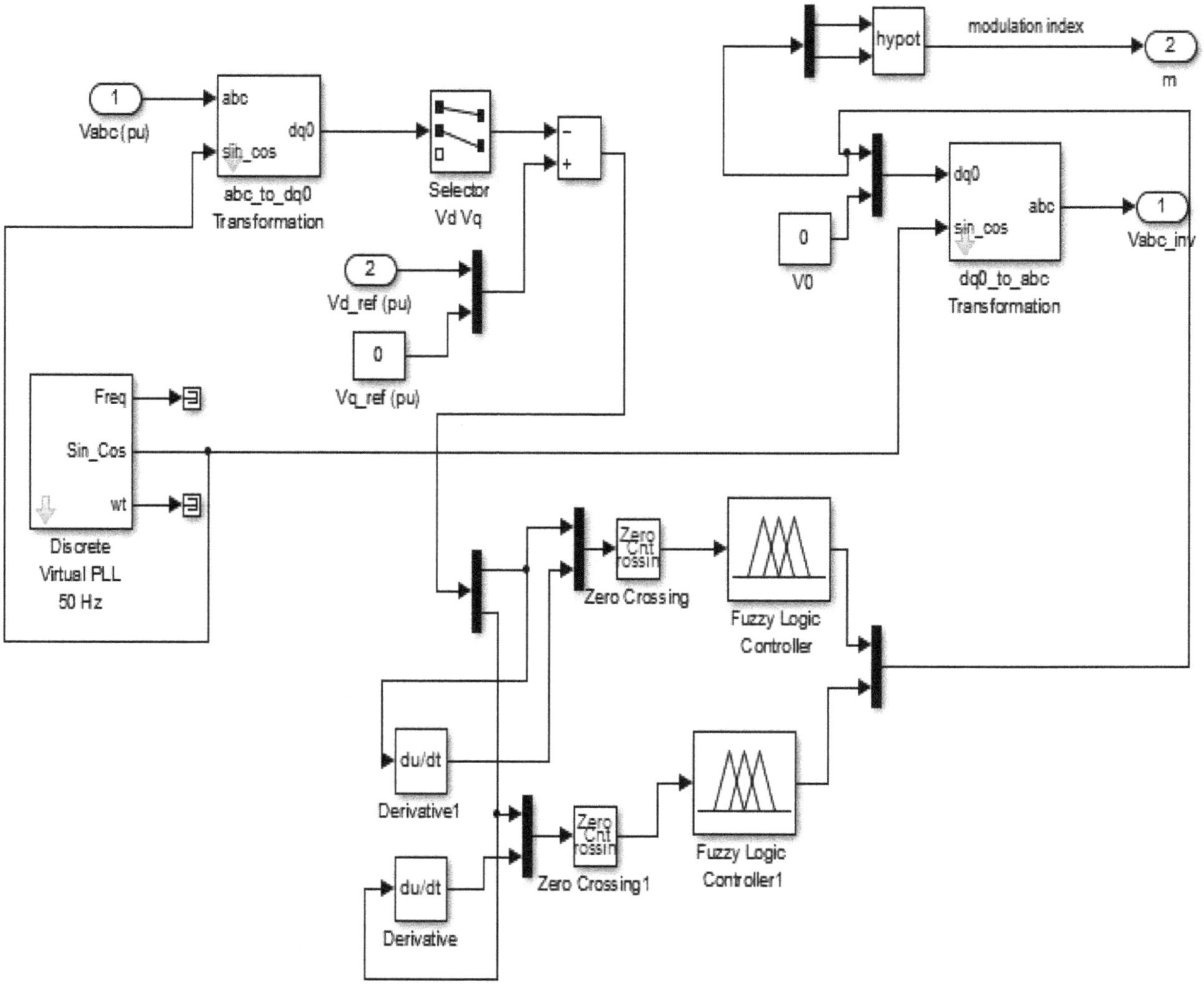

Figure 10. *Control Strategy of Fuzzy Controller.*

Table 1. *Fuzzy Rule Based System.*

"e"							
"ce"	**NL**	**NM**	**NS**	**Z**	**PS**	**PM**	**PL**
NL	L	L	L	M	Z	S	Z
NM	L	L	M	Z	Z	Z	S
NS	L	M	S	Z	Z	S	S
Z	M	S	S	Z	S	S	M
PS	S	S	Z	Z	S	M	L
PM	S	Z	Z	Z	M	L	L
PL	Z	S	Z	M	L	L	L

5.7. Hybrid PID-Fuzzy Based DVR

The hybrid PID-Fuzzy control scheme uses fuzzy as adjustor to adjust the parameters of proportional gain K_P, integral gain K_I and derivative gain K_D based on the error e and the change of error e [5]. PID-Fuzzy based Controller has been designed by taking inputs as error which is difference between measured voltage and reference voltage of DVR for voltage regulator and its derivative while ΔK_P, ΔK_I and ΔK_D as output for voltage regulator where K_P, K_I and

K_D are proportional gain, integral gain and derivative gain respectively [4] as shown in Figure11.

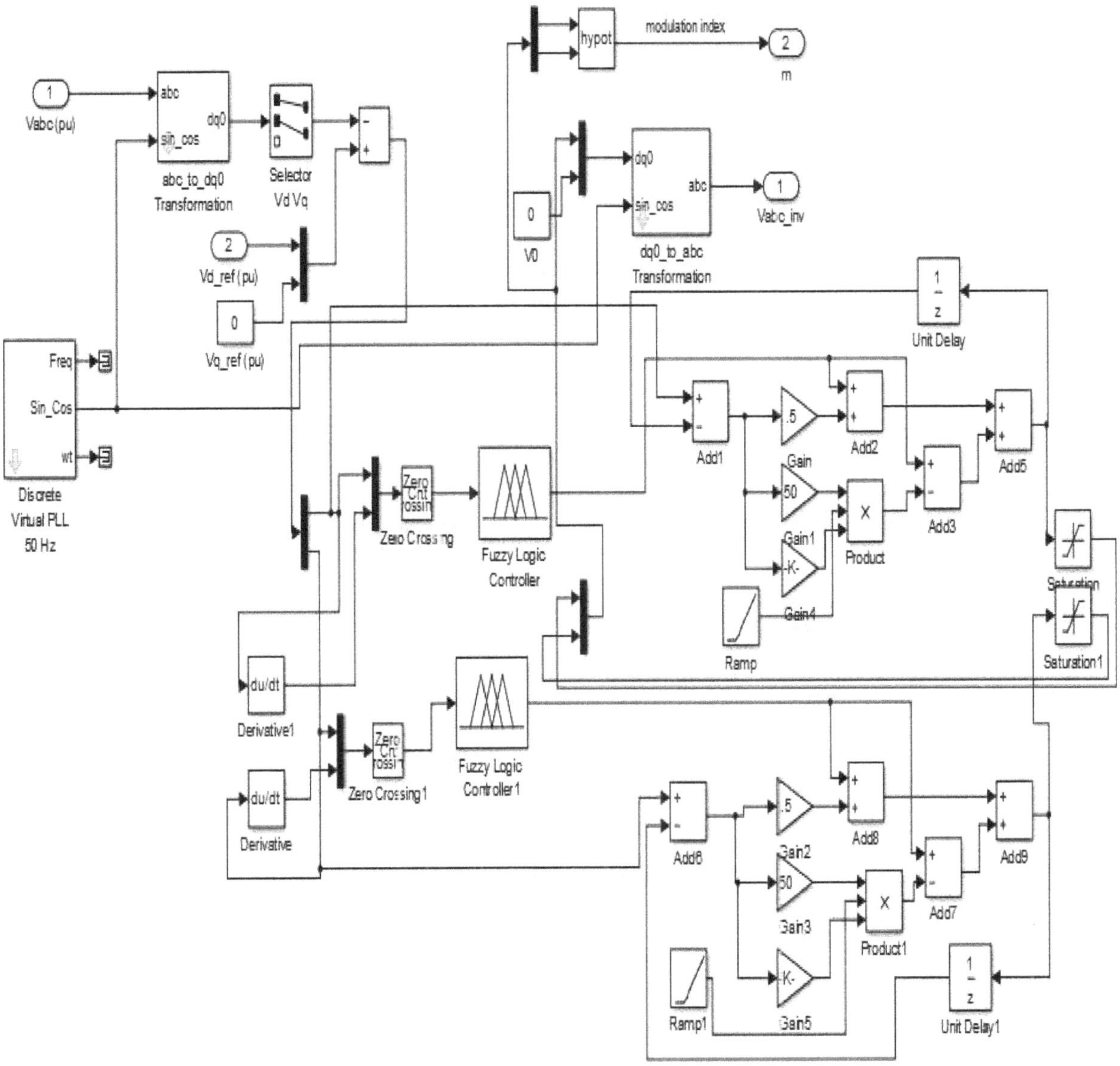

Figure 11. Control Strategy of PID FUZZY Controller.

6. Simulation Test Models

6.1. Introduction

In the SIMULINK test model, two feeders are drawn from the same supply using 3- winding transformer. One of the feeders is compensated using DVR while the other uncompensated. The parameters for the whole system model are explained in Table 2. These are further connected to identical loads so that their performances are fairly compared. The controllers PID, Fuzzy, hybrid PID-Fuzzy are employed step by step in the compensated feeder to compare their performances. The single line test diagram is shown in Figure12.

Table 2. System Parameters.

Sr.No.	System Quantities	Standards
1.	Source	3-Phase,13kV,50Hz
2.	Inverter Parameters	IGBT based,3arms,6 Pulse, Carrier frequency -1080Hz Sample time=50μsec
3.	PID Controller	K_P=0.5,K_I=50, K_D=100, sample time=50μsec
4.	RL Load	Active power = 1 KW Reactive power = 400 VAR
5.	Three Winding Transformer	Y/Δ/Δ 13/115/115kV
6.	TwoWinding Transformer	Δ/Y 115/11 kV

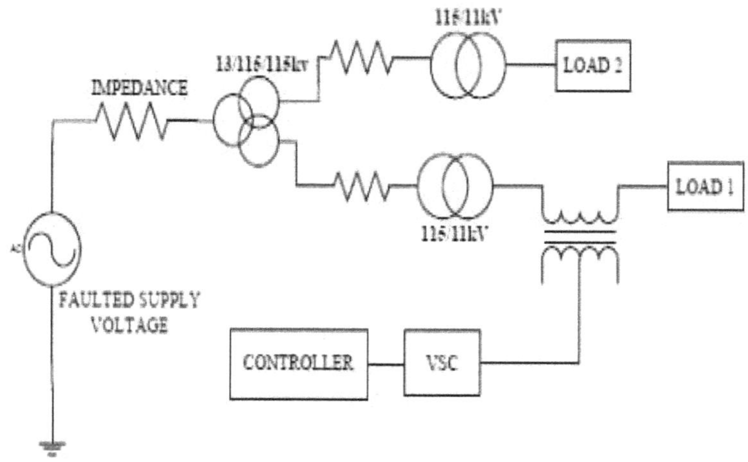

Figure 12. *Single Line Diagram of Test Circuit.*

6.2. Simulink Model of the Test System with PID Controller

In the SIMULINK model, the PID controller based DVR is investigated during a three phase fault to remove the voltage sag arises in output waveform due to the fault as shown in Figure13. In the simulation results, we study the output waveform to be uniformly sinusoidal after voltage magnification and the amount of harmonics reduced from during fault to post fault condition by using PID controller based DVR.

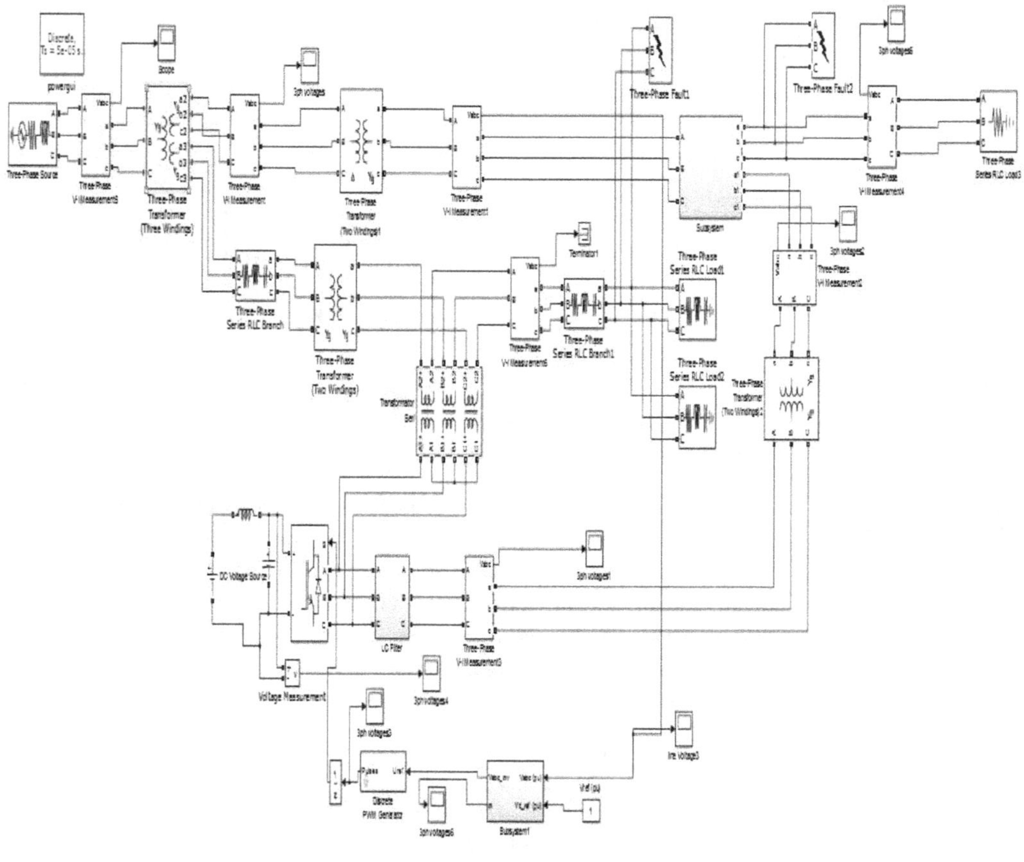

Figure 13. *Simulink Model of PID Controller Based DVR.*

6.3. Simulink Model of Test System with Fuzzy Controller

In this case, fuzzy logic controller is employed to compensate the uncompensated system shown in Figure14. It overcomes the disadvantage suffered with linear PID controller by providing better compensation and reducing the

THD level of uncompensated system. In here simulation results, we comparatively study its output waveform with previous result of PID controller based DVR, and also the variation in THD for both PID and Fuzzy controller based DVR.

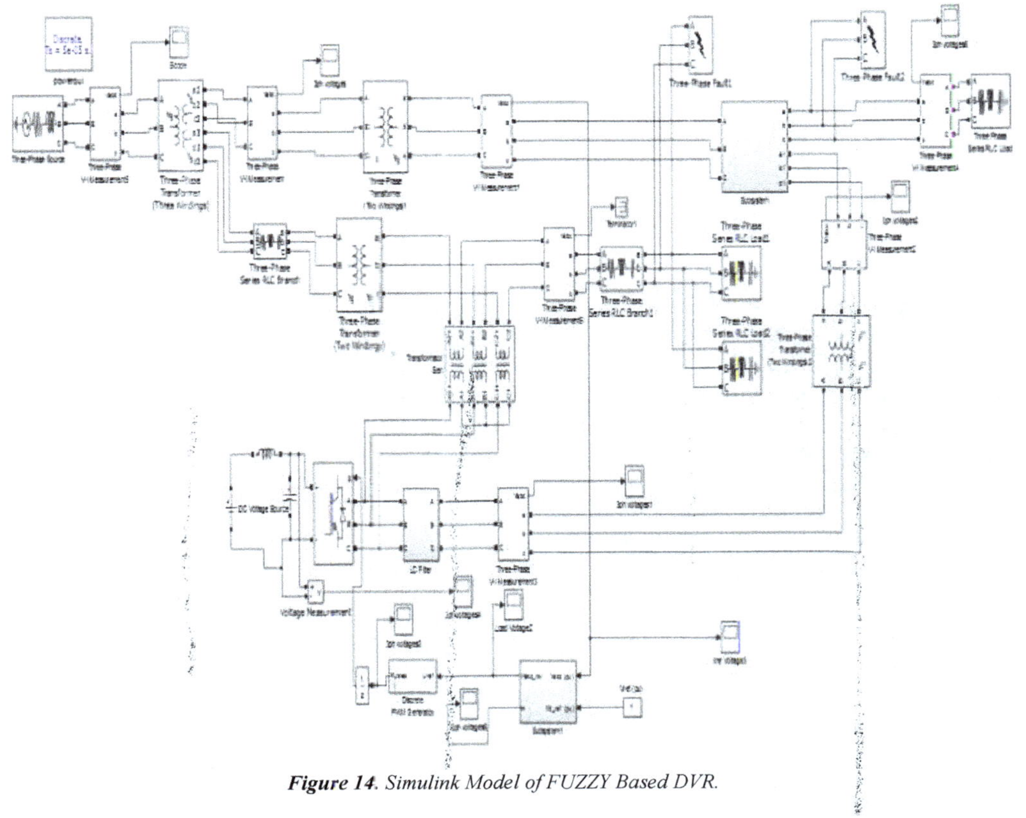

***Figure 14**. Simulink Model of FUZZY Based DVR.*

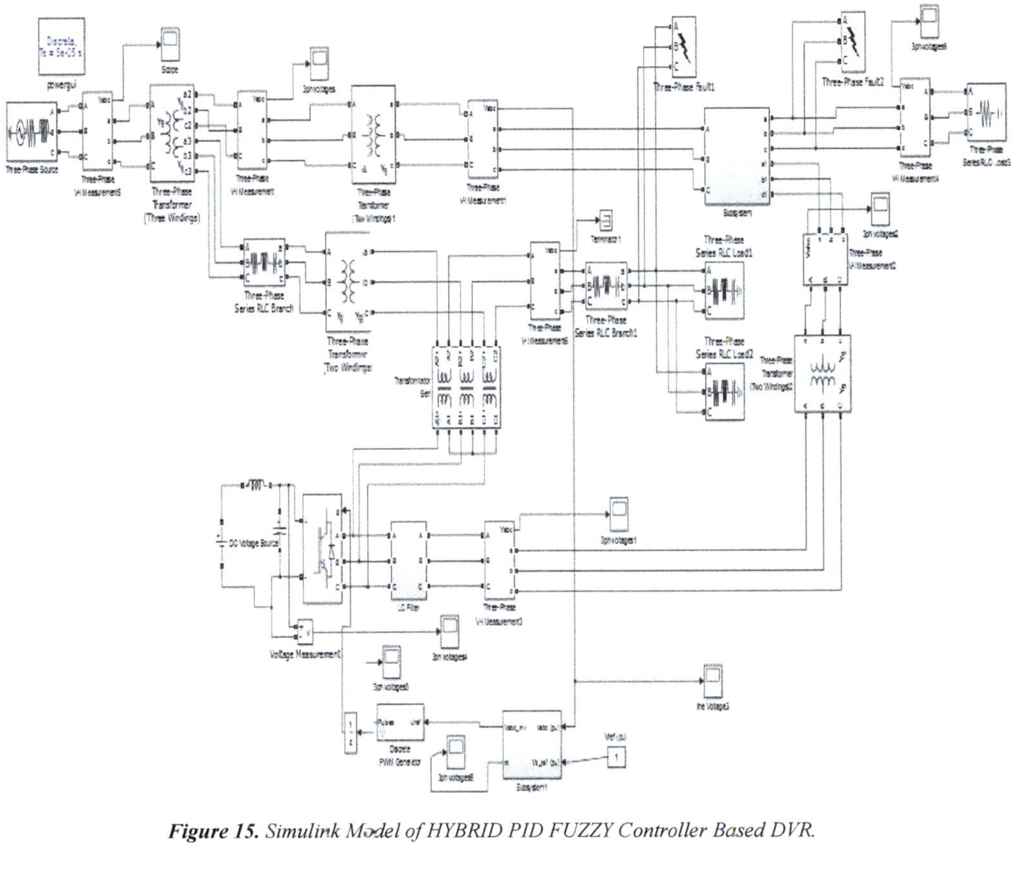

***Figure 15**. Simulink Model of HYBRID PID FUZZY Controller Based DVR.*

6.4. Simulink Model of Test System with Hybrid PID-Fuzzy Controller

In this case, hybrid PID-Fuzzy control scheme is employed and tested for a three phase fault in a distribution network. This hybrid controller adjusts the proportional, integral gains and derivative gain K_P, K_I and K_D of PID using the trapezoidal membership function and the rule base system for regulating the voltage of the system as shown in Figure11. The complete simulation test model for hybrid PID-Fuzzy controller based DVR is shown in Figure15.

7. Result

After simulating the test model during a three phase fault in a distribution network we get the waveform during fault as shown in Figure16. which clearly shows the voltage sag during a three phase fault i.e. from time 0.04 sec to 0.10 sec after which the fault is removed leaving behind the higher order harmonics which is shown in the Figure16. The frequency spectrum analysis shows the percentage of Total Harmonic Distortion during fault condition is shown in Figure17. The value for THD during a three phase fault for a given test system is 23.67%.

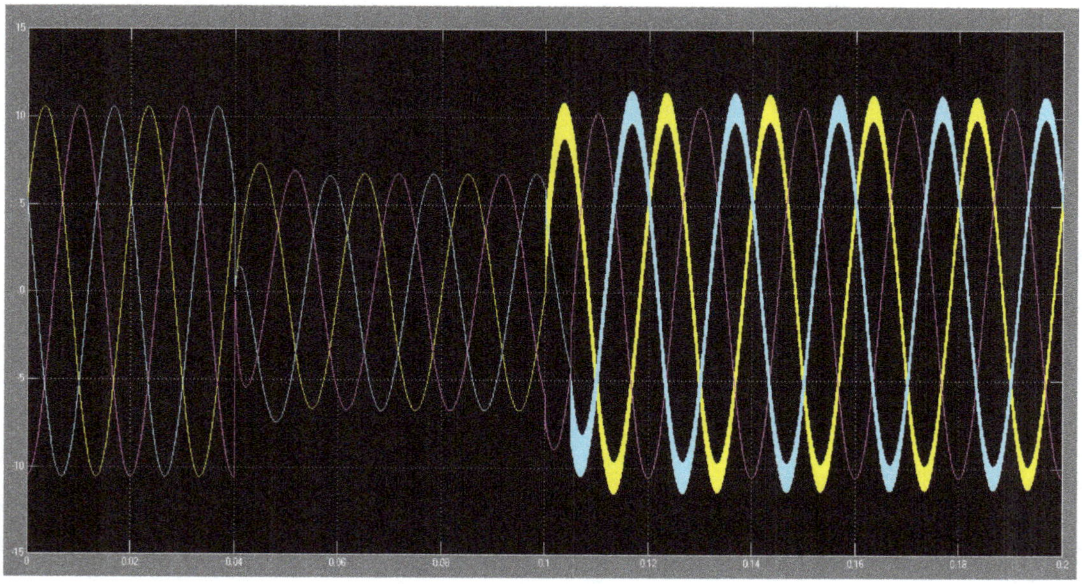

Figure 16. Load Voltage Waveform During Three Phase Fault.

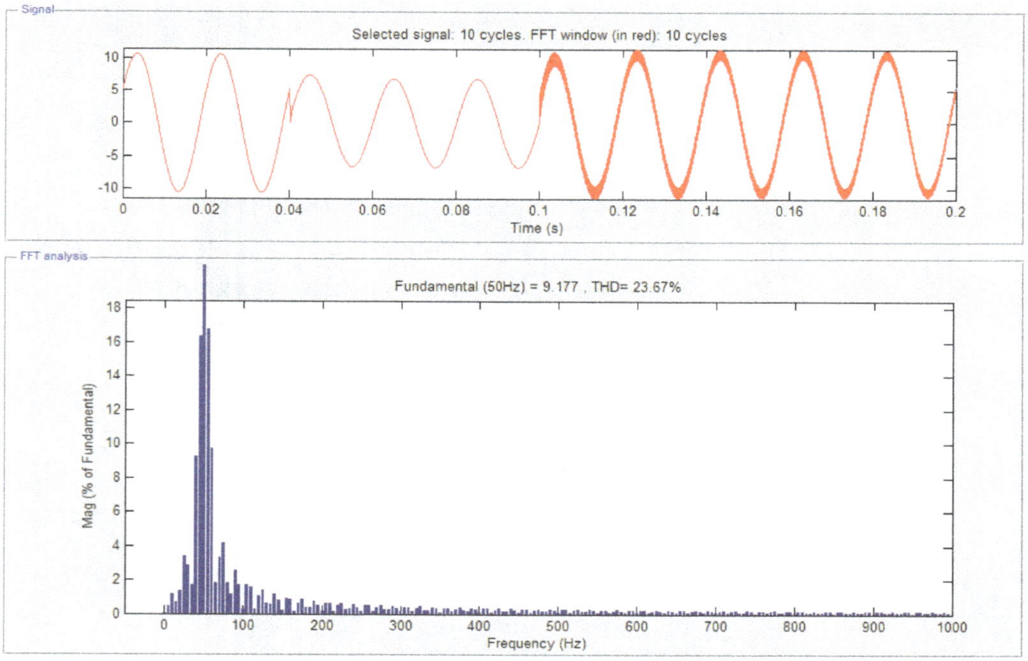

Figure 17. FFT Analysis of when Uncompensated During Three Phase Fault .

Now after simulating all the three controlling techniques i.e. PID, Fuzzy and Hybrid PID-Fuzzy controller based DVR respectively for the removal of a three phase fault, we get the following results:

Simulation Result for PID Controller Based DVR

The resultant output waveform after mitigation of fault in the network is shown in the Figure18. In the output waveform we have seen that the voltage sag is removed but a slightly distorted waveform will remain after fault clearing time as in Figure18.

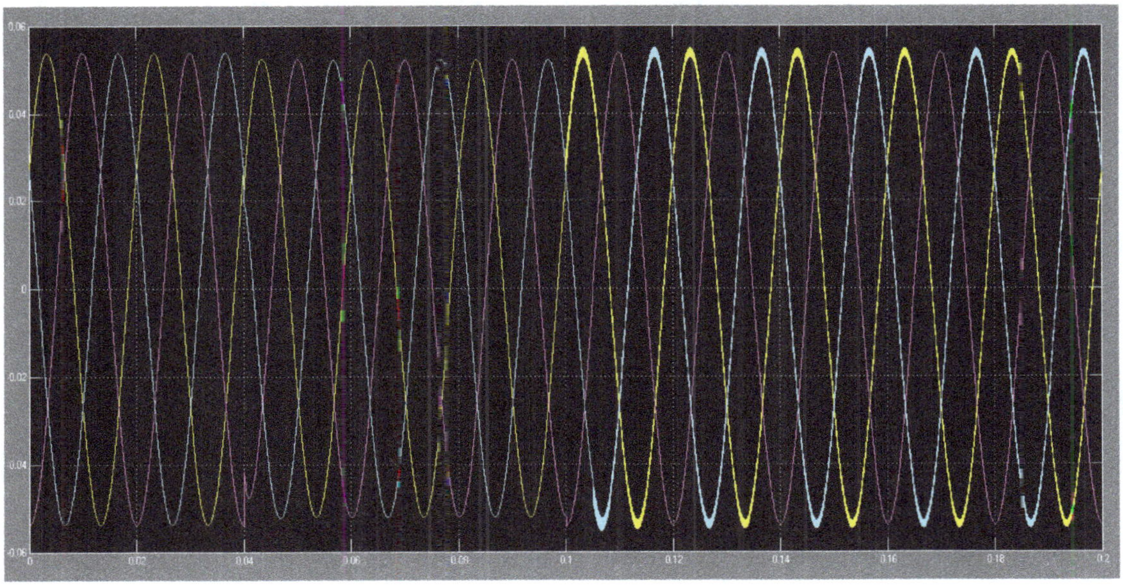

Figure 18. *Output Voltage Waveform when Compensated Using PID Controller Based DVR.*

Figure19.shows the frequency spectrum analysis showing the THD at a fundamental voltage level. Now after clearing the fault, PID controller reduces the THD from 23.67% (during fault) to 3.09% (by using PID compensator).

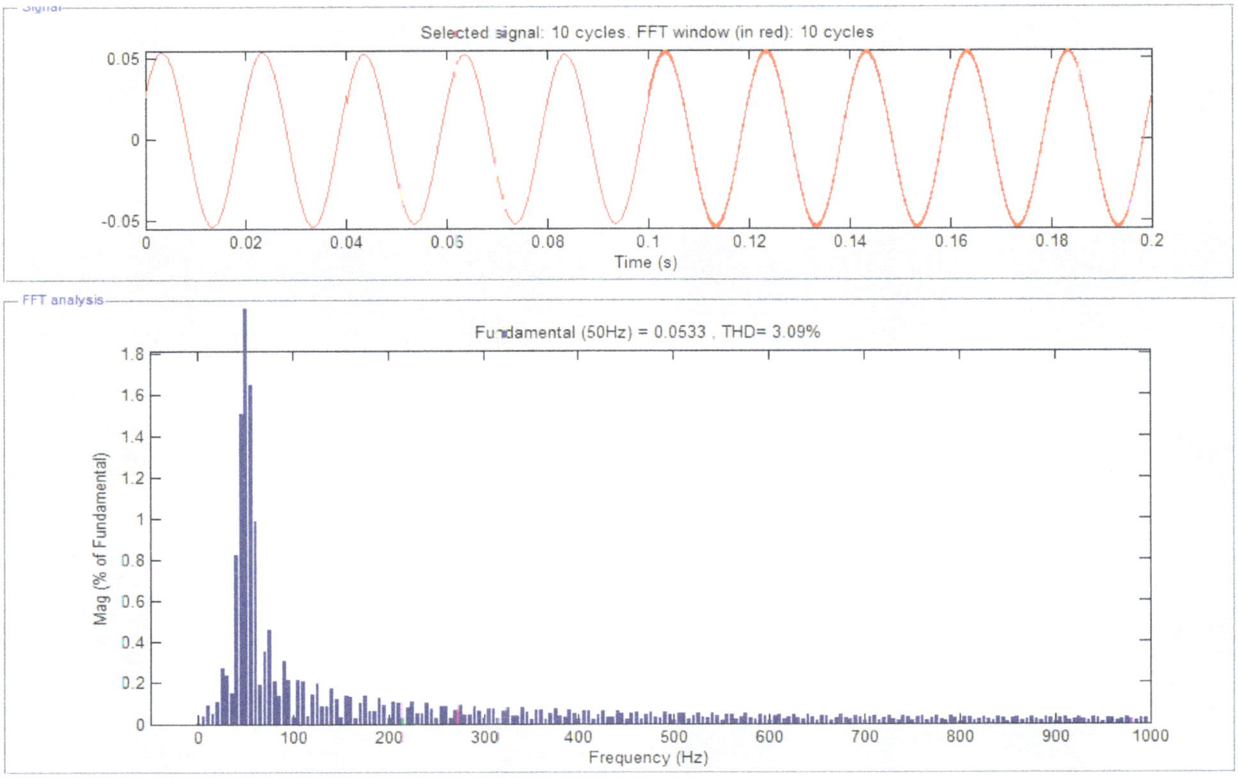

Figure 19. *Load Frequency Spectrum when Compensated Using PID Controller Based DVR.*

Simulation Result for Fuzzy Controller Based DVR

By using FLC, the output waveform will remain same only the distortion which is coming after the fault clearing time is minimized to further extent as shown in Figure20.

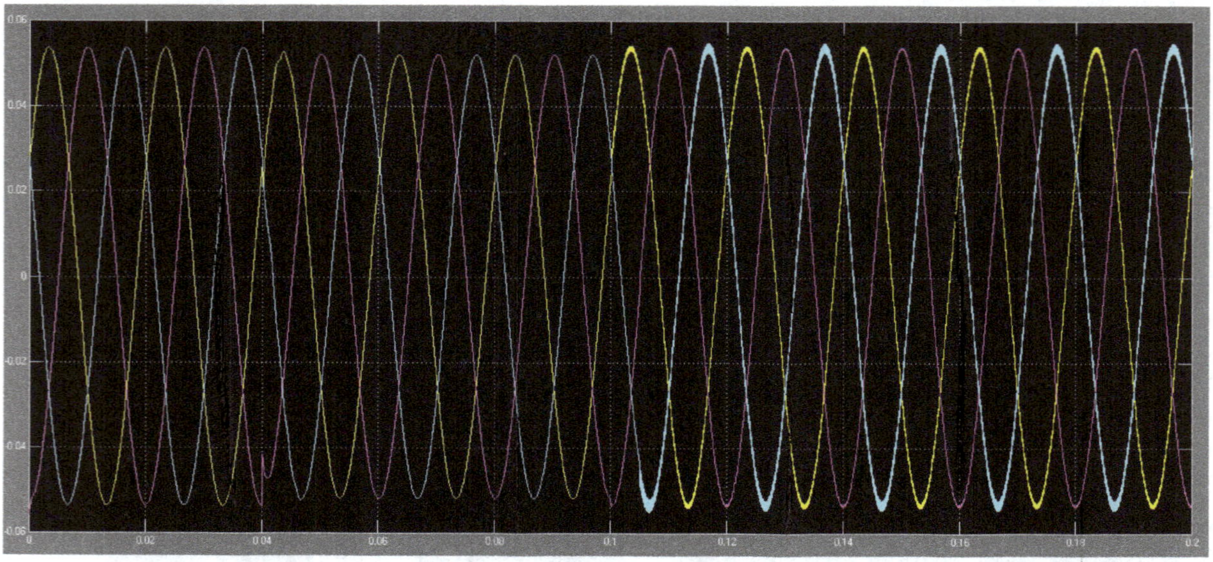

Figure 20. *Load Voltage Waveform when Compensated Using FUZZY Controller Based DVR*

Figure21. shows the FFT analysis for Fuzzy controller based DVR for compensation of a three phase fault. From the graph we can see that the THD is further more reduced from 23.67% to 2.58% while the percentage of THD was 3.09% for PID compensated system.

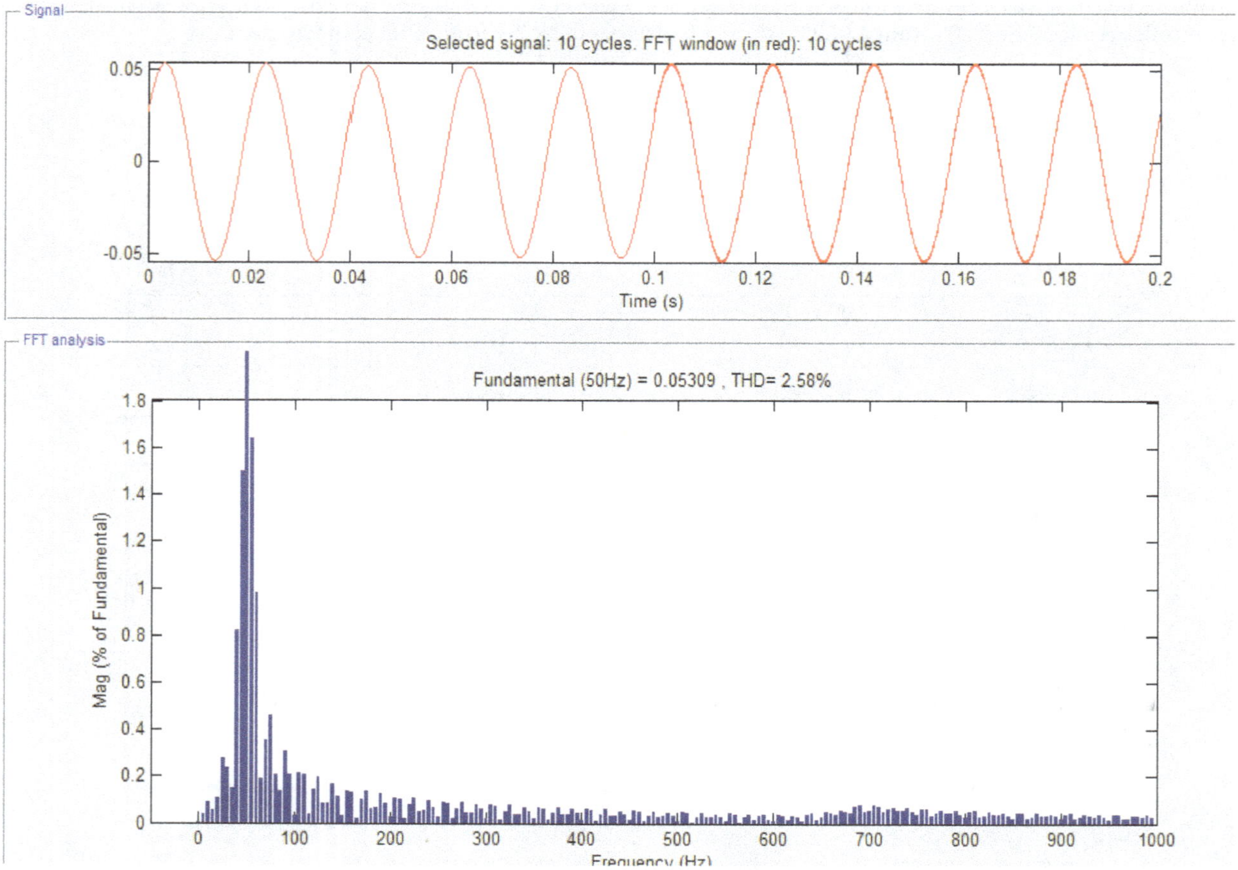

Figure 21. *Load Frequency Spectrum when Compensated Using FUZZY Controller Based DVR.*

Simulation Result for Hybrid PID-Fuzzy Controller Based DVR

As it is the combination of both the above explained compensation techniques, so the results for this controller are supposed to be good and the results are really much up to our expectations as shown in the Figure22. The harmonics in the output waveform are further reduced to certain level.

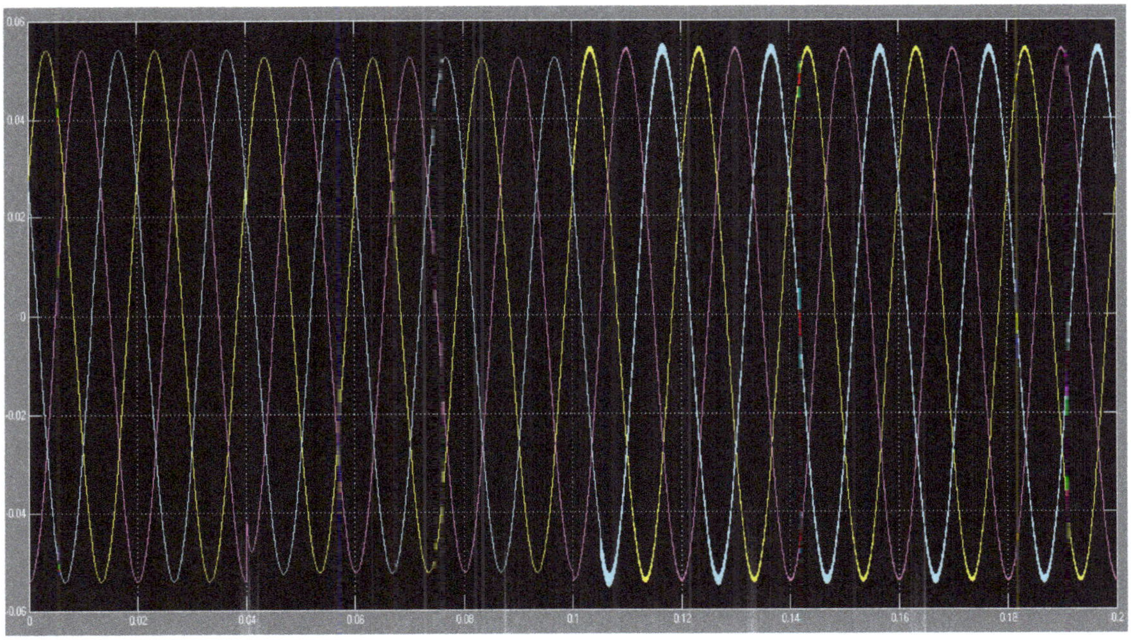

Figure 22. *Load Voltage Waveform When Compensated using PID FUZZY Controller Based DVR.*

The values for the reduced THD are shown in load frequency spectrum for hybrid PID-Fuzzy controller based DVR in Figure23. The percentage for THD is now reduced to a level of 2.42% which is really less than 3.09% of PID controller and as well as 2.58% of Fuzzy controller.

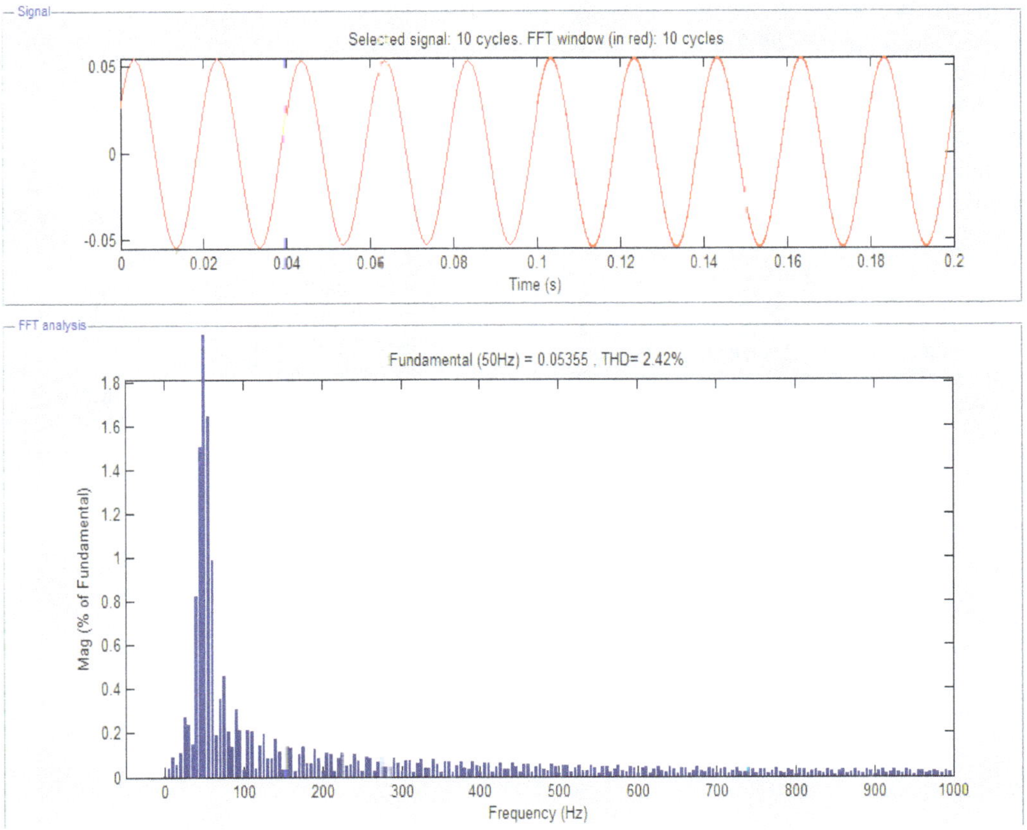

Figure 23. *Load Frequency Spectrum When Compensated using PID FUZZY Controller Based DVR.*

As given in the Table 3, all the three control strategies namely, PID controller based DVR, Fuzzy controller based DVR and hybrid PID-Fuzzy based DVR give the respective values as shown in the Table no. 3 during a three phase fault in the system. In which hybrid PID-Fuzzy gives better compensation by reducing the THD level of uncompensated system from 23.67% to a much reduced value of 2.42%% as compared to 2.58% with Fuzzy controller and 3.09% with PID controller.

Table 3. *THD Values for Different Control Strategies*

Controllers	Load voltage (50Hz)	THD (%)
PID	0.05333	3.09%
FUZZY	0.05359	2.58%
PID- FUZZY	0.05355	2.42%

8. Conclusions

In this work, the various control strategies are employed and tested for 13kV distribution system. The PID controller based DVR, fuzzy controller based DVR and PID-Fuzzy controller based DVR's are connected step by step in the compensated feeder to compare their performances. The effectiveness of different control techniques based DVRs for three phase fault have been investigated. As seen from the load voltage waveform and frequency spectrum of uncompensated system, the THD level is reduced effectively from 23.67% to a much less value of 3.09% in case of PID controller. Load voltage waveforms and frequency spectrum of Fuzzy control scheme for a three phase fault depicts that the harmonics are effectively reduced to a less value as compared to 2.58% with Fuzzy controller and 2.42% with Hybrid PID-Fuzzy controller. Simulation results indicate that Hybrid PID-Fuzzy compensation techniques provide better compensation to the system as compared to the linear PID technique or Fuzzy controller based DVR connected to the feeder during three phase fault.

References

[1] Rosli Omar, N.A.Rahim, Marizan Sulaiman, "Dynamic Voltage Restorer Application for Power Quality Improvement in Electrical Distribution System: An Overview", Australian Journal of Basic and Applied Sciences, vol.5, pp.379-396, December 2011.

[2] R. H.Salimin and M. S. A.Rahim, "Simulation Analysis of DVR Performance for Voltage Sag Mitigation", IEEE 5th International Conference iineering and Optimization, pp.261-266, June 2011.

[3] Ravilla Madhusudanl, G. Ramamohan Rao "Modeling and Simulation of a Dynamic Voltage Restorer (DVR) for Power Quality Problems Voltage Sags and Swells", IEEE-International Conference On Advances In Engineering, Science And Management (ICAESM -2012) March 30, 31,2012 442.

[4] Jurado, Francisco, Valverde, M, "Fuzzy Logic Control of a Dynamic Voltage Restorer", IEEE International Symposium on Industrial Electronics", vol. 2, pp.1047-1052, May 2004.

[5] A.Luo, C.Tang, Z.Shuai, J.Tang, X.Y. Xu, D.Chen, "Fuzzy-PID Based Direct-Output- Voltage Control Strategy for the Statcom Used in Utility Distribution systems", IEEE Transactions on Industrial Electronics, vol.56, pp.2401-2411, July 2009.

[6] S.Chauhan, V. Chopra, S.Singh, "Power System Transient Stability Improvement Using Fuzzy-PID Based STATCOM Controller", 2nd International Conference on Power, Control and Embedded Systems, pp.1-6, December 2012.

[7] M.Srivatsa, S. Srinivasa Rao "THD analysis with custom power devices at distorted load conditions: Implementation of UPQC using PI-Resonant controller" International Journal of Innovative Research in Science, Engineering and Technology Volume 3, Special Issue 1, February 2014 International Conference on Engineering Technology and Science-(ICETS'14)

[8] RAMANPREET KAUR, PARAG NIJHAWAN "Comparative Analysis Of The Pi, Fuzzy And Hybrid Pi-Fuzzy Controllers Based DVR", Thapar University.

[9] Harkesh Pathak, Mohd. Iliyas "Power Quality Improvement Of Distribution Networks Using Dynamic Voltage Restorer" Department of Electrical Enginering, AFSET.

[10] U.V Krishnan, M.Ramasamy, "An Enhancement Method for the Compensation of Voltage Sag/Swell and Harmonics by Dynamic Voltage Restorer", International Journal of Modern Engineering Research, vol.2, pp.475-478, March-April 2012.

[11] H.P.Tiwari and S.K.Gupta, "Dynamic Voltage Restorer Against Voltage Sag", International Journal of Innovation, Management and Technology, vol.1, pp.232-237, August 2010. [15] P.C Loh, D.M Vilathgamuwa, S.K Tang, H.L Long, "Multilevel Dynamic Voltage Restorer", International Conference on Power System Technology, vol.2, pp.1673-1678, November 2004.

Factors Affecting the Harmonics Generated by a Group of CFLs: Experimental Measurements

Muhyaddin J. H. Rawa, David W. P. Thomas

Electrical Systems and Optics Research Division, the University of Nottingham, Nottingham, NG7 2RD, UK

Email address:
mrawa@kau.edu.sa (Muhyaddin J. H. R.), dave.thomas@nottingham.ac.uk (David W. P. T.)

Abstract: The penetration of nonlinear loads on power systems increases the harmonics that can cause detrimental problems to power systems such as an increase in voltage distortion, equipment failure, system resonance, increase system losses and decrease system efficiency. Many factors affect the harmonics produced by nonlinear loads. Hence, the aim of this paper is to evaluate the impacts of system voltage, impedance, frequency and background voltage distortion on the harmonics generated by a group of Compact Fluorescent Lamps (CFLs). Experimental measurements are performed for this purpose.

Keywords: Distortion, Harmonics, Power Quality, CFLs, Nonlinear Loads, THD

1. Introduction

Recently, due to the widespread use of power electronics based nonlinear loads and equipment, distortion levels in power systems are increased. It has been estimated that by 2012, 60% of the loads on power systems will be nonlinear [1]. The proliferation of nonlinear loads increases the harmonic level in power system networks and can cause severe problems such as an increase in voltage distortion, system resonance, a decrease in Power Factor (PF), increase system losses, equipment failure, thermal effects on rotating machines and decrease in the overall system efficiency [2-7].

Lighting forms one of the most important loads in power systems. According to [8], about 25%-35% of the global generated power is consumed by lighting. Another study claims that lighting energy constitutes 15% of the total energy in residential buildings and 30% in commercial buildings [5].

Due to the advance in power electronics technology, Fluorescent Lamps (FLs) and Compact Fluorescent Lamps (CFLs) are becoming increasingly more popular for many reasons. First, the efficiency of FLs is three to four times the efficiency of Incandescent Lamps (ILs) for the same light output. FLs efficiency can also be improved by 20%-30% using a higher switching frequency (>25 kHz) [5, 9]. Second, lifetime of a typical CFL is two to three times that of an IL.

FLs and CFLs are discharge lamps; hence, they require a ballast to establish a high initial voltage across the lamp tube for proper lamp ignition and to limit the lamp current once the arc is established [5, 6, 9]. There are electronic ballasts and magnetic ballasts. Magnetic ballasts use iron-core transformers for stable operation; however, this might cause extra heat losses across the ballast [5, 9]. These transformers also make magnetic ballasts bulky and heavy. Magnetic ballasts operate at a line frequency of 50 or 60 Hz, which cause light flickering. Magnetic ballasts need a pre-heat or switch start circuit to heat up the tube to form the required arc across fluorescent tubes. A standard magnetic ballast current distortion is 15% [6]. Typically, lifetime of electronic ballasts is less than 8 years, while lifetime of magnetic ballasts is more than 30 years [10]. Magnetic ballasts will be phased out from the United Kingdom market by 2017 because their efficiencies are much lower than those of electronic ballasts [11].

Electronic ballasts use SMPSs to convert the power frequency to a very high switching frequency in the range of 25 to 40 kHz [5]. Due to the fact that Electronic ballasts employ SMPSs, their current distortion could be as high as 100% or even more [12-15]. Electromagnetic Interference (EMI) filters are used to minimize the high frequency conducted noise [5, 6]. In order to minimize their cost and size, most of the commercial CFLs nowadays do not include Power Factor Correction (PFC) circuit [9]. Hence, a typical FL with electronic ballast has a power factor of 0.6 [12, 15].

A number of studies have been conducted to investigate the factors that can affect the harmonics produced by nonlinear loads. In [16-22], attenuation, which refers to the interaction of voltage and current distortions due to shared system impedance, and diversity, which refers to the partial cancellation of harmonic currents due to phase angle diversity, are discussed. The effects of mixing single and three phase non-linear loads on harmonic cancellation are also quantified in [23]. Harmonic cancellation due to different single-phase nonlinear loads is illustrated in [24].

The influence of supply voltage distortion on the harmonic currents of a single-phase diode bridge rectifier is discussed in [25]. Effects of voltage distortion on the harmonics generated by household appliances are described in [26-29]. However, all previously mentioned experimental studies have not employed purely sinusoidal power supplies. Hence, harmonics quantification due to a particular factor is not fully assessed.

Notwithstanding, authors in [30] discuss the effects of voltage, impedance and frequency variations on the harmonics generated from a single Personal Computer (PC). The impact of linear loads on the harmonics generated from nonlinear loads is described in [31]. Different factors that can affect the harmonics generated by a cluster of PCs are discussed in [32, 33] including system voltage, impedance, frequency, background voltage distortion, percentage of linear loads, attenuation and diversity effects.

This paper aims to investigate the effects of source voltage, frequency, impedance and background voltage distortion on the harmonics generated by a group of CFLs. One of the key objectives of this research study is to evaluate experimentally the effects of different factors individually and precisely. This is achieved by controlling the input voltage magnitude, frequency, source impedance and background distortion using a programmable AC source Chroma™ 61511 [34]. The Chroma helps in isolating the test rig and filtering out harmonics and fluctuations from the mains. Also, a high accuracy KinetiQ PPA1530 Power Analyzer was used to monitor the input voltage and current of the device under test [35].

Due to the fact that the current drawn by a single CFL is too small, its voltage distortion is almost negligible. Therefore, variation in source voltage, frequency or impedance has a minor effect on voltage distortion. Hence, only current distortion is considered. However, the background voltage distortion affects voltage harmonics notably. Therefore, both THD_i and THD_v were evaluated.

Voltage, impedance, frequency and background voltage distortion variations are within the permissible limits according to BS 7671, IEEE 519, for National Grid and Central Networks in the UK [36-39]. One parameter was changed at a time while the other factors were kept constant at nominal values. Nominal values for the system were set to be 240V, 50Hz, 0.25Ω and zero background voltage distortion (a pure sine wave voltage source).

The paper is organized as follows: section II presents harmonic indices calculations based on IEEE 1459 [40]. Section III discusses the CFL Model. The factors that can affect the harmonics generated from the CFLs are introduced in section IV. Finally, Section V presents the summary and conclusions.

2. Harmonic Indices Calculations

The most commonly used harmonic quantification indices are the Voltage and Current Total Harmonic Distortion (THD_v and THD_i) [40].

The RMS voltage and current are [40]:

$$V_H^2 = V^2 - V_1^2 \tag{1}$$

$$I_H^2 = I^2 - I_1^2 \tag{2}$$

Voltage and current total harmonic distortion can be calculated as follows [40].

$$THD_v = \frac{V_H}{V_1} = \sqrt{\left(\frac{V}{V_1}\right)^2 - 1} \tag{3}$$

$$THD_I = \frac{I_H}{I_1} = \sqrt{\left(\frac{I}{I_1}\right)^2 - 1} \tag{4}$$

where

V, V_1 and V_H: total, fundamental and harmonic voltages, V

I, $I_1$and I_H: total, fundamental and harmonic currents, A

3. CFL Model

In general, a CFL consists of four main block diagrams, the EMI filter, rectifier, resonant inverter stage with its control system and lamp [8, 9, 12, 41-43]. According to [5], EMI filters provide large attenuation for high switching noise currents that are generated due to the rapid changes of di/dt and dv/dt at high switching frequency. The rectifier transforms the 50Hz ac voltage into a dc. The function of the resonant inverter is to convert the dc voltage to an ac voltage at a very high switching frequency suitable to drive the lamp. The harmonics generated by the CFLs are mainly caused by the ac/dc and dc/ac converters.

4. Factors Affecting the Generated Harmonics

In this section, different factors that may affect the harmonics generated by a cluster of CFLs are introduced through laboratory measurements.

4.1. Supply Voltage Variation

Supply voltages in power system can vary within prescribed limits. In order to evaluate the impact of voltage variation on the harmonics produced by a group of CFLs, the input voltage varies from 228V to 252V, ±5% on a 240V supply. It can be shown from Figure 1 that increasing supply voltage increases THD_i.

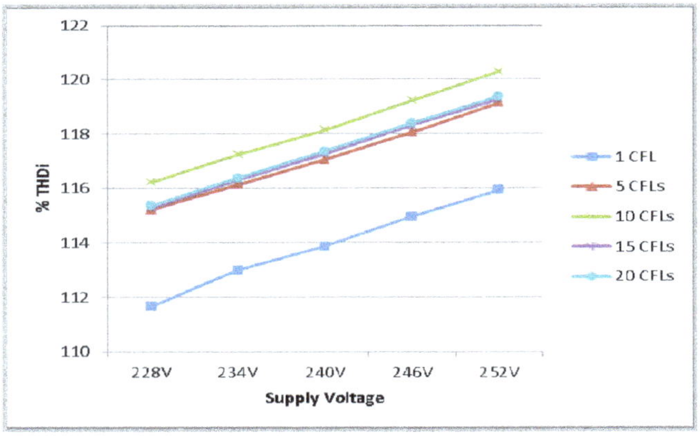

Figure 1. *THD$_i$ vs. supply voltage variation*

Figure 1 also shows that all curves have the same slope; therefore, the same percentage effect due to the supply voltage changes.

4.2. Source Impedance Magnitude and X/R Ratio Variations

Cable length and system loading affect source impedance magnitude and X/R ratio. Figure 2 illustrates that for a group of up to 20 CFLs supplied through a wide range of system impedance magnitudes, increasing the source impedance has almost no impact on the harmonics generated from the CFLs. In this case, the system impedance magnitude varies from 0.15Ω to 0.95Ω while the X/R ratio is 15°.

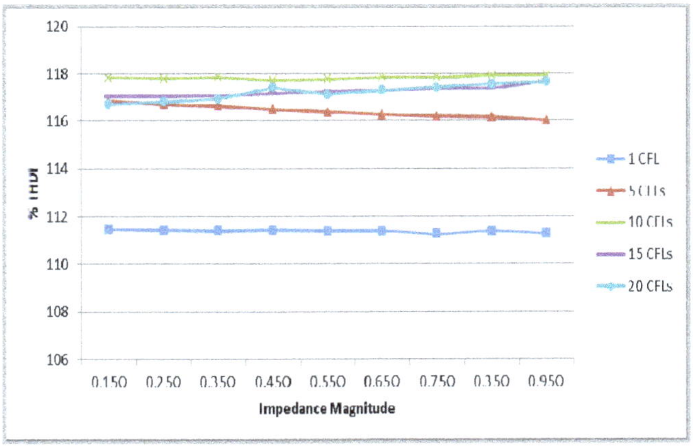

Figure 2. *THD$_i$ vs. source impedance magnitude variation*

X/R ratio in power networks is quite high. This is because the impedance of the generation and transmission equipment is mainly inductive. However, as cable sizes inside commercial offices and residential buildings are of small sizes, their X/R ratios are lower. Therefore, the source impedance seen by commercial and residential loads towards the source varies depending on their location within the building. Figure 3 proves that for a single CFL, increasing X/R ratio has almost no effect on the generated harmonics. However, when a group of 5 or more CFLs is considered, THD$_i$ increases with increasing the X/R ratio.

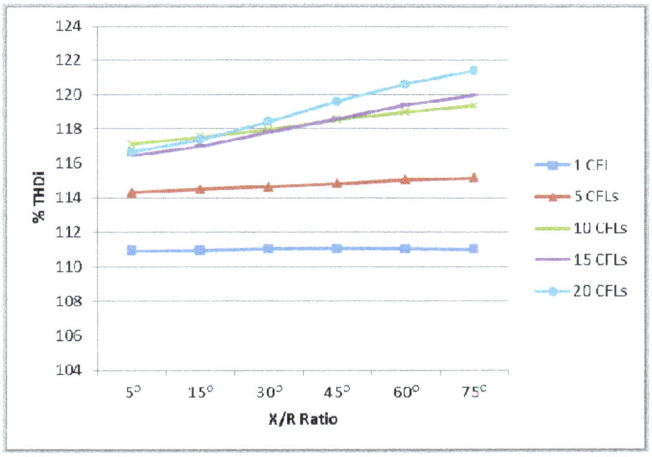

Figure 3. *THD$_i$ vs. source impedance X/R ratio variation (arc tan X/R)*

In this case, the system impedance X/R ratio varies from 5° to 75° while the impedance magnitude is set to be 0.25Ω.

4.3. System Frequency Variation

System frequency can change due to faults on bulk transmission systems or the disconnection of large loads [6]. When a group of up to 20 CFLs are connected simultaneously to a single supply, changing the system frequency from 49.5Hz to 50.5Hz has very minor impact on the generated harmonic currents as seen in Figure 4.

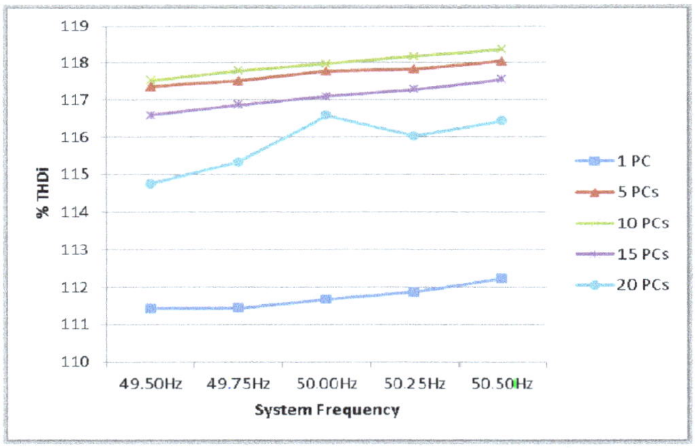

Figure 4. *THD$_i$ vs. system frequency variation*

4.4. Background Voltage Distortion Variation

Due to the ever increasing number of nonlinear loads in power systems, source voltages are becoming more distorted. Figure 5 shows the impact of background voltage distortion on the harmonics generated from CFLs. It can be shown from the Figure that increasing the background voltage distortion within the IEEE recommended limits [36] increases current distortion notably. For example, for a group of 20 CFLs increasing the voltage distortion from 0% (sinusoidal waveform) to 5% increases THD$_i$ from 117.4 to 125.9%.

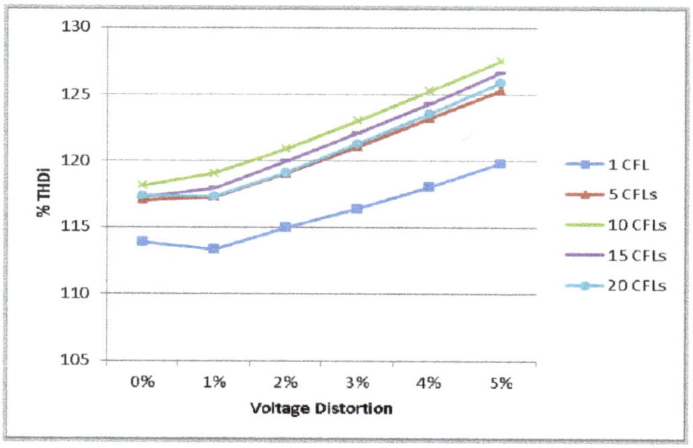

Figure 5. *THD_i vs. background voltage distortion variation*

In power system analysis, it is very important to assess the harmonics penetration and their impact on the networks and customers. For this reason IEEE 519 not only recommends the THD_i limits, but also set maximum permissible limits for individual harmonics. For this purpose, assessing the effects of background voltage distortion on individual harmonics is important.

Figure 6 shows that increasing voltage distortion to 5% decreases the magnitudes of the 3rd harmonic currents. The percentage decrements tend to be more when more CFLs are considered.

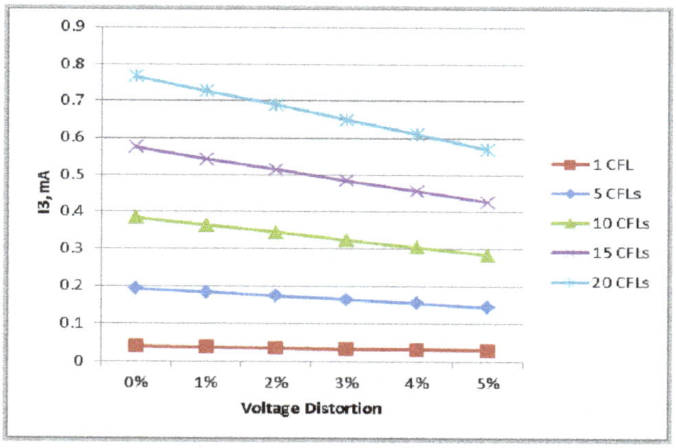

Figure 6. *3rd harmonic currents vs. background voltage distortion variation*

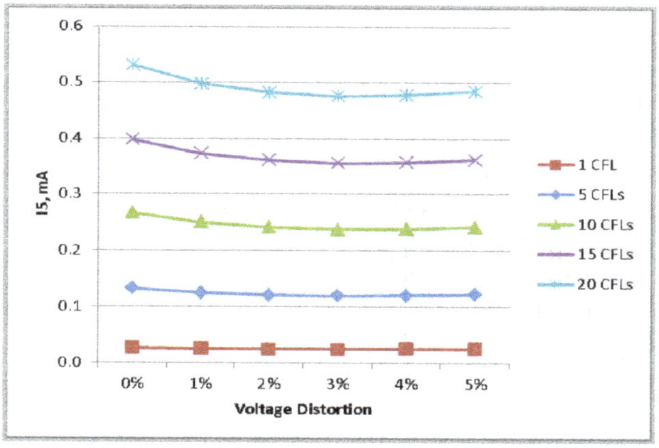

Figure 7. *5th harmonic currents vs. background voltage distortion variation*

Increasing the background voltage distortion has a minor impact on the 5th harmonic currents when a cluster of 10 or more CFLs are turned on simultaneously as shown in Figure 7. Notwithstanding, it has no effect when 5 or less CFLs are considered.

However, increasing the background voltage distortion from 0 to 5% increases the 7th harmonic currents as shown in Figure 8.

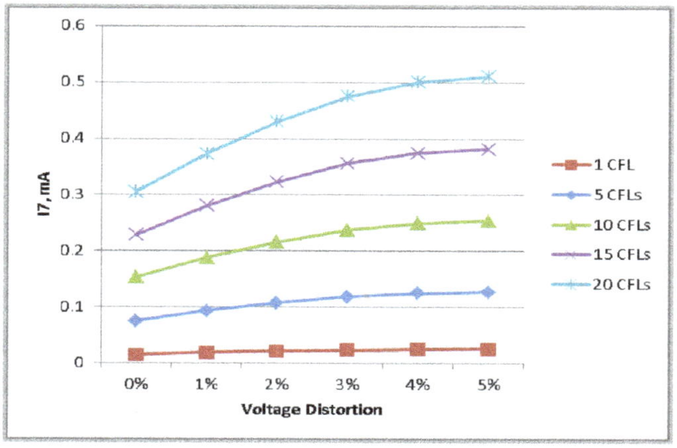

Figure 8. *7th harmonic currents vs. background voltage distortion variation*

As the background voltage distortion increases from 0 to 5%, THD_v and individual harmonic voltages increase sharply as shown in Figures 9 to 12.

As the background voltage distortion is mainly produced by the Chroma power supply and the voltage distortion caused by the CFLs are very small compared to the former, the curves shown if Figure 9 are almost identical. The same is also true for all curves of Figures 10 to 12.

It should be noted that the different background voltage distortion levels were obtained by adding individual voltage harmonic sources in series with the 50Hz fundamental voltage source which distorts the supply voltage. The Chroma power supply can be programmed to give voltage distortion up to 5%.

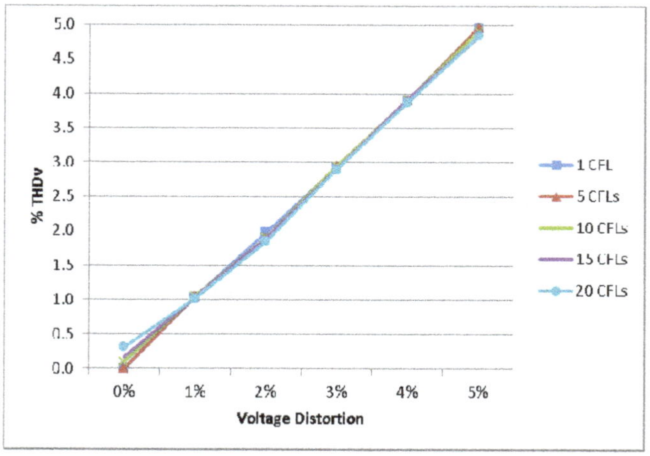

Figure 9. *THD_v vs. background voltage distortion variation*

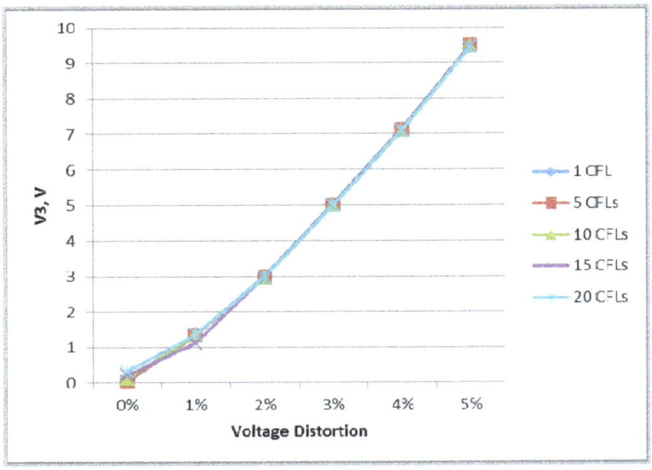

Figure 10. 3rd harmonic voltages vs. background voltage distortion variation

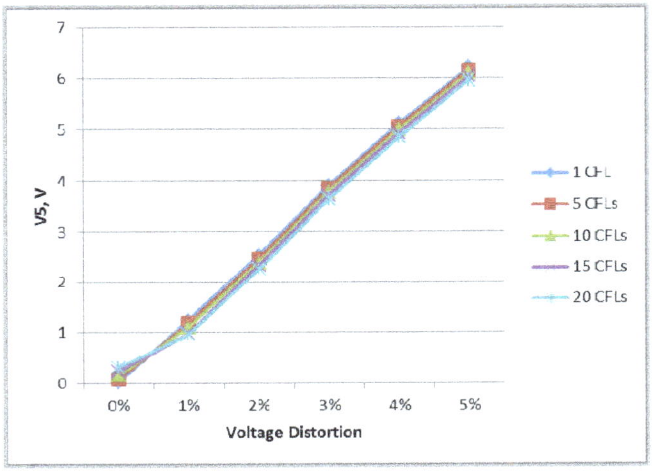

Figure 11. 5th harmonic voltages vs. background voltage distortion variation

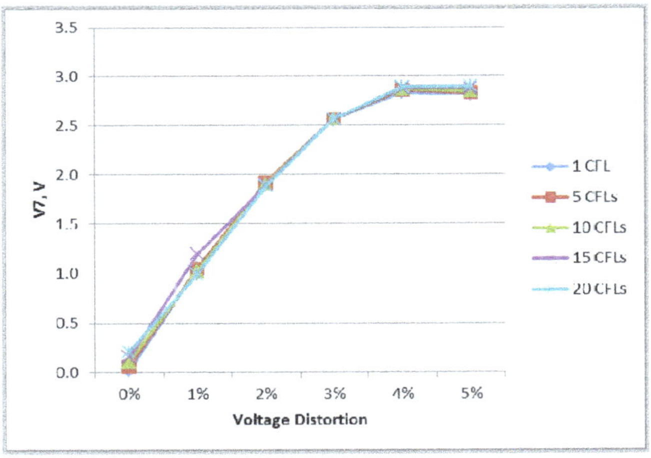

Figure 12. 7th harmonic voltages vs. background voltage distortion variation

5. Conclusion

Power systems harmonics are continuously varying due to many factors. In this paper the impacts of source voltage, impedance, frequency and background voltage distortion variation on the harmonics generated from a group of CFLs have been evaluated for the normal variation of parameters within a power supply. Although system voltage, impedance X/R ratio and frequency affect these harmonics slightly, system impedance magnitude has a very minor impact on the produced harmonics.

Notwithstanding, the background voltage distortion affects THD_i, THD_v and individual harmonic currents and voltages significantly when the background voltage distortion varies within the IEEE 519 recommended limits. Individual harmonic currents tend to behave differently with increasing the background voltage distortion as the 3rd harmonic currents decrease, 5th harmonic currents decrease then increase whereas the 7th harmonic currents increase.

Moreover, these factors are not isolated from each other. Rather, they occur simultaneously and their accumulative effects can be effectively high or some kind of harmonic compensation may occur.

Acknowledgment

The authors would like to acknowledge King Abdulaziz University for their financial support of this PhD study.

References

[1] J. C. Das, "Power System Analysis: Short-Circuit Load Flow and Harmonics", Marcel Dekker, 2002.

[2] E. F. Fuchs and M. A. S. Masoum "Power Quality in Power Systems and Electrical Machines", Academic Press, 2008.

[3] B. W. Kennedy, "Power Quality Primer", McGraw-Hill, New York, 2000.

[4] F. C. De La Rosa, "Harmonics and Power Systems", CRC Press, 2006.

[5] N. Mohan, T. M. Undeland, and W. P. Robbins, "Power Electronics: Converters, Applications and Design", 3rd ed., New York: John Wiley & Sons, 2003.

[6] R. C. Dugan, M. F. McGranaghan, S. Santoso, and H. Wayne Beaty, "Electrical Power System Quality", McGraw-Hill Professional, 3rd ed., 2012.

[7] A. Ghosh and G. Ledwich, "Power Quality Enhancement Using Custom Power Devices", Kluwer Academic, Dordrecht, 2002.

[8] Shrivastava, A.; Singh, B.; , "PFC Cuk converter based electronic ballast for an 18 W compact fluorescent lamp," Industrial and Information Systems (ICIIS), 2010 International Conference on , vol., no., pp.393-397, July 29 2010-Aug. 1 2010.

[9] Lam, J.C.W.; Jain, P.K.; , "A High-Power-Factor Single-Stage Single-Switch Electronic Ballast for Compact Fluorescent Lamps," Power Electronics, IEEE Transactions on , vol.25, no.8, pp.2045-2058, Aug. 2010.

[10] Wei Yan; Tam, E.; Hui, S.Y.; , "A Semi-Theoretical Fluorescent Lamp Model for Time-Domain Transient and Steady-State Simulations," Power Electronics, IEEE Transactions on , vol.22, no.6, pp.2106-2115, Nov. 2007.

[11] "Guidance on current and forth coming legislation within the lighting sector,": The Institution of Lighting Professionals, 2011.

[12] Ying-Chun Chuang; Chin-Sien Moo; Hsien-Wen Chen; Tsai-Fu Lin; , "A Novel Single-Stage High-Power-Factor Electronic Ballast With Boost Topology for Multiple Fluorescent Lamps," Industry Applications, IEEE Transactions on , vol.45, no.1, pp.323-331, Jan.-feb. 2009.

[13] Jing Yong; Liang Chen; Nassif, A.B.; Wilsun Xu;, "A Frequency-Domain Harmonic Model for Compact Fluorescent Lamps," Power Delivery, IEEE Transactions on , vol.25, no.2, pp.1182-1189, April 2010.

[14] Cunill-Sola, J.; Salichs, M.; , "Study and Characterization of Waveforms From Low-Watt (25 W) Compact Fluorescent Lamps With Electronic Ballasts," Power Delivery, IEEE Transactions on , vol.22, no.4, pp.2305-2311, Oct. 2007.

[15] Ying-Chun Chuang; Hung-Liang Cheng; , "Single-Stage Single-Switch High-Power-Factor Electronic Ballast for Fluorescent Lamps," Industrial and Commercial Power Systems Technical Conference, 2006 IEEE , vol., no., pp.1-7, 0-0 0.

[16] A. Mansoor, W. M. Grady, A. H. Chowdhury, and M. J. Samotyj, "An investigation of harmonics attenuation and diversity among distributed single-phase power electronics loads," IEEE Trans. Power Delivery, vol. 10, pp. 467–473, Jan. 1995.

[17] A. Mansoor, W. M. Grady, P. T. Staats, R. S. Thallam, M. T. Doyle, and M. J. Samotyj, "Predicting the net harmonic currents produced by large numbers of distributed single-phase computer loads," IEEE Trans. Power Delivery, vol. 10, pp. 2001–2006, Oct. 1995.

[18] P. J. Moore and I.E. Portugués, "The influence of personal computer processing modes on line current harmonics", IEEE Transactions on power delivery, vol. 18, no. 4, October 2003.

[19] A. I. Maswood and J. Zhu, "Attenuation and Diversity Effect in Harmonic Current Propagation Study", Proc. 2003 IEEE Power Engineering Society (PES) General Meeting, Toronto, Canada, 2003.

[20] Nassif, A.B.; Acharya, J., "An investigation on the harmonic attenuation effect of modern compact fluorescent lamps," Harmonics and Quality of Power, 2008. ICHQP 2008. 13th International Conference on, vol., no., pp.1-6, Sept. 28 2008-Oct. 1 2008.

[21] Nassif, A.B.; Wilsun Xu;, "Characterizing the Harmonic Attenuation Effect of Compact Fluorescent Lamps," Power Delivery, IEEE Transactions on , vol.24, no.3, pp.1748-1749, July 2009.

[22] Nassif, A.B, "Modeling, Measurement and Mitigation of Power System Harmonics", PhD Thesis, University of Alberta, 2009.

[23] S. Hansen, P. Nielsen, and F. Blaabjerg, "Harmonic cancellation by mixing nonlinear single-phase and three-phase loads" IEEE Trans. Ind. Applicat., vol. 36, pp. 152–159, Jan./Feb. 2000.

[24] W. M. Grady, A. Mansoor, E. F. Fuchs, P. Verde and M. Doyle, "Estimating the Net Harmonic Currents Produced by Selected Distributed Single-Phase Loads: Computers, Televisions, and Incandescent Light Dimmers", Power Engineering Society, Winter Meeting, IEEE, 2, 1090-1094, 2002.

[25] A. Mansoor, A.; Grady, W.M.; Thallam, R.S.; Doyle, M.T.; Krein, S.D.; Samotyj, M.J.; , "Effect of supply voltage harmonics on the input current of single-phase diode bridge rectifier loads," Power Delivery, IEEE Transactions on , vol.10, no.3, pp.1416-1422, Jul 1995.

[26] Paulo F. Ribeiro, "Time-varying waveform distortions in power systems", John Wiley & Sons, 2009.

[27] M. H. J. Bollen And I. Y. H. Gu, "Signal Processing of Power Quality Disturbances", John Wiley & Sons, 2006.

[28] Blanco, A.M.; Stiegler, R.; Meyer, J., "Power quality disturbances caused by modern lighting equipment (CFL and LED)," PowerTech (POWERTECH), 2013 IEEE Grenoble, vol., no., pp.1,6, 16-20 June 2013.

[29] Prudenzi, A.; Grasselli, U.; Lamedica, R., "IEC Std. 61000-3-2 harmonic current emission limits in practical systems: need of considering loading level and attenuation effects," Power Engineering Society Summer Meeting, 2001, vol.1, no., pp.277, 282 vol.1, 2001.

[30] Rawa, M.J.H.; Thomas, D.W.P.; Sumner, M.; Chin, J.X., "Source voltage, frequency and impedance variation effects on the harmonics generated from a Personal Computer," Power Electronics, Machines and Drives (PEMD 2012), 6th IET International Conference on , vol., no., pp.1,6, 27-29 March 2012 doi: 10.1049/cp.2012.0210.

[31] Rawa, M.J.H.; Thomas, D.W.P.; Sumner, M., "Harmonics attenuation of nonlinear loads due to linear loads," Electromagnetic Compatibility (APEMC), 2012 Asia-Pacific Symposium on , vol., no., pp.829,832, 21-24 May 2012.

[32] Rawa, M.J.H.; Thomas, D.W.P.; Sumner, M., "Factors affecting the harmonics generated by a cluster of personal computers," Harmonics and Quality of Power (ICHQP), 2014 IEEE 16th International Conference on , vol., no., pp.167, 171, 25-28 May 2014.

[33] Rawa, M.J.H.; Thomas, D.W.P.; Sumner, M., "Background voltage distortion and percentage of nonlinear load impacts on the harmonics produced by a group of Personal Computers," Electromagnetic Compatibility (EMC Europe), 2014 International Symposium on , vol., no., pp. 626,630, 1-4 Sept. 2014.

[34] Programmable AC Source 61511/61512 User's Manual (Chroma), Chroma ATE INC., Version 1.1, 2009, P/N A11 001293.

[35] PPA1500 KinetiQ User Manual, Newtons4th Ltd, 2010.

[36] IEEE Standard 519-1992, "IEEE Recommended Practices and Requirements for Harmonic Control in Electrical Power Systems".

[37] BS 7671:2008+A1:2011, "Requirements for electrical installations". IET Wiring Regulations. 7th edition.

[38] The Grid Code, N. G. E. T. plc, CC.6.1.2, (2011).

[39] (31 August 2011). Central Networks Supply Information. Available: http://www.eonuk.com/distribution/electricians.aspx

[40] "IEEE Standard Definitions for the Measurement of Electric Power Quantities Under Sinusoidal, Nonsinusoidal, Balanced, or Unbalanced Conditions," IEEE Std 1459-2010 (Revision of IEEE Std 1459-2000) , volume, no., pp.1-40, March 19 (2010).

[41] Lam, J.C.W.; Jain, P.K.; Jain, P.K.; , "A Dimmable Electronic Ballast With Unity Power Factor Based on a Single-Stage Current-Fed Resonant Inverter," Power Electronics, IEEE Transactions on , vol.23, no.6, pp.3103-3115, Nov. 2008.

[42] Chan, S. S. M.; Chung, H. S.-H.; Lee, Y.-S.;, "Design and Implementation of Dimmable Electronic Ballast Based on Integrated Inductor," Power Electronics, IEEE Transactions on, vol.22, no.1, pp.291-300, Jan. 2007.

[43] Chang-Shien Lin;, "Low power 60 kHz electrodeless fluorescent lamp for indoor use," IPEC, 2010 Conference Proceedings , vol., no., pp.682-686, 27-29 Oct. 2010.

7

Reduction of Losses and Capacity Release of Distribution System by Distributed Production Systems of Combined Heat and Power by Graph Methods

Parsa Sedaghatmanesh[1, *], **Mohammad Taghipour**[2]

[1]Electrical Power Engineering, Islamic Azad University of Saveh, Markazi, Iran
[2]Industrial Engineering, Science & Research Branch of Islamic Azad University, Tehran, Iran

Email address:
Parsa.sm.86@gmail.com (P. Sedaghatmanesh), mohamad.taghipour@srbiau.ac.ir (M. Taghipour)

Abstract: Formulation of long term program of optimization of energy sector has positive effect on economy of country and improving the role of Iran in global energy markets. One of the results of optimization of energy supply sector is improvement of efficiency and reduction of environmental pollutants of energy generation. There are various optimization solutions in energy supply as combined power and heat generation at proper location of distribution network. This thesis is aimed to locate combined generation source via integrated graph algorithm with sensitivity analysis to reduce electric power loss and release capacity and increase economic productivity. The capacity is determined based on applying restrictions of voltage and available levels of candidate locations in the studied networks. The results of simulation are presented in standard 30-bus IEEE network to evaluate efficiency of the above method.

Keywords: Combined Generation System, Distribution Networks, Placement, Graph Algorithm, Sensitivity Analysis

1. Introduction

Now, due to economic issues, power plants are in big sizes with 150-1000 MW capacity. Indeed, such great power plant needs great economic investment, employees, space and strong and long transmission lines and this leads to increase of cost, design time, installation, operation and maintenance [1]. This great structure leads to vulnerability of system against natural events. Besides environmental problems, these issues increase the tendency to new generation methods. In distributed generation as called little generation at the point of consumption [2], applies distributed generators in distribution system to improve quality of power and eliminate technical problems. Some of the positive effects of DGs activity in network are as follows: Reduction of loss of lines, improvement of voltage profile, reduction of pollutant gases, increase of total return of energy, improving security and reliability of network, improvement of power quality, capacity release of distribution and transfer systems, delay of investment for network development, productivity improvement, reduction of treatment and health costs to reduce pollutants emission and increase of security in sensitive loads [3], [4]. In this study, combined heat and power/CHP connected to network is evaluated as one of DG technologies and their optimal place and capacity in radial distribution network can be determined by new method based on a combination of graph algorithm and sensitivity analysis and with the aim of reduction of losses of power and improvement of profile of voltage. Optimum state of network equipment with DG units is as besides fulfilling designer goals can be economical or they can have economic justification for energy suppliers. Based on high cost of installation and launching combined generation in CHP placement, the goal of minimizing costs is not acceptable. The capital return is similar at the worst and best state and it is better to investigate objective function except the cost. Thus, objective function of loss reduction and voltage profile modification are on priority of placement projects. The capacity of combined generation systems is computed based on the available levels of installation candidate locations in network and voltage restraint and finally placement and sizing algorithm is presented with definite goals of sample network: 30-bus IEEE standard and the results of studies are presented to evaluate the efficiency of above method. Normally, the required electricity of industrial units, commercial and residential buildings is supplied via the major power plants of our country. The thermal need of all of them is generated in the same site. Another method receiving much attention is combined heat and power generation as combined power and heat generation by a system. Thus, besides electric or mechanic power by the system, the loss heat of generator or engine is used as thermal energy. IN this system, the heat loss resources include emitting bases from initial place, cooling cycle and oil of lubrication and by putting thermal convertors, lost heat is recovered as heat.

Distributed/Dispersed Generation (DG) is an old idea referring to the end of 19th century [1]. The required power plants are divided into three in terms of type of resources : Water (using energy of flowing water of river or water stored in reservoir of dams), thermal (using energy of fuels as oil, gas or coal) and nuclear (e.g. atomic energy) to the end of 20th century, other generation technologies were unconventional.

Table 1-1 shows the share of each of above technologies in energy generation over the world to the recent century.

Generally, any technology of electric energy generation as combining in distribution system or it is connected to network via consumer of measurement device [5], can be called distributed generation. DG systems are introduced as modular systems [6] with capacity of less than 100 mvA [7] and less than 10MW [8]. It is predicted that in future, about 25% of new manufacturers of electric power in power networks are dedicated to distributed source [11].

Table 1. *The share of different technologies in energy generation in GW to the end of 20th century [2].*

Region /Technology	Thermal	Water	Nuclear	Others/ Renewable	Total
US	642	176	109	18	945
Southern America	64	112	2	3	181
Western Europe	353	142	128	10	633
Eastern Europe and Russia	298	80	48	0	426
Middle East	94	4	0	0	98
Africa	73	20	2	0	95
Asia and Oceania	651	160	69	4	884
Total	2.175	694	358	35	3.262
Percent	66.6	21.3	11.0	1.1	100

2. Distributed Generation Technologies

Distributed generators are divided into renewable and non-renewable (fossil fuel technologies) including these systems [2]:

Table 2. *Different distributed generation generators [2].*

Technology	Capacity range	Relationship with network
Internal combustion engine	Hundreds of kilowatt to ten megawatts	synchron generator or AC/AC convertor
Combined cycle	Ten megawatt to hundreds of megawatts	
Combustion turbine	Some Megawatts to hundreds of Megawatts	synchron generator
Microturbine	Tens of kilowatts to some megawatts	AC/AC convertor
Fuel cell	Ten kilowatts to ten hundreds megawatts	DC/AC convertor
Wind	Some hundreds watts to some megawatts	Synchron generator
Solar, photolytic	Some watts to some hundreds of kilowatts	DC/AC convertor
Geothermal	Some hundreds of kilowatts to some megawatts	Synchron generator
Ocean	Some hundreds of kilowatts to some megawatts	4-polar synchron generator

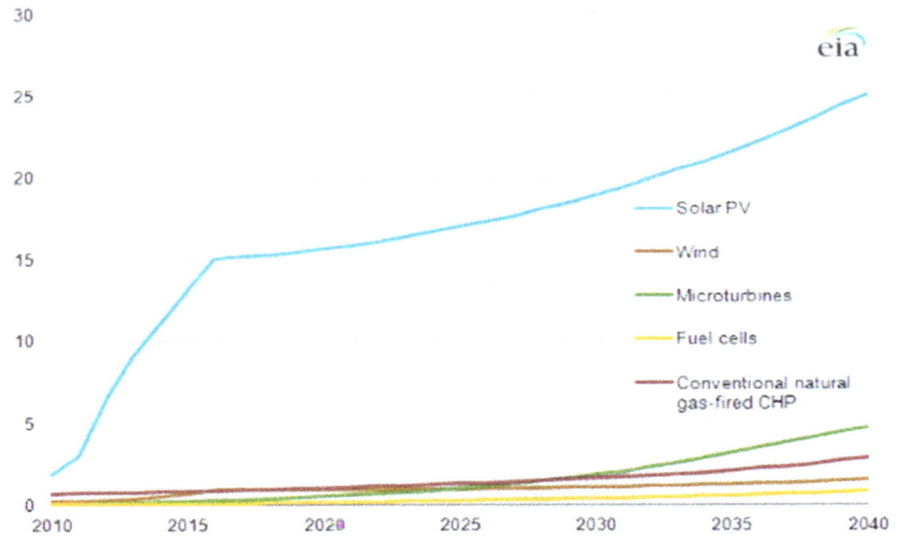

Figure 1. *Estimation of energy generation by distributed generation resources (GW).*

The share of distributed generation technologies in power generation market

DG share in electric energy generation is increased considerably as in the early 90s; Non- Utility Generators (NUGs) dedicate 5% of total generation in US. At the end of this decade, it reached 20% of total generation. In other words, it is increased from 40Gigawatt to more than 150 Gigawatts [2]. Fig 1 shows the annual increase of distributed generation capacity in Gigawatts from 2010 to 2040 [13].

Based on increasing utilities of DG generators in energy supply, future power networks are fed from various distributed resources making general power network and despite the complexity of power supply, the existence of a center as leadership of network and power control can be necessary.

3. CHP and its Division Based on Technology of Incentives [16]

CHP systems are based on separated components of initial incentive (thermal engine), heat recovery convertors and electric connections consider as a unified system. As the combined generation is based on power generation, these systems are classified based on its electric generator. The type of equipment providing the movement force of entire system determines the type of CHP system as including:

- Steam Turbine
- Gas Turbine
- Micro Turbine
- Reciprocating Engines
- fuel cell

Classification of CHP systems

Classification of CHP systems is defined in three levels and based on capacity of generated electric power as in following Table. We cannot consider this division as absolute.

Table 3. *Classification of CHP based on capacity and use.*

Use	Power range (kilowatt)	Capacity
Domestic-administrative	1-100	Small/low
Administrative-Commercial-Industrial	100-10000	Average
Industrial	Above 10000	Big/High

The application and influence of project is based on required heat of the building in terms of qualitative and quantitative aspects. If hot water is required at low scale, commercial piston engines are the best choice. For each kilowatt/h energy, the heat of device can increase water temperature as 500 to 900 kg/°C. Water heating is one of the most common needs of all types of use and the difference is in quality and quantity. In case of the need to steam, turbines are the best choice. The need to heat to be used in ventilation systems is changed over the year and temperature changes. The best places to stabilize the domestic and administrative heat are those with low temperature changes over the year. Briefly, we can summarize the benefits and disadvantages of different CHP s and their capacity in the following Table:

Table 4. *Summary of CHP technology.*

Type of system	Benefits	Disadvantages	Existing capacities
Gas Turbine	High reliability Low pollution Heat with high quality Without the need to condenser	The high use of gas or gas compressor Low return in case of production with low capacity Reduction of generation with the increase of temperature e	500 kilowatts to 250 Mw
Micro turbines	Low and small parts Compressed size and low weight Low air pollution and noise pollution Without condenser Long maintenance distance	High investment cost Low mechanic return Limited to application at low temperature Complexity of connection to electricity network	30kw to 250 kw
Piston engines	High efficiency Flexibility in performance of a part of capacity Rapid launching Relative low investment cost Being used as independent from network Regulation of total generated load Regulated at site with normal operators Function with low pressure gas	High maintenance cost Limited to application at low temperature Relatively high air pollution It should be cold permanently; even output heat is not used. Noise pollution	Spark ignition For local consumption less than 5mw-with high speed 1200 lower than 4mw Compression ignition Low speed To 75mw
Steam turbine	Total high return Variety of fuel Heating more than one terminal High reliability and life The change in force to heat ratio	Long launching Low force to heat ratio	50kw to 250mW
Fuel cell	Low noise pollution High return for various generated loads Separated design	High price Low continuance and force density The need to the process on fuel (except pure hydrogen)	5kw to 2mw

4. Graph Algorithm

Graph algorithm can create each of above algorithms as a part of itself and it is a general algorithm and by selection of a dimensional state compared to the present state is based on constraints and conditions and it is the most flexible algorithm against speed and precision change as the algorithm designer based on easy use of the rest of algorithms can make the speed or precision as dominant. Convergence in graph algorithm is an unavoidable issue as at the worst states, after long time, convergence is occurred, even if convergence is not found. In case of correct design of algorithm, graph algorithm can direct trial and error to the test of all states and this is a specific feature in all existing algorithms. Regarding the escape from relative optimal responses, graph algorithm is one of the strongest algorithms as easily by creating a loop in tree algorithm can escape from relative optimum value. This escape is unique among the algorithms and it is the best escape.

4.1. Explanation of Graph Algorithm

Here, we define the application of graph theory in optimization of an objective function and determination of optimum response. Assume that the system has the following state vector as n state x1 to xn:

$$X = [X_1, \dots, X_n] \tag{1}$$

Also, objective function has the regulation

We try to find a trend to optimize F function determining optimum response as one of the states of X vector. In graph algorithm, we assume that we have n vertices, each vertex refers to a definite state. The objective function in each state as the relevant vertex refers is called the weight of vertex. It means that if n=3, we achieve three vertices as: Inside each vertex, the referred state is mentioned and number of each vertex is written under it. Thus, we have:

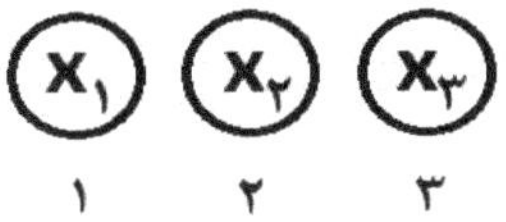

Figure 2. *View of three vertices.*

$$X = [X_1, X_2, X_3] \tag{2}$$

The equation of vertices weight vector:

$$W = [W_1, W_2, W_3] \tag{3}$$

Indeed, optimal response is vertex or minimum weight, at ideal state, investigation of n states and finding the weight of n vertices can direct use to optimum response. The number of industrial system states is high and investigation of all of them requires much time and this leads to solutions with absolute precision. Absolute precision is completely exact responses. At first, we consider a state as initial state. The selection of initial state can be performed after a process on system. For example, in a system, strong relations between system parameters can be selected and this leads to initial state.

Rank 1 is given to the initial state. This value is called the index of number of replications and it is called rank and is denoted by m. For example if ith state is selected as initial state, to find the next replication or next rank, we can move along (n-i) other vertices and investigate their algorithm.

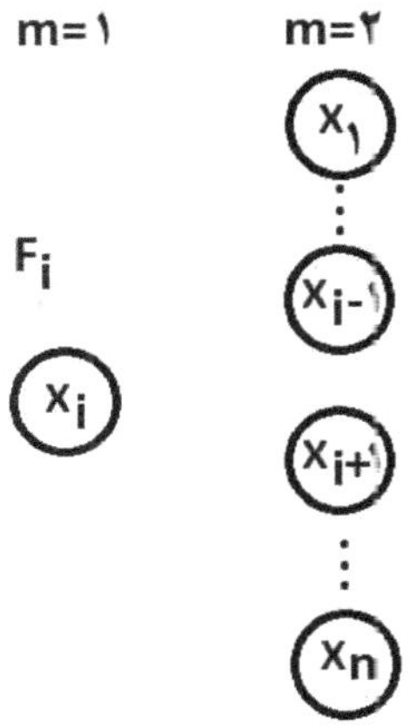

Figure 3. *Ranking vertices of graph.*

This method is the investigation of all states as time-consuming. Thus, a solution should be thought to avoid the investigation of all states of rank 2. At first, some concepts are defined and the mentioned method is defined finally.

a. Edge: The line between two vertices and it exits if the difference of rank of two vertices is one.
b. Edge weight: Edge weight is zero or one. If edge weight is zero, dimension state is not investigated but if edge weight is one, the dimension state is investigated and objective function is evaluated.

After determining the weight of all mentioned edges between vertices rank 1 and vertex rank 1, objective function is considered one for rank two vertices connected to edges with weight 1. If the weight of kth vertex is the lowest, the vertices of all states except XK is eliminated and is used in scale m=3.

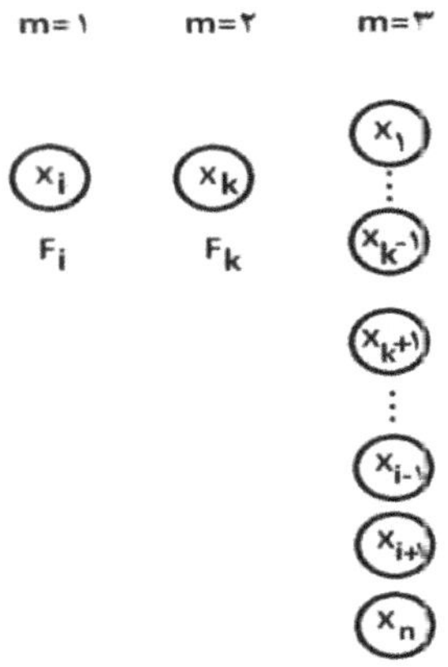

Figure 4. *Formation, scale m=3 graph.*

Now we deal with n-2 vertices. As before, the edges are defined and weight of edges is determined. Then, the weight of vertices connected to edge is calculated with weight 1 or objective function of referred state and this trend is continued to convergence.

4.2. Convergence Condition

If the difference of weight of each vertex with the previous vertex is ignored, we achieve convergence and an optimum state is achieved and it is local or absolute. To escape from local optimas, a design is thought.

4.3. Escaping from Local Optimas

To escape from local optimas, we can use loop in in oriented graph algorithm. For example, Graph: 0(7) is considered.

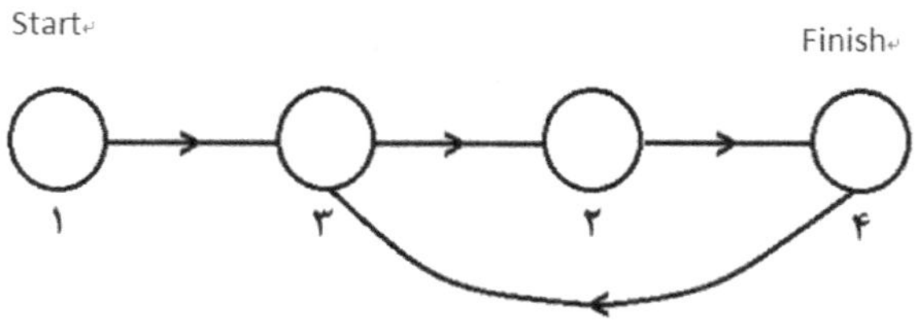

Figure 5. *Graph algorithm with loop of escape from relative optimas.*

In creation of this loop, we can return from final vertex to one of the previous vertices. If vertex 4 is not local optima, after vertex 3, we go to vertex 2, then vertex 4. If vertex four is local optima, as the system is changed, from state 3, we can go to another state. Selection of a vertex from algorithm path from which local optima to loop is created, it depends the function selected by the designer to escape from local optimas. This function is effective on changing responses close to absolute optima and it is one of the key functions of graph algorithm and based on its application is determined by algorithm designer. The mentioned function is called mutation function.

4.4. Algorithm Finish

Decision in algorithm finish is dependent upon the view of designer and type of function and it is after some continuous convergences to a value.

4.5. Conclusion

As it was said, the designer of graph algorithm should determine the followings based on the type of graph algorithm application:

Identification and determining of various system states

Initial state: The stages of initial state are determined as suitable initial selection helps us in absolute optimal solution.

Edge discriminant function: The determination of this function is based on decision making regarding the selection of states and compromise between speed and precision.

Selection of mutation function to escape from local optima's Algorithm finishing method.

5. Objective Functions in Placement of Combined Generation Systems in Distribution Network

The experts and planners of distribution systems aim to use distributed generations namely combined generation resources, growth of economic indices and increase of satisfaction of consumers to reduce environment pollution, reduction of losses, and improvement of reliability and increase of power quality. The benefits of distributed generation are achieved if they are in proper place and optimal capacity in network. The optimal allocation and placement of distributed generation besides mentioned benefits have other benefits and they are not ignored. These benefits include improvement of profile of voltage of load points and placement of voltage of busses of network in allowable range and reduction of power from network feeders due to compensation of losses and compensation of a part of power of network load and this reduces stress on feeders of network namely in output feeders from distribution post of network and this increases life of equipment. The objective function in this study is to evaluate losses reduction of distribution system and improvement of network voltage profile. To evaluate the saving of using CHP in distribution network, installation cost of system is also considered in objective function and a combination of goals of loss reduction and minimum cost is used for the designer.

5.1. Economic Saving

If the dedicated capital to project of CHP application in distribution network is limited, capital return time is considered by experts of network and the cost of installation of CHP systems is used as one of the most effective factors on costs of placement project and establishment of devices is applied in objective function as minimizing of total cost function includes the difference of value of generated energy by combined generation generators from the sum of energy loss costs in year and establishment of combined generation generators is the goal of placement and capacity determination project. The objective function is considered as (4):

$$\mathrm{Cost} = CHP_C + \mathrm{C_{loss}} - \mathrm{CHP}_i \qquad (4)$$

Where $\mathrm{CHP_C}$ is installation cost of CHP and $\mathrm{C_{loss}}$ annual loss cost and $\mathrm{CHP_i}$ is annual generation energy CHP. This function is cost function that is minimized and the most important factor is high cost of CHP. This cost is not compensated over a year with saving cost of loss reduction and annual generation energy. Also, combined generation is not established only to be used in the first year and its generation is considered in life service of system in economic evaluation. It is better to define another objective function as the number of years of capital return by which a comparison is made between life service of system and capital return period and economic aspect of purchase of CHP systems is investigated. The capital return function is introduced as (5).

$$Y = \frac{CHP_C}{CHP_L + \Delta I_L} \quad (5)$$

Where, Y is the number of capital returns years, C is CHP_C installation cost of CHP, CHP_L annual generation energy value CHP and ΔI_L as the cost of compensated loss. This function should be minimized to do capital return at short period and profit of CHP placement is achieved earlier. The installation cost of CHP systems as it was said for per kilowatt generation capacity is averagely 2420dollars and price of per kilowatts hour electric energy in entire electricity network of Iran without subsidy is 0.1$. In calculation of capital return period of each of (9-3) equations is generalized by mentioned values and (6) is achieved:

$$Y = \frac{P_{CHP} \times \sum_{i=1}^{K} CHP_i}{(\sum_{i=1}^{k} CHP_i + \Delta L) \times P_{kw}} \quad (6)$$

Where, P_{CHP} is price of per kilowatt capacity of CHP system, CHP_i capacity of combined generation system ith, K is the number of CHP systems in network, P_{kw} is the value of per kilowatt hour electricity in network, ΔL is total compensated losses.

5.2. Losses Reduction

Losses reduction is followed via various methods as installation of generated generators including combined generation systems in distribution network. Among all possible states for configuration and capacity of CHP systems, optimal state is selected in terms of power loss value in network lines as the study result. This objective function is defined by (7):

$$Loss = 3 \sum_{i=1}^{N} Re(V_i I_i^*) \quad (7)$$

Where, V_i, I_i are voltage and current of ith line of distribution network and N number of network lines

5.3. Modification of Voltage Profile

Combined generation systems can be useful in modification of fluctuation and weakness of distribution network voltage. Based on this reality, the second objective function as considered in another stage of project studies is the deviation function of mean of busses voltage from nominal voltage (Vb) in distribution network and is defined as:

$$V = \sqrt{\frac{\sum_{i=1}^{M} (V_i - V_b)^2}{M}} \quad (8)$$

Where M is number of busses, V_i is bus ith voltage, Vb is voltage of basis of network.

By minimizing this function, we can achieve a condition for capacity and place of CHPs as optimal where busses voltage has the least difference with nominal value 1p.u. If the goal is reaching voltage of all buses to definite limit, this condition is added beside objective function and when this condition is satisfied, CHP systems enter the network.

6. Problem Solving of Determining Place and Optimal Capacity of Combined Generation Systems in Distribution Network

After determining the conditions of distribution system, feasibility of CHP installation based on load type, installation site and introduction of objective functions in project, it is required to define the general trend of problem solving in the form of stages and steps of start to end to have accurate perception of total studies. The problem of placement and suitable capacity of systems is based on goals. The first step is placement as by graph algorithm with high speed in problem solving compared to most of common algorithms and methods [5], we try to determine placement of combined generation generators. In this algorithm, for each of studied scenarios, this question is raised which capacities of systems are required by designer and which combined generation systems should be placed. Thus, we enter sizing field and require the size of systems for placement scenario from sizing block. The result of this block depends upon the calculations of load distribution in sample network. Thus, we use load distribution block to respond the higher block of sizing and the response achieves placement block and place scenario is analyzed and the result is stored. Then, another scenario and then the best state are determined in terms of objective function. In explanation of project performance, we go from smaller block as load distribution to main block as placement of system and problem solving method is defined.

6.1. Determination of Optimal Capacity of Combined Generation Systems in Distribution Networks

In this section, by knowing the system installation location, its capacity is determined to achieve highest profit of CHP system and escape from disorder in network performance. One of the important features of network is voltage of busses as kept in the defined limit. CHP generation power control can avoid voltage excess in network. Generally, in CHP sizing, definite conditions are used:

a. Distribution networks are used as radial without loop and placement of CHP is limited to future networks.
b. Constraint of voltage excess is considered. Busses voltage is modified by injection of output power of CHP but voltage excess is not occurred. Thus, in sizing CHP systems, each time with load distribution, voltage of busses is checked and in case of voltage exiting from the defined range 0.01lower or upper than 1 per unit, even if the losses show the minimum value, the state is eliminated from possible states that CHP installation doesn't make any problem in its performance.
c. The higher the power of CHP system, or condition with lack of violation of problem constraints, power losses are lower.

If maximum generation power of CHP is considered as Pmax and sum of load of other busses of CHP with local consumption of bus of CHP is considered as L, optimal power of CHP production in terms of objective function is called P_{opt} and each step as determined by precision of problem solving is assumed as Ps and we achieve P_{opt} in which objective function is minimized. Flowchart 5 shows work trend.

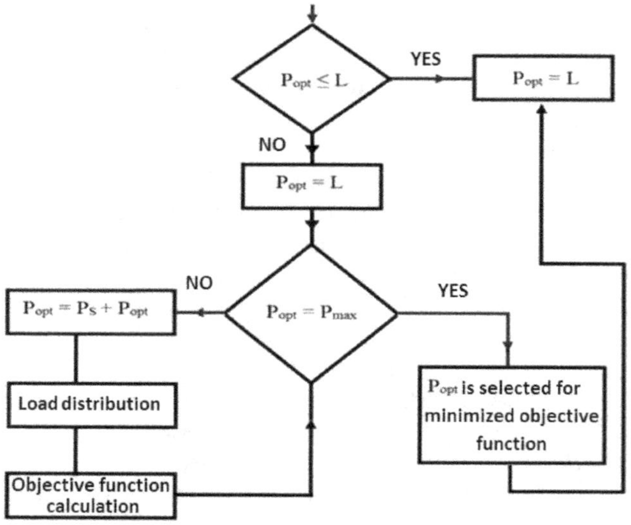

Figure 6. *CHP sizing flowchart.*

6.2. Determining Optimal Location of Combined Generation Systems in Distribution Networks

IN power system studies, if a state is selected among a series of initial states and the number of initial states are high, as all of them are not evaluated in terms of speed and memory, an algorithm is selected among various problem solving algorithms as genetics, neural networks, fuzzy, elimination and graph to define the movement between states based on proposed method of algorithm and without investigation of all of them, the best state is defined for designer.

Among various algorithms, graph algorithm has suitable speed and high efficiency in solving locating problems in power network [6]. This algorithm needs the definition of edge discriminant function as guiding the movement among various states. This study applies sensitivity analysis and sensitivity index in two types of studies of loss reduction and voltage profile modification as edge discriminant function of graph algorithm. Each movement is directed based on sensitivity and the effect of candidate bus to achieve the goal of reduction of losses and mean deviation of voltage to the next state.

6.3. Sensitivity Analysis [7]

In solving optimization problems, we are faced with two following methods [8]:

a. Objective function has soft trend guaranteeing the partial derivatives.
b. The equation system of objective function is solved as analytic

If C is objective function dependent upon X1, Xa variables, $C=f(X_1,\ldots\ldots,X_a)$, we can define partial differential vector n variable dx and gradient vector as Equation 14:

$$\nabla C = \begin{bmatrix} \frac{\partial C}{\partial X_1} \\ \cdot \\ \cdot \\ \cdot \\ \frac{\partial C}{\partial X_n} \end{bmatrix} \quad dx \triangleq \begin{bmatrix} dx_1 \\ dx_2 \\ \cdot \\ \cdot \\ \cdot \\ dx_n \end{bmatrix} \tag{9}$$

General differential of objective function is as Equation (10):

$$dC = \nabla C^T . dx \tag{10}$$

As the internal product of two vector is maximum that they are parallel, we can say if the changes are along the gradient vector, we have the highest change in objective function, we can present the following algorithm to search the optimal response:

The start point of xn is selected as well.

The gradient value is computed in that point and we take a differential step in that direction to achieve the next point. The previous stage is repeated as a good measure guarantees we approached adequate to optimal point. In placement of combined generation systems, like other optimization issues, step by step movements in problem solving algorithm are based on definite criterion as directing the movement to the defined destination. In placement among the candidate points of CHP installation, sensitivity index is a guidance from one step to another one and based on defined goals of problem, loss reduction and voltage profile modification, two indices of voltage sensitivity and losses are introduced by which the most sensitive candidate busses are selected in placement with the definite goals, the searching algorithm is started from those points and further selections are based on lower sensitivities.

6.4. Sensitivity Index with the Goal of Reduction of Network Power Losses

The main goal of placement of combined generation systems in distribution network is reduction of losses and release of capacity of distribution system. Thus, sensitivity index in losses reduction as called losses sensitivity index is used as guidance of directing in problem solving process. Losses sensitivity index is introduced with the Equation:

$$L_s = \frac{\Delta loss}{loss} \tag{11}$$

Where, L_s is losses sensitivity index and $\Delta\ loss$ is the difference between the losses before and after installation of CHIP and loss, losses of network before CHP installation.

Sensitivity index is important in differential movement. Thus, a small part of CHP system (e.g. 1 kw) is added to the network without CHP and loss changes are evaluated. The sensitivity of each bus indicates its importance. Thus, for each bus, differential change in CHP capacity is applied from zero to small component and the effect in changes of network loss is evaluated.

6.5. Sensitivity Index with the Aim of Modification of Voltage of Network Buses

Another goal of placement in this section is modification of voltage profile. Thus, sensitivity index is computed with the goal of voltage modification for each of busses and its value is defined in effect on voltage of network busses. The voltage sensitivity index is defined based on the mean of squares of changes of voltage of busses of network before and after installation of small part of CHP system as each time in a candidate bus.

$$V_S = \sqrt{\frac{\sum_{i=1}^{a}(V_{i\,with\,chp} - V_{i\,whit\,out\,chp})^2}{n}} \quad (12)$$

Where, $V_{i\,whit\,out\,chp}$ و $V_{i\,with\,chp}$ are bus ith voltage before and after installation of combined generation generator in one of the busses of network and n is the total number of busses.

7. Searching Optimal Response by Graph Algorithm

Now, we can discuss graph algorithm structure in placement solution of combined energy and heat systems. At first, we form a vector of elements of candidate points as $X=[x_1,x_2,\ldots,x_n]$ in which n is the number of candidate points, X is state vector in which x1s are selection condition of candidate point ith. If it is one, it means the selection of the candidate point and zero indicates the lack of selection of candidate point. Thus, we have 2n state vector from zero to one and only one of them is optimal. Thus, we attribute a graph node to each of 2n states and in each repetition from one node to another node, if these movements are directed, we are guided to response node. If each repetition is denoted by edge connecting two nodes, we achieve the graph showing repetition stages to response.

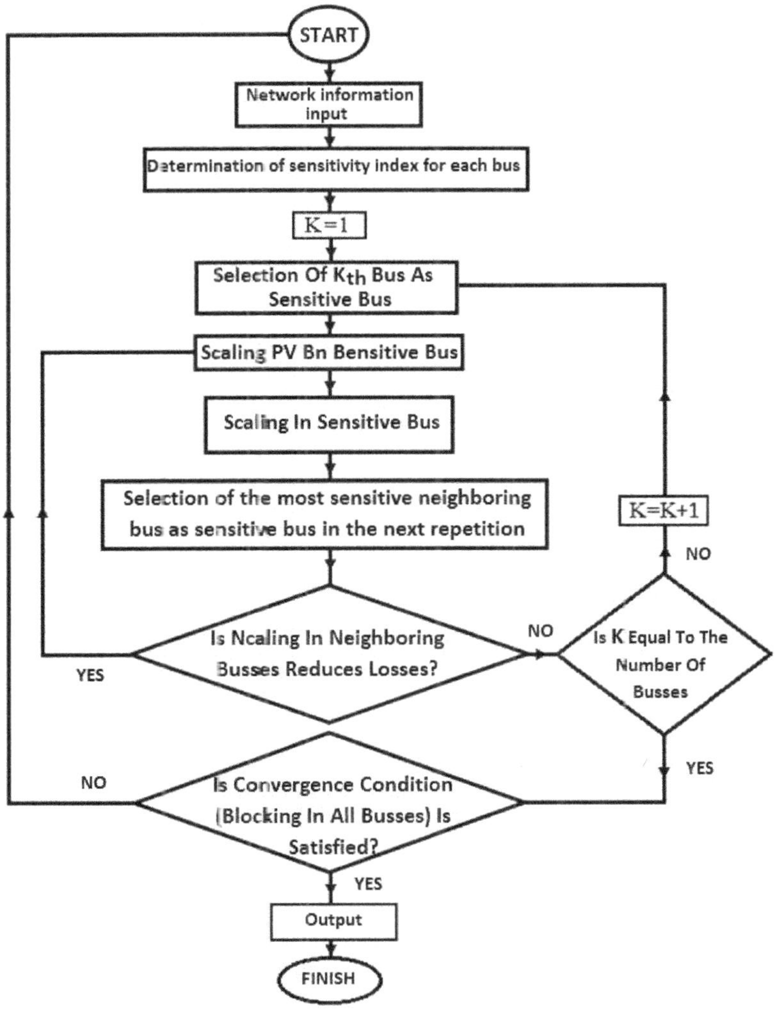

Figure 7. General flowchart of graph algorithm.

As it was said, to find optimal state vector, the following four stages are used:

a. Selection of initial state: In CHP placement, initial state to start algorithm is selected by analysis of sensitivity and defined sensitivity indices.

b. Selection of edge discriminant function: According to graph. According to graph algorithm assumptions, movement of problem state in two continuous repetitions is as only CHP installation is change in one or two busses. The changes in this algorithm are limited to neighboring busses. Thus, edge discriminant function is selected as the most sensitive neighboring bus and to find the next state in one repetition by putting CHP in neighboring busses, we select the bus in which generator installation has the highest loss and we put CHIP in it.

c. Relative convergence condition: Convergence is occurred when the sensitivity of all buses is equal and by putting CHIP in neighboring busses, based on precision of problem solution, the losses of lines are not reduced compared to the previous state.
d. The loop of escape from local optima: To escape from local optimal, after convergence in an optima, we order the busses on priority, higher to less, the most valuable bus reducing losses is the start point of algorithm.

Flowchart of proposed method of locating combined generation systems in distribution networks

The results of studies in this study are presented in proposed method for placement of combined generation systems in distribution networks as shown briefly in Flowchart 7. The sizing block is observed in the body of algorithm and the load distribution block among the blocks is used repeatedly.

8. Numerical Studies and Results of Analysis

In this section, we evaluate the proposed algorithm and analysis of results of placement of CHP systems in distribution network. First, the functions of economic and electric goal including cost goal, capital return, losses and voltage are compared and the result is a good function for placement of CHP generators. Thus, the first stage of studies directs total project and main goal. In this section, 30-bus IEEE standard is investigated. Various researches have been conducted regarding the effect of two factors of capacity and installation place to achieve the required goals in this type of load distribution and it had also interesting results. The initial capital to purchase CHP system as one of the terms of capital return functions and costs is an exception in this regard and performance of system has no effect on its establishment cost. The calculations in first stage to study the objective functions are performed with optimistic view of equal generation with capacity or efficiency 100%. The results showed that capital return period is higher than life service of combined generation systems. Thus, economic goals are outside of the list of followed goals in placement of combined generation generators in existing conditions. Thus, if efficiency is considered in two economic functions, better response is not achieved and capital return period is increased.

8.1. The Study and Comparison of Four Objective Functions in Placement of Combined Generation Systems in Distribution Networks

At first, 30-bus distribution networks are considered as sample network with fixed loads. CHP generators should be installed and used in feeders as commercial, administrative and industrial loads at peak time. In this thesis, CHP source is tested in load busses in network test. In this stage, as initial stage of placement studies of CHP systems, the network load is fixed. The goal of execution of this stage of studies is determining a suitable objective function. Selection is the best way.

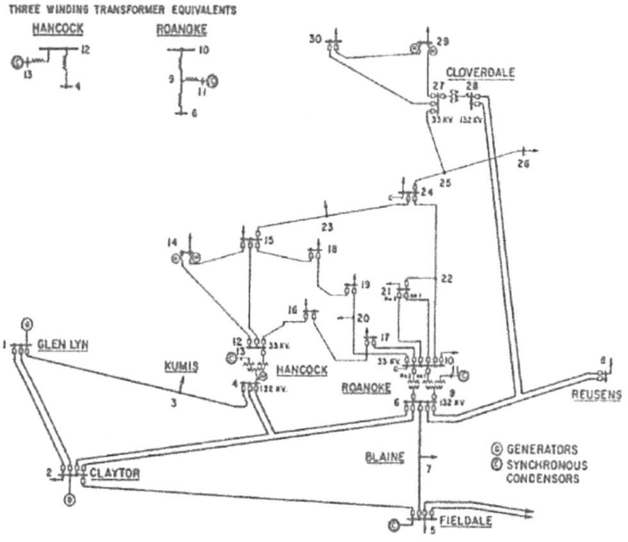

Figure 8. *30-bus distribution network IEEE.*

8.2. Features of Test Network

Table 5. *General features of test network.*

General features of test network	
Number of busses	30
Bus number PV	2-5-8-11-13
Bus number PQ	24
Bus NO. SW	1
Number of lines	41

Table 6. *Features of lines of test network.*

Line	From Bus	To Bus	R(p.u.)	X(p.u.)	Rating(p.u.)
1	1	2	0.0192	0.0575	0.0300
2	1	3	0.0452	0.1852	0.0300
3	2	4	0.0570	0.1737	0.0300
4	3	4	0.0132	0.0379	0.0300
5	2	5	0.0472	0.1983	0.0300
6	2	6	0.0581	0.1763	0.0300
7	4	6	0.0119	0.0414	0.0300
8	5	7	0.0460	0.1160	0.0300
9	6	7	0.0276	0.820	0.0300
10	6	8	0.0120	0.420	0.0300
11	6	9	0.0000	0.2080	0.0300
12	6	10	0.0000	0.5560	0.0300
13	9	10	0.0000	0.2000	0.0300
14	9	11	0.0000	0.1100	0.0300
15	4	12	0.0000	0.2560	0.0650
16	12	13	0.0000	0.1400	0.0650
17	12	14	0. 1231	0.2559	0.0320
18	12	15	0.0662	0.1304	0.0320

Line	From Bus	To Bus	R(p.u.)	X(p.u.)	Rating(p.u.)
19	12	16	0.0945	0.1987	0.0320
20	14	15	0.2210	0.1997	0.0160
21	16	17	0.0824	0.1932	0.0160
22	15	18	0.1070	0.2185	0.0160
23	18	19	0.0639	0.1292	0.0160
24	19	20	0.0340	0.0680	0.0320
25	10	20	0.0936	0.2090	0.0320
26	10	17	0.0324	0.0845	0.0320
27	10	21	0.0348	0.0749	0.0300
28	10	22	0.0727	0.1499	0.0300
29	21	22	0.0116	0.0236	0.0300
30	15	23	0.1000	0.2020	0.0160
31	22	24	0.1150	0.1790	0.0300
32	23	25	0.1320	0.2700	0.0160
33	24	25	0.1885	0.3292	0.0300
34	25	26	0.2544	0.3800	0.0300
35	25	27	0.1093	0.2087	0.0300
36	28	27	0.0000	0.3960	0.0300
37	27	29	0.2198	0.4153	0.0300
38	27	30	0.3202	0.6027	0.0300
39	29	30	0.2399	0.4533	0.0300
40	8	28	0.0639	0.2000	0.0300
41	6	28	0.0169	0.0599	0.0300

Table 7. *Voltage of test network.*

Bus	1	2	3	4	5	6	7	8	9	10	11	12	13	14	15
Voltage(KV)	132	132	132	132	132	132	132	132	33	33	11	33	11	33	33
LOAD (MW)	0.0	21.7	2.4	67.6	34.2	0.0	22.8	30.0	0.0	5.8	0.0	11.2	0.0	6.2	8.2
Bus	16	17	18	19	20	21	22	23	24	25	26	27	28	29	30
Voltage(KV)	33	33	33	33	33	33	33	33	33	33	33	33	1322	33	33
LOAD (MW)	3.5	9.5	3.2	9.5	2.2	17.5	0.0	3.2	8.7	0.0	3.5	0.0	0.0	2.4	10.6

Table 8. *The reactive power range of test network.*

Bus	1	2	3	4	5	6	7	8	9	10	11	12	13	14	15
Q MIN(P.U)	-0.2	-0.2			-0.15			-0.15			-0.1		-0.15		
Q MAX (P.U)	0.0	0.2			0.15			0.15			0.1		0.15		
Bus	16	17	18	19	20	21	22	23	24	25	26	27	28	29	30
Q MIN(P.U)		-0.05	0.0					-0.05				-0.05			
Q MAX (P.U)		0.05	0.055					0.055				0.055			

8.3. Evaluation of Economic Objective Functions

At first, economic goals are followed including cost objective and capital return. The defined cost function in (19) is based on 0.1 dollar for per kWh of electricity without using subsidy and 2420 dollars capita for each kW capacity of combined generation and (12) is achieved.

$$cost = 2420 * \sum chp + 0.1 * (loss - \sum chp) \qquad (13)$$

$\sum$CHP is the sum of CHP capacities in network and loss electric loss of distribution network. Another economic objective function is capital return as with high meaning compared to cost function as cost function puts reduction of losses of system in a year beside great capacity to purchase CHP. Thus, CHP generation capital return in continuous years and loss reduction is compared with initial capital and required duration to achieve CHP income is computed to the maximum dedicated cost. Inflation rate is not considered for each of investment costs and generated electricity value and reduced losses and value of money is fixed and based on defined equation (10), capital return is shown in (14).

$$Y = \frac{\sum CHP * 2420 * 10^{6}}{0.1 * (\sum CHP + (LOSS_{WO/chp} - LOSS_{W/CHP}))} \qquad (14)$$

Where, $\sum$CHP is the sum of CHP capacities installed in network in the studied state and $LOSS_{(W/CHP)}$ is electric losses of network and $LOSS_{WO/chp}$ is losses at base state without installation of CHP in distribution network.

8.4. Evaluation of Electric Objective Functions

Second view is electric one and two objective functions of loss and voltage profile can be included. Loss objective function is defined with (15) and minimizing active power loss is the goal following by this function. Voltage profile objective function is observed in (16), the mean of deviation of busses voltage is shown from ideal voltage of network (V_i-V_b):

$$LOSS = \sum_{i=1}^{N} Re\,(V_i I_i^*) \qquad (15)$$

$$V = \sqrt{\frac{\sum_{i=1}^{M}(V_i - V_b)^2}{M}} \qquad (16)$$

Where, V_i is bus ith voltage in the studied state, N is the number of busses; M is the number of lines of network.

8.5. The Results of Objective Function Evaluation

To compare simulation state, without and with CHP, at first load distribution is done in the studied network and basic electric losses are computed and then placement is done with the minimizing goal of objective functions.

8.6. Placement with the Goal of Minimizing Power Loss Function and Voltage Profile

8.6.1. The Evaluation of Power Losses

At first, by load distribution in test network, power losses are achieved. These losses are shown on graph network in Figure 9 as weight of each line.

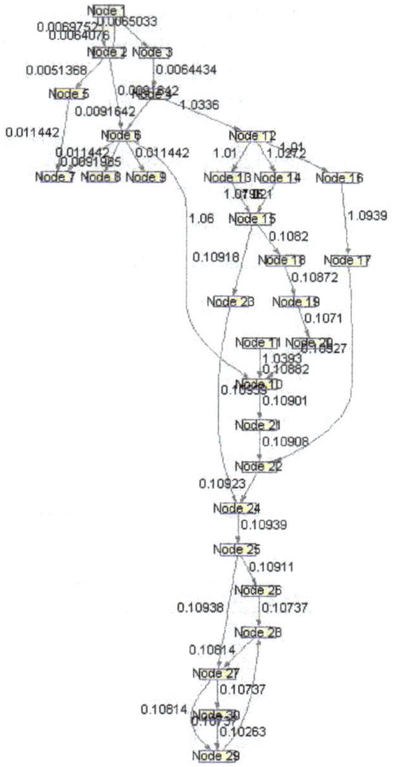

Figure 9. Graph of power losses in entire test network in each line.

Figure 10 shows the lines with highest losses in mw.

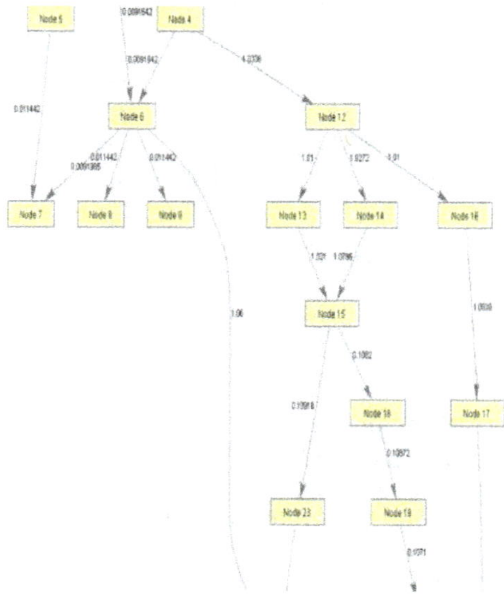

Figure 10. Graph of power loss in a part of test network.

The sum of power losses in entire test network without CHP is 11.23MW.

8.6.2. The Evaluation of Voltage Profile

At first, by load distribution in test network, voltage profile is achieved. As shown in Figure 11, busses 21-30 show the highest voltage reduction in test network.

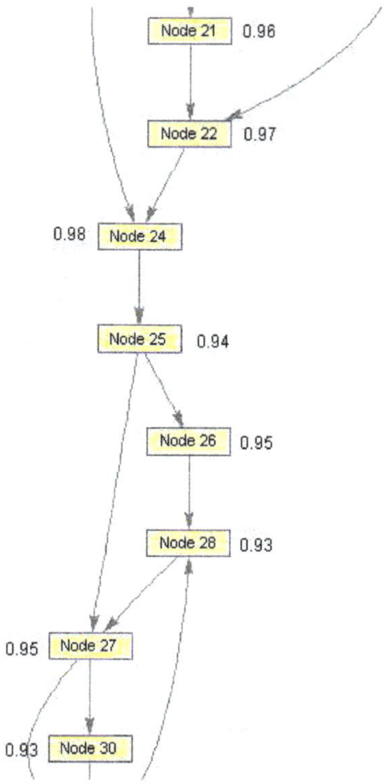

Figure 11. Graph of voltage size in some of busses of test network.

Table 9. Voltage of busses of test network.

Bus NO.	Bus voltage (p.u)	Bus No.	Bus voltage (p.u)
1	1.03	16	1.00
2	1.03	17	0.93
3	1.01	18	1.01
4	0.97	19	1.02
5	1.04	20	1.04
6	1.01	21	0.96
7	1.02	22	0.97
8	1.01	23	1.00
9	1.03	24	0.98
10	1.02	25	0.94
11	1.03	26	0.95
12	1.02	27	0.95
13	0.98	28	0.93
14	1.00	29	0.93
15	1.00	30	0.93

8.6.3. Putting CHP Source in Test Network

To put CHP source in test network, at first by sensitivity analysis, we find a bus to have the lowest changes with the highest change in sum of electric power losses. Indeed, the goal of using sensitivity analysis is increasing convergence speed of graph algorithm. By this method, the best state of start is achieved.

(i). Formation of Edge Discriminant Function Matrix

To define the movement of CHP source among busses o network, as shown in Chapter 3, edge discriminant function is used and in neighbourhood of selected bus by sensitivity analysis can be applied.

Table 10. Matrix of edge discriminant function.

1	2	3	4	5	6	7	8	9	10	11	12	13	14	15
0	1	1	0	0	0	0	0	0	0	0	0	0	0	0
0	0	0	1	1	1	0	0	0	0	0	0	0	0	0
0	0	0	1	0	0	0	0	0	0	0	0	0	0	0
0	0	0	0	0	1	0	0	0	0	0	1	0	0	0
0	0	0	0	0	0	1	0	0	0	0	0	0	0	0
0	0	0	0	0	0	1	1	1	1	0	0	0	0	0
0	0	0	0	0	0	0	0	0	0	0	0	0	0	0
1	0	0	0	0	0	0	0	0	0	0	0	0	0	0
0	0	0	0	0	0	0	0	0	0	0	0	0	0	0
0	0	0	0	0	0	0	0	0	0	0	0	0	0	0
0	0	1	0	0	0	0	1	0	0	0	0	0	0	0
0	1	0	0	0	0	0	0	0	0	0	0	1	1	0
0	0	0	0	0	0	0	0	0	0	0	0	0	0	0
0	0	0	1	0	0	0	0	0	0	0	0	0	0	1
0	0	0	0	1	0	0	0	0	0	0	0	0	0	0
0	0	0	0	0	0	0	0	0	0	0	0	0	0	0
0	0	0	0	0	0	1	0	0	0	0	0	0	0	0
0	0	0	0	0	0	0	0	1	0	0	0	0	0	0
0	0	0	0	0	0	0	0	0	0	0	0	0	0	0
0	0	0	0	0	0	0	0	1	0	0	0	0	0	0
0	0	0	0	0	0	0	0	0	1	0	0	0	0	0
0	0	0	0	0	0	0	0	0	0	1	1	0	0	0
0	0	0	0	0	0	0	0	0	0	0	0	0	0	0
0	0	0	0	0	0	0	0	0	0	0	0	0	1	1
0	0	0	0	0	0	0	0	0	0	0	1	0	0	0
0	0	0	0	0	0	0	0	0	0	0	0	0	0	0
0	0	0	0	0	0	0	0	0	0	0	0	0	0	0
0	0	0	0	0	0	0	0	0	0	0	0	0	0	0
0	0	0	0	0	0	0	0	0	0	0	0	0	0	0

16	17	18	19	20	21	22	23	24	25	26	27	28	29	30
0	0	0	0	0	0	0	0	0	0	0	0	0	0	0
0	0	0	0	0	0	0	0	0	0	0	0	0	0	0
0	0	0	0	0	0	0	0	0	0	0	0	0	0	0
0	0	0	0	0	0	0	0	0	0	0	0	0	0	0
0	0	0	0	0	0	0	0	0	0	0	0	0	0	0
0	0	0	0	0	0	0	0	0	0	0	0	0	1	0
0	0	0	0	0	0	0	0	0	0	0	0	0	0	0
0	0	0	0	0	0	0	0	0	0	0	0	1	0	0
0	0	0	0	0	0	0	0	0	0	0	0	0	0	0
0	1	0	0	1	1	1	0	0	0	0	0	0	0	0
0	0	0	0	0	0	0	0	0	0	0	0	1	1	0
0	0	0	0	0	0	0	0	0	0	0	0	0	0	0
0	0	0	0	0	0	0	0	0	0	0	0	0	0	1
0	0	0	0	0	0	0	0	0	0	0	0	0	0	0
0	0	0	0	0	0	0	0	0	0	0	0	0	0	0
0	0	0	0	0	0	0	0	0	0	0	0	0	0	0
0	0	0	0	0	0	0	0	0	0	0	0	0	0	0
0	0	0	0	0	0	0	0	0	0	0	0	0	0	0
0	0	0	0	0	0	0	0	0	0	0	0	0	0	0
0	0	0	0	0	0	0	0	0	0	0	0	0	0	0
0	0	0	0	0	0	0	0	0	0	0	0	0	0	0
0	0	0	0	0	0	0	0	0	0	0	0	0	0	0
0	0	0	0	0	0	0	0	0	0	0	0	0	0	1
0	0	0	0	0	0	0	0	0	0	0	0	0	0	0
0	0	0	0	0	0	0	0	0	0	0	0	0	0	
0	0	0	0	0	0	0	1	0	0	0	0	0	0	0
0	0	0	1	0	1	0	0	0	0	0	0	0	0	
0	0	0	0	1	0	0	0	0	0	1	0	0	0	0

Graphic chart of edge Discriminant and CHP movement in neighboring busses

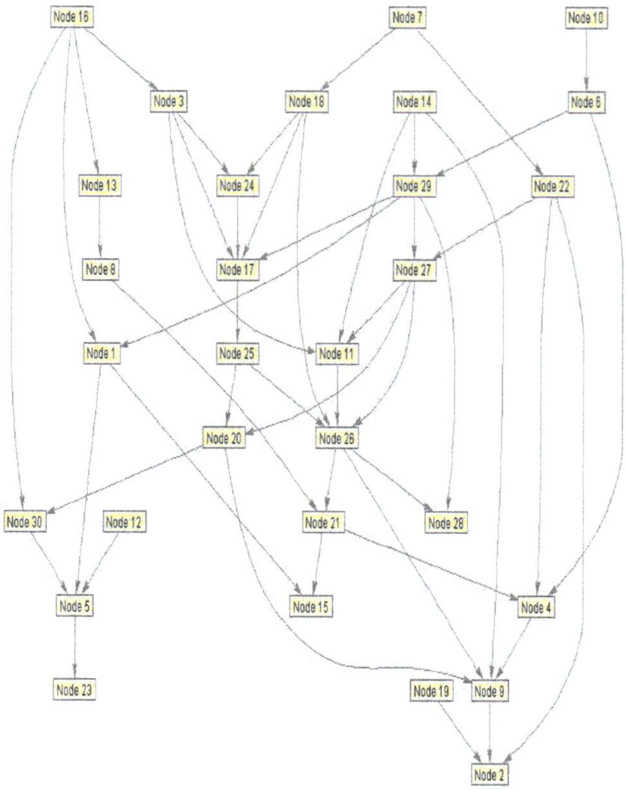

Figure 12. Graphic view of edge discriminant.

Graphic chart of Figure 13 shows a path in test network with the lowest changes in size of losses and profile of voltage. In this path, graph algorithm achieves convergence. This path includes busses 21, 22, 24, 25.

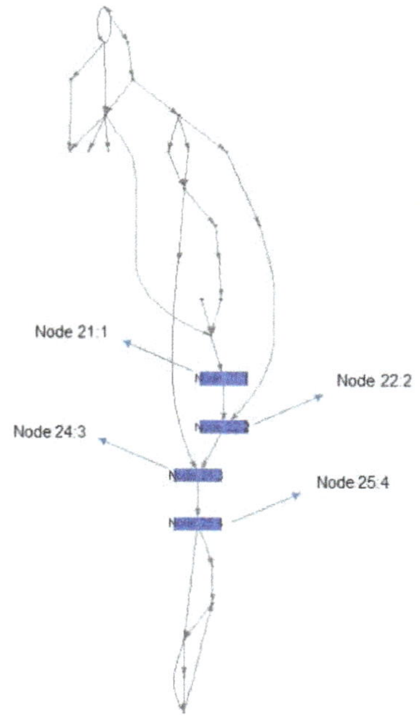

Figure 13. Convergence path based on sensitivity analysis.

(ii). CHP Installation Site and Its Capacity

Among 4 busses with numbers 21, 22, 24, 25 busses. NO. 25 is with capacity 2.14Mw.

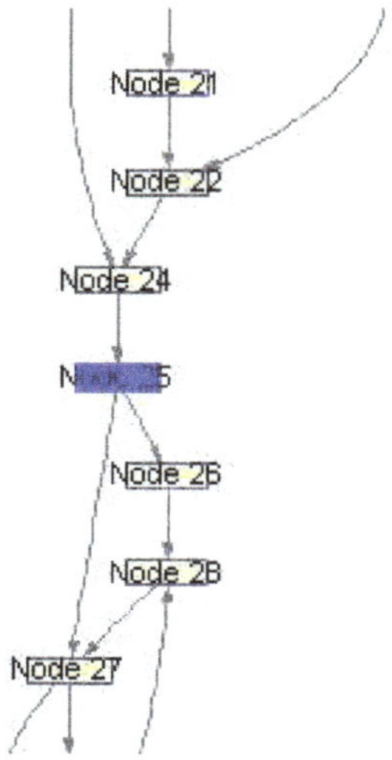

Figure 14. CHP installation site

8.7. Changes of Power Loss after CHP Installation

The sum of power losses in entire test network in CHP installation time is 7.24mw. In (31), the change of power losses is shown in some high loss lines of test network.

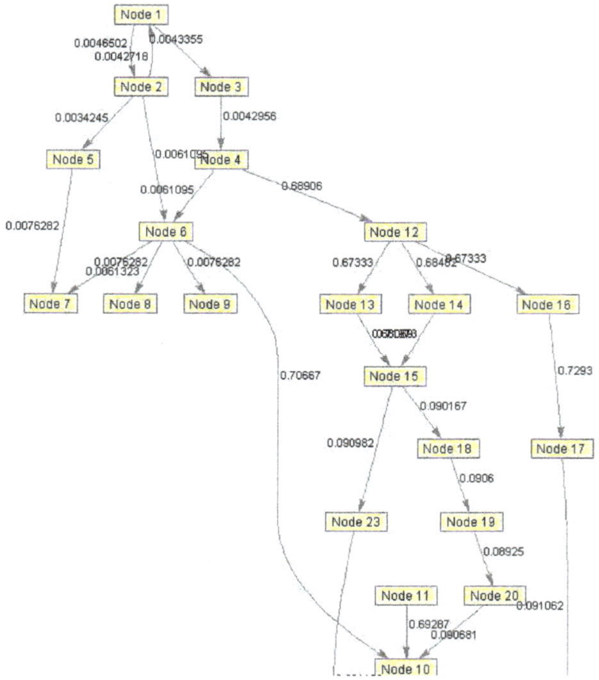

Figure 15. The power losses in test network in case of CHP source installation.

8.8. The Changes of Voltage Profile in Test Network

In 16 values of changes in voltage size after CHP installation in some busses, we can say the voltage size is improved in these busses.

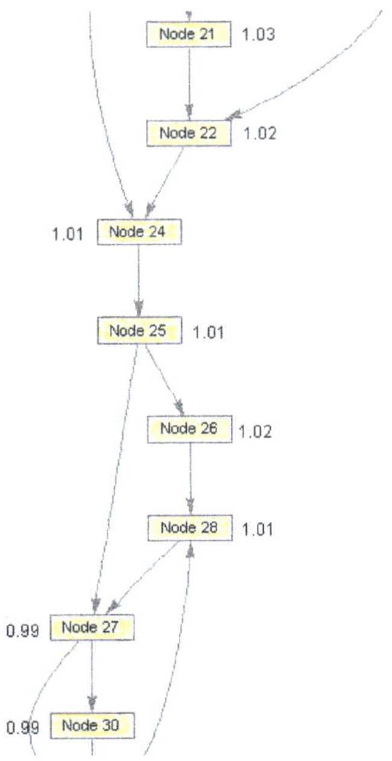

Figure 16. *The profile changes of voltage after CHP installation*

Table 11. *The changes of voltage profile after CHP installation.*

Bus NO.	Bus voltage (p.u)	Bus No.	Bus voltage (p.u)
1	1.03	16	1.01
2	1.03	17	0.95
3	1.01	18	1.01
4	0.98	19	1.03
5	1.04	20	1.05
6	1.01	21	1.03
7	1.02	22	1.02
8	1.01	23	0.99
9	1.01	24	1.01
10	1.01	25	1.01
11	1.02	26	1.02
12	1.01	27	0.99
13	0.97	28	1.01
14	1.00	29	0.99
15	0.98	30	0.99

8.9. The Evaluation of Placement with the Goal of Minimizing Capital Return Function

Placement is considered with the aim of minimizing capital return function, installation of a CHP system with capacity 2.23mw in bus-25 as the best state with 34.34 years of capital return and the number of capital return years is almost 34. It can be said that putting local power plants in busses with high sensitivity to the generators with the aim of reduction of losses has short capital return. By analysis of gradient, we can say the most sensitive non-ending bus to capital return is bus 25 and in placement analysis with graph algorithm, the same bus is selected as a proper site to install CHP generator. It can be said that in none of above analyses in optimization of network from electric aspects, no considerable success is achieved. Under these conditions, placement with economic goal is not useful. The life service of combined generation systems is estimated as about 15-20 years. The capital return in the best state for our studied network is more than 34 years and the revenue of CHP systems are not equal with the initial capital. Indeed, this condition is the result of some realities in country. Electric energy is not valuable compared to other economic goods and investment to generate non-utility energy is not economical and combined generation technology is achieved of other industrial countries and combined generation system costs are high in our country due to its non-local nature. In addition, renewable systems don't have spiritual and materialistic support in country. Thus, when combined generation technology is not localized by internal experts, or suitable rules are not executed to support combined generation industry of energy and non-utility electricity generation as distributed units by policy makers, construction and operation cost from CHP systems is high as power generation is not economical and not accepted by capital owners.

Placement is achieved with electric goals with the aid of two objective functions of loss reduction and voltage modification in both cases and similar answer is achieved. For optimized reduction of loss and optimized increase of voltage range to good value, CHP generators are sized with highest capacity in all candidate busses. These losses are reduced as 0.36, 41% and average voltage of buses is increased 1.74v. The lowest voltage in buss no. 25 is increased from 382.724v at basic state to 385.748 v in network equipped with CHP generator. The placement of CHP in distribution networks can reduce losses distribution considerably and this increases busses voltage range. By the calculations and results of load distribution in distribution networks, we can say placement of CHP systems in distribution network can reduce losses considerably and increase bus voltage range. CHP generators with low production for occupied area are suitable for distribution networks and can be used in weak pressure levels. The price of CHP system for each kw is very high as capital long-term return is achieved. Thus, CHP placement in distribution networks is not useful with economic analyses.

9. Conclusion

In this study, a new method is used for placement and optimal capacity of combined generation systems connected to utility with the aim of reduction of losses and release of capacity and modification of voltage profile in radial distribution networks. In placement of CHP systems, combined graph algorithm with sensitivity analysis is applied and their capacity is determined based on applying constraints of network voltage and available levels of candidate locations in the studied network. The simulation results are presented in sample network to support efficiency of above method.

The numerical study section of this project is full of various analytic studies with valuable results. 30-bus IEEE network as sample network is studied to prove efficiency of proposed method in placement of combined generation systems in distribution network with the goal of reduction of losses and modification of voltage profile and the results show the accuracy of placement by combination of graph algorithm with sensitivity analysis. Finally, these studies in addition of providing accuracy of proposed method in locating CHP systems show the benefits of using CHP generators in distribution network in terms of reduction of losses and improvement of voltage of network and shows a new way to cope up with the problems of distribution networks for operation from distributed systems namely energy generators with renewable source and combined generation generators for designers and planners of power network. Also, it proposes that we should actualize the new thoughts to avoid energy crisis or destruction of human environment.

References

[1] Philipson L., Willis H. L., 1999. Understanding Electric Utilities and De-Regulation, Marcel Dekker, New York.

[2] Puttgen H. B., MacGregor P. R., Lambert F.C., 2003. Distributed Generation: Semantic Hype or The Dawn of a New Era, IEEE Trans. On Power and Energy, 1(1): 22-29.

[3] Brown R. E., 1996. Reliability Assessment and Design Optimization in Electric Power Distribution Systems, Ph.D. Dissertation, University of Washington, WA.

[4] Lamarre L., 1993, The Vision of Distributed Generation, EPRI Journal.

[5] Ackermann T., Knyazkin V., 2002, Interaction Between Distributed Generation and the Distribution Network: Operation Aspects In: Proc.Of the IEEE/PES T&D Conference and Exhibition, vol. 2, Asia Pacific.

[6] Griffin T., Tomsovic K., Secrest D., Law A., 2000. Placement of Dispersed Generation Systems for Reduced Losses. In: Proc. of the 3rd Annu. Hawaii Int. Conf. Systems Sciences, Maui, HI.

[7] Hegazy Y. G, Salama M. A., Chikhani A. Y., 2003. Adequacy Assessment of Distributed Generation Systems Using Monte Carlo Simulation, IEEE Trans. On Power Systems.

[8] Brown R., Pan I., Feng X., Koudev K., 2001 Siting Distributed Generation to Defer T& D Expansion. of the IEEE T&D Conference.

[9] L. R. Mattison, "Technical Analysis of the Potential for Combined Heat and Power in Massachusett," University of Massachusetts Amhers, Massachusetts, May 2006.

[10] Devender Singh, R. K. Misra, and Deependra Singh, "Effect of load models in Distributed Generation planing," IEEE Transaction on Power system, vol. 22 no 4, Nov. 2007.

[11] Dugan, R.C., and McDermott, T.E., "Distributed generatio," IEEE Industry Applications Magazin, vol. 8. No 2, pp. 19-25, 2002.

[12] Ramakumar R., Chiradeja P., 2002, Distributed Generation and Renewable Energy Systems, of the Intersociety Energy Conversion Engineering Conference.

[13] U.D.o. Energy. Modeling Distributed Generation in the Buildings Sectors. Washington DC. 2013.

[14] R. R. Chiradeja P, An Approach to Quantify the Technical Benefits of Distributed Generation," IEEE Trans. on energy Conversion, vol. (4)19, 2004.

[15] Fraser P., 2002, the Economics of Distributed Generation, International Energy Agency, Energy Prices and Taxes.

[16] Complete guidance of heat and power combined generation. Ministry of energy. 2009. Productivity improvement and power and energy economy.

[17] Seyed Mehrdad Hosseini, Mohammad HOssein Javidi. 2011. Placement and sizing combined generation of heat and power by PSO algorithm.

[18] B. Zhao, C. X. Guo, B. R. Bai, and Y. J. Cao, An improved particle swarm optimization algorithm for unit commitment, Int. J. Elect. Power Energy Syst., 28, (September (7)), 2006, 482-490.

[19] C. C. Kuo. "A Novel Coding Scheme for Practical Economic Dispatch by Modified Particle Swarm Approach". IEEE Trans. Power Syst., vol. 23, no. 4, pp. 1825-1835, Nov. 2008.

[20] Holland J.H., 1975, Adaptation in Natural and Artificial Systems, Ann Arbor, the University of Michigan Press, Michigan.

[21] Caire R., Retikre N., Morin N., Fontela M., Hadjsaid N., 2003, Voltage Management of Distributed Generation in Distribution Networks, IEEE.

[22] Celli G., Pilo F., 2001, Optimal Distributed Generation Allocation in MV Distribution Networks, pp. 81-86in Proc. of IEEE PICA Conf., Sydney, NSW, Australia.

[23] Willis H. L., Scott W. G., 2000, Distributed Power Generation, Marcel Dekker, New York.

[24] Willis H. L., 1997, Power Distribution Planning Reference Book, Marcel Dekker, New York.

[25] Carpinelli G., Celli G., Pilo F., Russo A., 2001, Distributed Generation Siting and Sizing Under Uncertainty, pp. 376-401in Proc. IEEE Powertech Conf., vol. 4, Porto, Portugal.

[26] B, Defino, Modeling of the Integration of Distributed Generation into the Electrical System",IEEE Power Engineering Society Summer Meeting ,vol 10, pp. 17.

[27] El-Khattam W., Hegazy Y., Salama M. M. A., 2005, An Integrated Distributed Generation Optimization Model for Distribution System Planning, IEEE Trans. On Power Systems, 20(2).

[28] El-Khattam W., Bhattacharya K., Hegazy Y., Salama M. M. A., 2004, Optimal Investment Planning for Distributed Generation in a Competitive Electricity Market, IEEE Trans. On Power Systems, 19(3).

[29] Derek W. A., Fletcher W., Fellhoelter K., 2003, Securing Critical Information and Communication Infrastructures through Electric Power Grid Independence, IEICE/IEEE INTELEC'03, pp: 19-23.

[30] Caisheng W., Nehrir, M.H., 2004, Analytical Approaches For Optimal Placement of Distributed Generation Sources in Power Systems, IEEE Trans. On Power Systems, 19(4): 2068 – 2076.

[31] Chang SK, Marks GE, Kato K (1990) optimal real-time voltage control. IEEE Trans. on Power Systems, vol 5, no 3, pp 750-756.

[32] Hollenstein W, Glavitch H (1990) Linear programming as a tool for treating constraints in a Newton OPF. Proceedings of the 10th Power Systems Computation Conference (PSCC), Graz, Austria, August 19-24.

[33] Karmarkar N (1984,) a new polynomial time algorithm for linear programming, Combinatorica 4, pp 373-395.

[34] Lu N, Unum MR (1993) Network constrained security control using an interior point algorithm. IEEE Transactions on Power Systems, vol 8, no 3, pp 1068-1076.

[35] Irisarri GD, Wang X, Tong J, Mokhtari S (1997) Maximum loadability of power systems using interior point nonlinear optimisation method. IEEE Trans. on Power Systems, vol 12, no 1, pp 167-172.

[36] Wei H, Sasaki H, Yokoyama R (1998) an interior point nonlinear programming for optimal power flow problems within a novel data structure. IEEE Trans. on Power Systems, vol 13, no 3, pp 870-877.

[37] Torres GL, Quintana VH (1998) an interior point method for non-linear optimal power flow using voltage rectangular coordinates. IEEE Transactions on Power Systems, vol 13, no 4, pp 1211-1218.

[38] Zhang XP, Petoussis SG, Godfrey KR (2005) Novel nonlinear interior point optimal power flow (OPF) method based on current mismatch formulation. IEE Proceedings Generation, Transmission & Distribution, to appear.

[39] El-Bakry S, Tapia RA, Tsuchiya T, Zhang Y (1996) On the formulation and theory of the Newton interior-point method for nonlinear programming. Journal of Optimisation Theory and Applications, vol 89, no 3, pp 507-541.

[40] Dashti R., Haghi, Qam. R., 2004. Re-configuraiton of distribution networks to reduce losses by graph theory. MA thesis. Tarbiat Modarres University. Tehran.

[41] Greatbanks J. A., PopoviC D. H., BegoviC M., Pregelj A., Green T. C., 2003, On Optimization for Security and Reliability of Power Systems with Distributed Generation, in Proc. Of IEEE Bologna PowerTech Conference, Bologna, Italy.

[42] Bayegan M., 2001, A Vision of The Future Grid, IEEE Power Eng. Review. 21:10-12.

Research About China's Electricity Market Reform Based on Hall's Three Dimensions Structure Model

Jun Dong, Rong Li

School of Economics and Management, North China Electric Power University, Beijing, China

Email address:
lirong_huadian@163.com (Li Rong)

Abstract: To promote the development of the national economy, how to improve the electricity market and establish effective competition mechanisms becomes the focus of the new round of electricity market reform in China. Referring to the basic thoughts of Hall's three dimensions structure theory of system engineering discipline, and analyzing from three levels of logic dimension, time dimension and knowledge dimension, this paper provides recommends on the implementation of electricity market reform in China. Through the compared analysis of two reform programs in the logic dimension, we obtain the modified electricity reform program for current China that is keeping the grid company having the whole business of transmission and distribution, establishing electricity sale companies, establishing power trading center and promoting power direct trading.

Keywords: Hall's Three Dimensions Structure, Electricity Market Reform, Implement Suggestion

1. Introduction

Electricity industry as the basic pillar industry of the national economy, whether it can keep healthy, orderly and efficient development gets much attention from the national Government and various parties. Electricity system reforms in China started from the 70's of the last century which have been trying to establish an electricity market suitable to China's national conditions and economic development. The reforms have experienced several key points [1]: During 1978-1985, when the national Governments implemented the policy of "Funding to establish plants", the electricity industry pioneered to use the bank loans in order to broaden the channels of funding and began to cultivate the market pricing mechanism. During 1987-1998, the electricity industry was separated from the Governments, and the electricity industry ministry was withdrawn with the establishment of the national power company. In 2002, the State Council issued the "Fifth Document", and started the electricity reform. The former State power grid company was split into two major power grid companies, five major power generating groups and four auxiliary groups. In 2003, the State Electricity Regulatory Commission was set up to achieve "Separating Government and Supervision"; then the "Electricity Price Reform Scheme" enacted by the State Council divided the electricity price into four forms including Internet access price, transmission price, distribution and sales price. In 2004, the National Development and Reform Commission issued the "Provisional management approaches of direct power-purchase pilots between electricity users and power generation enterprises", as well as the "Views on the establishment of coal-electricity price linkage mechanism". In 2010, as there were many controversies in the verification of transmission and distribution prices and direct power purchase for high energy-consuming enterprises and so on, several Central Government ministries stopped the local "direct purchase" pilots. First half of 2014, more than 10 provinces such as Anhui, Jiangsu, Jiangxi reset "direct purchase" pilots, and in November, the "pilot reform program for Shenzhen power transmission and distribution price" issued by the National Development and Reform Commission launched a new round of transmission and distribution price reform. In March 2015, the CPC Central Committee and the State Council promulgated the electricity reform "Ninth File"—"Some opinions on further deepening the reform of electricity system" ([2015]9), which officially opened the second round reform of electricity market.

The existing studies on the electricity market reform in China are mainly focused on the design of electricity market patterns and the selection of reform programs. Yao Jiangang discusses the models suitable to China's electricity market, and proposes the "contracts and spot market" form is the best way to address the problems of interests of all parties in the reform [2]. Shang Jincheng and Fu Shuguo propose the related reform models of electricity market by analyzing a regional power market from transaction types, market structures, competitive modes as well as function divisions, combined with the demand of electricity market development in China[3-4]. By discussing a series of changes of pricing mechanisms in China, Lin Boqiang proposes the concentration of state-owned enterprises are the root of many fundamental problems of the power industry, and points out that a successful reform program must fully take into account electricity prices, coal prices, state-owned enterprises, energy agencies and other supporting reforms[5]. Song Xiaosong and Zhang Jianhua, who analyze the kinds of power market operation modes and status of electricity system reform in China, present a gradual reforming program for China's power market[6]. Li Han and Li Guoren by comparing the electricity industry reform processes of United Kingdom, the United States and Japan, propose that rationally determining the price formation mechanisms is the key to the success of China's electricity market reform[7]. Chen Weiyong proposes the bilateral trading and centralized trading patterns for large purchasing users, and two approaches for cross-subsidies, including users paying additional fees and taking into account transmission and distribution fees[8]. However, these existing studies are still confined to one or two aspects of designing or researching the electricity market, but not from a system perspective to comprehensively study how to select and implement a reform program. But it is a systemic project to set up a sound and orderly electricity market involving various aspects. Thus this paper uses one of the classical system engineering theories—Hall's three dimensions structure—to systematically and comprehensively analyze the selection and implement of new round of electricity market reform in China from time dimension, logical dimension as well as knowledge dimension, and then puts forward related suggestions, based on combing the past several rounds of changes in China.

2. Hall's Three Dimensions Structure

Hall's three dimensions structure is a system engineering methodology proposed by United States engineering expert A.D. Hall in 1969, which provides a systematic thinking method for planning, organization, management of large and complex systems, and has been widely used all over the world. Hall's three dimensions structure divides a whole system engineering process into 7 stages and 7 steps closely linked with each other, and meanwhile it also considers various required professional knowledge and skills for the completion of these phases and steps. The structure is formed by three dimensional space structure including time, logic and knowledge dimension [9]. Among them, the time dimension represents the whole process of system engineering activities from start to finish according to the chronological order, which is divided into 7 time stages including planning, programming, development, production, installation, operation, updating. Logical dimension refers to the work contents and thought process that should be followed during each time stages, including defining the problem, identifying the objectives, system integration, system analysis, system optimization, decision and implementation, these 7 logical steps. Knowledge dimension means those required knowledge synthesis when solving complex problems, including engineering, medicine, architecture, business, law, management, social sciences, arts and other knowledge and skills. The three dimensions structure vividly describes the framework of system engineering, and each step at any stage can be further expanded, finally forming a hierarchical structure system.

This paper refers to the thoughts of Hall's three dimensions structure theory, from the perspective of the time dimension to analyze the key steps of an electricity reform program implementation, from the viewpoint of the logical dimension to propose and verify a viable power market reform program, from the perspective of knowledge to analyze the conditions required before and in the process of program implementation.

3. Application of Hall's Three Dimensions Structure in the Electricity Market Reform

3.1. Logical Dimension

3.1.1. Defining the Problem

In the previous round of reform launched in 2002, despite the fact that the separations of governments and enterprises, plants and grids, main and auxiliary service, have been achieved, the "transmission and distribution separated" and "bidding on grid" these two major reforms identified in the file have been stalled for a long time. In order to realize the strategic goals proposed in the eighteenth National Congress of the CPC which is "By 2020, GDP and income of urban and rural residents are more than quadrupled than 2010 ", electricity as a fundamental industry must be first developed towards a higher level. But China's current power market at generation side and sale side does not form a good competitive environment and both the electricity on-grid price and sale price do not realize the goals set by the market reform. How to form a competitive pricing mechanism is the next important issue faced by the electricity reform.

3.1.2. Identifying the Objectives

The Grid company's monopoly in transmission-distribution and scheduling is the key factor hindering market-oriented pricing. In order to stimulate the enthusiasm of electricity consumers, market competition in sale side need to be brought in firstly to break the status quo that users can only purchase electricity from the grid companies and then a number of alternative electricity sale sources should be provided, so as to establish competition mechanisms of the sale price.

3.1.3. Forming System Schemes

Reform for the electricity sale side mainly has two executable programs. One is based on the "Fifth File" goals, "transmission and distribution separated" and "bidding on grid", to continue to deepen power market reform. Separate the transmission and distribution business of grid companies; establish more than one distribution companies and electricity sale companies, and achieve power generating enterprises bidding on grid. Another program is not to split the transmission and distribution business of grid companies, but to stripping their part of the electricity sale business, that is to say based on the former big users' electricity straight purchase pilots, further form the power direct transaction model.

3.1.4. Schemes Analysis

The first scheme mainly contains the contents that split the Grid companies' transmission and distribution business; establish independent distribution companies, opening distribution side and allowing social capital access into new distribution network; set up independent electricity companies, allowing private capital to enter. These measures, in theory, can effectively break the Grid company's monopoly whose profit mode will also turn from benefiting from price differentials between on-grid electricity price and sales price to charging net wheeling fees. For the second scheme, it ensures the integrity of the companies' transmission and distribution business, only spinning off their electricity sales business, and allowing the user to directly purchase electricity from electricity trading and generation companies. The grid corporations only charge net-using fees authorized by the Government during consumers directly dealing with the power generators. Compared to the previous large consumers' direct-purchasing, electricity direct trading mode emphasizes more on the user's own option, allowing users and generation companies trading on a unified platform. It realizes generators bidding to sale electricity and users can select with more choices or bidding to purchase, which will gradually achieve the market prices rather than the Government price. From the analyses of program one and program two, it can be found that the main difference between the two is that whether or not to separate the Grid company's transmission and distribution business and whether to set up an independent electricity companies.

3.1.5. Schemes Evaluation

Aimed at the points proposed in the program one that "separation of the Grid's transmission and distribution business", most experts hold a negative opinion. This paper also believes that splitting power grid pays too much, and is not conducive to the construction of China's power market. There are four main reasons: first of all, transmission and distribution functions are similar. Power transmission and distribution business both have the network properties and similar functions as the logistics behavior of electricity. Secondly, the transmission and distribution network is not easy to split. Power distribution network of 220kv and below at present also bears close-range transmission function, so it is difficult to precisely dividing the transmission and distribution from the assets interface. In addition, power transmission and distribution network has significant scale economies. Their huge construction costs determine the properties of natural monopoly. Finally, transmission and distribution as a whole is more conducive to power supply safety. Thus splitting the transmission and distribution is not the main point of the next reform.

Whether to set up independent electricity companies, power generation companies, Grid companies, and users hold uniform views. At the international level, this issue is also a great dispute. Texas of the United States conducted the similar reform in 2002, hoping to enhance competition in the market and lower retail prices and industrial electricity prices by bringing in independent selling companies [10]. But at last the reform effect does not achieve as expected. Its sale electricity side competition has no enhancement and electricity price is also not reduced [11]. Scholars Joskow and Jean Tirole in their articles have done demonstrations on whether electricity sale side should take complete competition. The results display that in the situation of without installation intelligent meters, market information is not complete, so social suboptimal choice will be the same as monopoly market equilibrium, that is, the competition reform of electricity sale market could reduce the whole market's efficiency instead. Competitive mechanisms may be more invalid than monopolistic mechanisms [12]. Tirole's research has a specific reference meaning, but we should also see that the conclusion is based on the transmission and distribution costs set to zero and derived under the assumption of full market monitoring. In the present electricity market in China, the practice in line with the actual is to form monopoly competition on the sales-side and then to reform gradually. Through the construction of a smart grid, smart meters installed on the user-side, the market will gradually translate to be with the full competition.

3.1.6. Decision

Decision making stage is mainly based on the analysis and evaluation of the steps above, to identify an optimal solution. Through the analysis of all aspects of the two different programs, this paper argues that the electricity market reform program most suitable to current China is to keep the transmission and distribution business of grid companies as a whole, and to establish an electricity trading center, allowing direct transactions between power plants and large consumers. In addition, grid companies still keep the selling business, but they need to purchase power through the trading center.

3.1.7. Implementation

This stage is to implement the selected program, namely decision execution phase. The electricity reform can be implemented in pilot regions first. If the implementation is successful, then it can be revised and perfected to be promoted in the whole country. However, often there will be new problems in practice. Thus it needs to take remedial actions or tracking regulatory and continue modification and perfection, depending on the actual situation of implementation and feedback.

3.2. Time Dimension

This dimension refers to the 7 activity development stages of electricity market reform program from the planning stage to the improvement stage, in accordance with the order of time. It can generally be divided into the following seven stages, which are interconnected with the 7 logical stages above, with no clear line in the system.

(1) Planning phase. It is the electricity market reform program planning and preparation stage at which government departments and various stakeholders should put forward proposals for the future fast development on the basis of the full investigation, and according to their own characteristics and actual environmental situation.
(2) Programming phase. Design and propose the system structure of the electricity market reform program, including the structures of subsystems.
(3) System analysis phase. According to the concrete architecture of the program, analyze the system and from different views, study what's change and influence the reform program may bring.
(4) Pilots implementation phase. Enact the determined reform program in the pilots with a sound electricity industry, laying the foundation for the further promotion.
(5) Debugging phase. Based on the implementation results of the pilots, adjust the problems exposed during the implementation of the reform program, and then put the modified program into pilot operation again until getting the improved scheme.
(6) Promotion phase. The improve program should be promoted in regions gradually, and continue to track some important indicators of it in the implementation process, ongoing monitoring whether the reform effect achieves the design requirements.
(7) Upgrading stage. According to the new problems that have emerged in the practice of regional promoting, improve the existing program based on the views of various parties, and transform reasonable reform paths to find the reform direction that can achieved the maximum benefits of the whole community.

3.3. Knowledge Dimension

Electricity market reform is a complex and comprehensive process, which needs to be actualized step by step. To complete each stages and steps above, it needs extensive knowledge and expertise. Its construction and running involves the knowledge of electricity systems, law, business, management, finance, social sciences, engineering technology and other social science and natural science, and each step and stage depending on different issues will involve the different knowledge. This requires that investigate fully during the program formulation and modification process; listen to experts' opinions, and follow the macro-control laws and step-by-step implement principle.

This dimension should not only consider the expertise and technologies involved in the development of reforms, but also the necessary precedent conditions before implementing the electricity market reform program. First, relevant laws should be sound. The electricity industry, as a basic pillar industry of the whole national economy, must be introduced a series of laws as the conduct codes to guarantee the effective reform application. Second, the Government should publish related policies including pricing rules, grid transmission and distribution requirements. Finally, establish sound market supervision and trading mechanisms to ensure that the reform can be effectively implemented. Hall's three dimensions structure of the power market reform in shown in Figure 1.

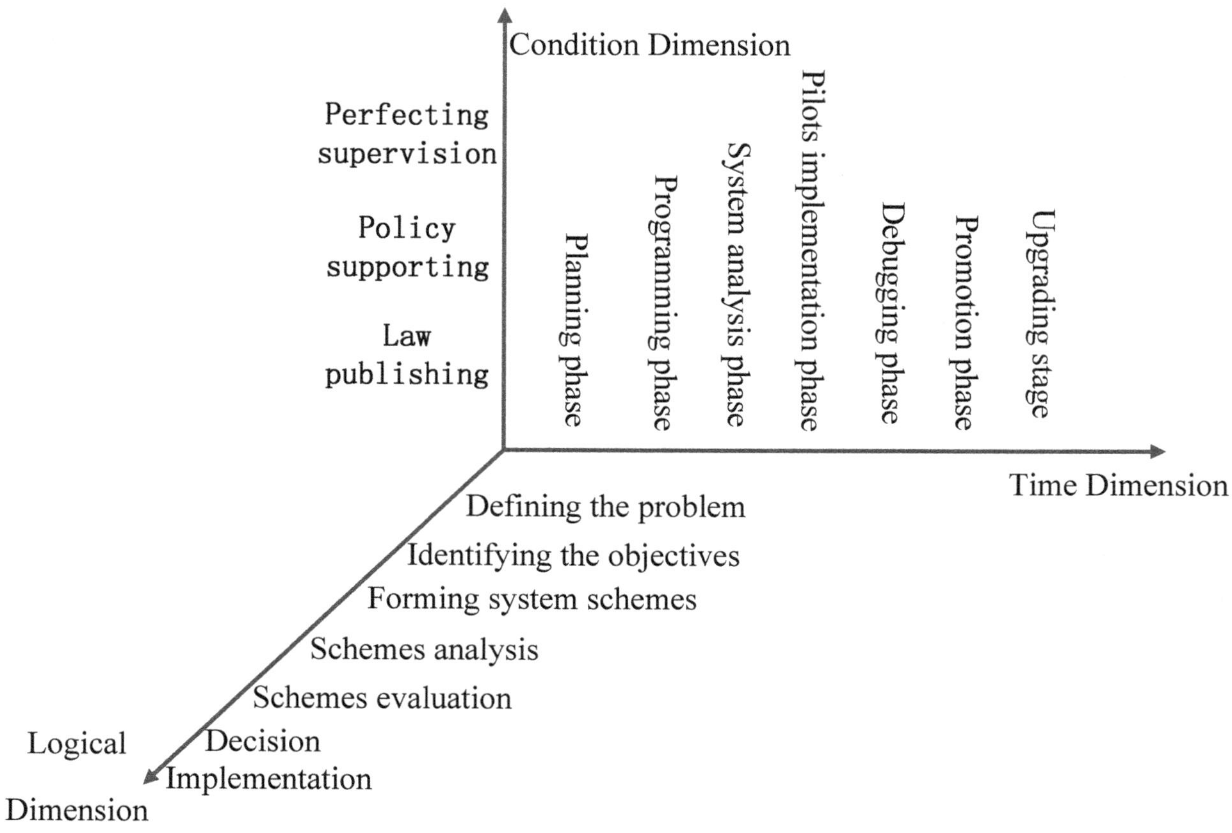

Figure 1. *Hall's three dimensions structure of the power market reform.*

4. Conclusion

Based on the theory of Hall's three dimensions structure, this paper builds an electricity market reform program adapted to the development of China. From logical dimension, time dimension and knowledge dimension, make a stepwise analysis of the planning and implementation of the reform scheme to provide a reference and guidance for policy makers and market entities about how to carry on the next step. But it should be also recognized that in theory, the logical dimension, time dimension and knowledge dimension of Hall's three dimensions structure are closely linked and interrelated, which also constructs an interacted and interdependent three-dimension structure of the power market reform as a complete system. To establish effective competition of power market should be based on the scientific planning mechanism, combine with the regional actual situations in China, and not just stuck with some theoretical dogmas. At the same time, the reform decision mechanisms should be also perfected, which needs related subjects participation and to combine the Government decision with expert arguments. Make Hall's three dimensions structure this systems engineering methodology play a real guide role in the power market reform, so as to promote the healthy, harmonious and sustainable development of China's economy and society.

References

[1] ZHOU Jun. Discussion on the Reform Issues of the Electric Power System in China[J], Journal of Guangdong Polytechnic Normal University, 2009(1):50-54.

[2] YAO Jiangang. Discussion on the structure and bidding modes of regional electricity market[J], Automation of electric power system, 2003,27(22):23-25.

[3] SHANG Jincheng, ZHANG Zhaofeng, HAN Gang. Study on transaction model and mechanism of competitive regional electricity market part one transanction model and mechanism, participation mode for hydroelectricity participants and power system security checking mechanism[J], Automation of electric power system, 2005,29(12):7-14.

[4] FU Shuguo, BAI Xiaomin, ZHANG yang et al. Regional electricity market mode and operation mode research[J], Automation of electric power system,2003,27(9):1-5.

[5] LIN Baiqang. The growth of China's electrical industry: the reform process, and reform that fit each other[J], Management World,2005(8):65-79.

[6] SONG Xiaosong, ZHANG Jianhua, LIU Zongqi. Analysis on the reform modes of electricity market in China[J], Electricity Education in China,2007, Management review and educational research Special Issue:245-248.

[7] LI Guoren, LI Han. Enlightenment from Electricity Industry Reforms in UK , US and Japan[J], SINO-GLOBAL ENERGY,2009(14):21-26.

[8] CHENG Haoyong, ZHANG Senlin, ZHANG Yao. Research on transaction mode of direct power purchase by large consumers in electricity market[J]. Grid Technology,2008,32(21):85-90.

[9] HALL AD. Three-dimensional morphology of system engineering [J], IEEE Transactions on System Science and Cybernetics. 1969: 5(2).

[10] HE Xueming. Electricity retail reform in Texas, China Electric Power Newspaper(N), 2005-04-12(008).

[11] XU Zizhi, ZENG Ming. Analysis on Electricity Market Development in US and Its Inspiration to Electricity Market Construction in China[J], Power System Technology, 2011, 35(6):161-166.

[12] ZENG Ming. Enlightenment for China's electricity sales side reform based on the theory of Tirole, China Energy News, 2015-01-19(005).

9

Development of Prototype Protection Setup for Standalone Solar Power System

Titu Bhowmick, Dharmasa

Department of Electrical and Computer Engineering, Caledonian College of Engineering, AL Hail, Oman

Email address:
titu11500@gmail.com (T. Bhowmick), rdharmasa@gmail.com (Dharmasa)

Abstract: This paper presents development of an electrical protection scheme using *Arduino* microcontroller for a prototype solar power system. First, literature review is carried out on numerical relays for the protection scheme having solar panel output of 12V DC, which is then converted into 220V AC power output using an inverter. Specifically two types of faults are demonstrated neglecting the fault impedance: (i) Over Current fault; (ii) Differential Overcurrent fault. Subsequently, as per the magnitude of the fault current, an Inverse-Definite Time (IDMT) characteristics are studied to *pre-set* the operational features of a relay. Afterwords, the relay signals are programmed using an *Arduino Microcontroller*. The proposed setup consists of 115V/15V transformer; 50W variable rheostat; electro-mechanical relay and low burden electronic current sensors (ACS712) to measure fault current. Then 220V AC auto-transformer is used to tap the voltage to a level of 115V as per the opted transformer. The work identifies the difference between the magnitudes of input-output of currents of a transformer and if magnitude is more than the pre-set value in AC section, finally a tripping signal will operate to disconnect the abnormal part. Further, setting of differential relay is investigated to find the efficient operation of the solar power system. Then validation of prototype model is done by creating an intentional fault using variable rheostat as load. This work investigates efficiently to obtain accurate results on both internal and external faults. In total the proposed scheme consumes less power, which is suitable to develop a prototype protection scheme.

Keywords: Arduino, Differential Relay, Overcurrent Relay, Inverter, ACS712 Current Sensor, Rheostat

1. Introduction

Present power system is in need of a high level of redundancy and reliable protection devices along with the periodic maintenance to keep the system healthy. For that, the relays must be pre-set to an accurate time of interval operation for the system under faulted condition and also monitor continuously scan to keep the system healthy. As per [1] Microcontroller Based Protection Relays (MBPR) have a real time embedded system and in case of abnormal variation of parameters in the system, then [MBPR] must operate in precise with programmed time-delays In these circumstances, there is need of an efficient protection scheme. Recently as per [2] MBPRs are programmed to monitor the health of DC trip coil circuits. MBPR will also help to validate signal status of inputs and output circuits. Further MBPR can be used to monitor the status of transformer breaker, auxiliaries and ambient environmental conditions using efficient communication links.

1.1. Problem Definition

Usually, power failures cause either short or long duration power cut in an electrical power distributed area. Which indicates that fault can be triggered due to numerous causes such as: short circuit; overloading; faults in electric generation, transmission and distribution.

The *Quality of Power* is an important issue for certain utilities such as: mines; hospitals and telecommunication systems. For these utilities the power outage should be minimized to keep the system in continuous operation. In general, voltage events are classified as: Under voltages-long term variations lasting more than 1 minute and if less than 90% of nominal voltage; Over voltages:-long term variations lasting more than 1 minute and if greater than 110% of nominal voltage; sags:- short term variations of duration less than 1 minute and the voltage is between 10% and 90% of nominal; Swells:- short term variations of duration less than 1 minute and the voltage is greater than

110% of nominal; Voltage unbalance: voltage on each phase conductor is different. The power interruption is more serious in a "Standalone Prototype Solar Power System". The updated, IEEE guide reported in 2011 presently available [3] – includes significant changes and additions like: Differential Relay – Primary Phase Fault Protection; Distance Relay – Backup for System; Generator Zone Phase Faults; for Generators 100 % Stator Ground Fault Protection and Time-Overcurrent Ground Relay.

1.2. Types of Power Failures

To achieve the power quality, it is necessary to understand failure classifications and impact of duration of fault on system:

i. Voltage reduction in the network, it can be visually observed by the dimming of the lights. If this fault persists then it will cause damage to the health of the system and also leads to certain equipment to malfunction, due to rise in current magnitude.

ii. Voltage rise occur mainly due to switching or lightning phenomena in the system. This instantaneous effect results in malfunctioning of the equipment and damage to certain machines.

iii. Blackout is the extreme cause of power failures. It isolates a whole network from electricity for a long duration of time, until the fault is rectified. It is mostly caused due to tripping in the power stations. It is still challenging to solve the issue.

1.3. Need of Protection Scheme for Power System to Reduce Power Failures

In a power system, all the equipment should operate efficiently without any interruptions and meeting demand of power. In case of sudden increase in demand the power station would not be able to provide the required power and it could cause overloading. Conventionally, solid state protection relays and fuses are used in numerous places to detect these odd effects and cut off the system to avoid any chance of damage mainly due to over current. The simulation study of injection of renewable energy such as solar power into the power grid is reported [4] considering occurrence of *overvoltage* in distribution feeder. If large amount of solar power is injected at low power demand, then the role of operation of overcurrent relays becomes more critical due to reverse power operation at 33 kV bus. In case of small-sized power grid bi-directional inverters on AC buses in charging battery banks and adjusting the relay current settings.

In [5] authors are reported simple MPBR approach, which operates as per strict-time based protection deadlines such as i) Overloading of electrical equipment, it is not always necessary to have sudden operation of protective relaying, for such conditions definite real time operations are suitable. ii) In case of long duration short circuit fault to avoid damage of the equipment, sudden removal is necessary.

In recent years good- quality research is reporting, in this topic to improve the reliability of the protection devices. Failures can occur in any part of the system and in any equipment, protection devices are aimed to protect the assets and guarantee continued supply of power, even under sudden inrush of current.

1.4. Overcurrent Relay

The most severe type of fault in a system is over load fault; where it can develop heat [I^2Rt] in the equipment and then damage the system. In case of Short Circuit, disconnection of the power system for high magnitude over current (I_{sc}) due to improper setting of disconnection time. Hence, there is a greater risk of blackout of electrical power supply area. As per IET regulations, in case of distribution system graded current range selected for the current carrying conductor as per equation (2):

$$I_b \leq I_n \leq I_z \leq I_{sc} \quad (1)$$

Where,

I_b = Design Current
I_n = Fuse setting Current [Nominal value]
I_z = Perspective current
I_{sc} = Short Circuit current

$$Iz = I_n/C_aC_gC_iC_h \quad (2)$$

Where, C_a, C_g, C_i, C_r are the correction factors used to design a current carrying conductor as shown in equation (2).

C_a= Conductor under ambient temperature
C_g = Grouping of cables having conductors
C_i= Thermal insulation of cable covering conductor
C_h = Presence of Harmonic in the current

The conventional Electro-Mechanical induction overcurrent relays are used as an energizing coil through which the current passes [6]. In such case, a circuit must be completely ON or OFF with minimal on-state voltage drop. In such case an Electro-Mechanical Relay [EMR] is the only choice. EMRs also are best, if heavy surge currents or intermittent voltages are anticipated from the load side. In normal conditions, the current is not sufficient to produce magnetic field to move the disc. In abnormal conditions when the current is very high, the coil produces large magnetic fields that move the disc and trigger the tripping of the breaker. The work considers solid -state relays (SSR)s relays as electronic devices. Usually, an SSR's input consists of an Opto-isolator, while its output is a Triac, SCR, or FET. One negative aspect of an SSR is that semiconductors are never completely on or off. In the on-state, substantial resistance is present, which can lead to significant heat generation when current is flowing. So SSRs must be placed on heat sinks, often several times the relay's weight. SSR sensitive to ambient heat and must be de-rated if used in hot environments country like Oman in the Gulf region.

Improved and advanced versions of the overcurrent relay is embedded in the digital relays, which determine in precision the abnormality of the system conditions.

Overcurrent protection motor [7] protection setup is as shown in Fig. 1 for three phase network. In this arrangement, an instrument transformer is used such as current transformer

(CTs), which take the current reading and sends analog values to the controller/processor in voltage from. First, controller converts the digital and then compares them to a preset value. The preset value is the current ratings of the system during normal condition.

In case of abnormal conditions the current reading will exceed the preset value and then the controller will send the tripping signal. Inverse Definite Minimum Time (IDMT) over current protection is used as the relay settings. This has an advantage of dealing with the severity of the fault current. If the fault current is high, the controller will send the tripping signal fast otherwise after a time delay. The time delay is set using standard equation (1):

$$t = TMS \times \left\{ \left[\frac{k}{\left(\frac{I}{Is}\right)^{a} - 1} \right] + c \right\} \quad (3)$$

Where,

t = operating time; I = energizing current;

I_s= overcurrent setting; TMS = Time Multiplier Setting.

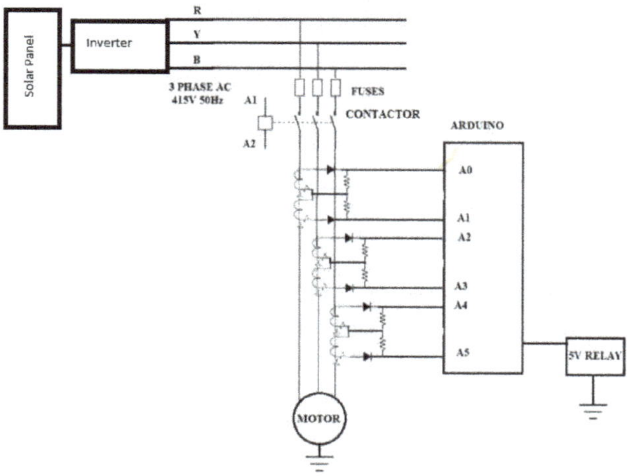

Fig. 1. Overcurrent motor protections circuit.

1.5. Differential Relay

This relay is basically unit-type or for a specific equipment protection. It functions to detect any internal faults in its zone of protection. It takes the current readings and checks the differential in the current entering and leaving as per simple KCL technique. Current Differential relays are basically applied to transformer giving rating upto 5 MVA and above (on transformers which are critical to operate) Transformers have Buchholz relay installed in them that monitors the internal health of the transformer using the insulating oil, but it cannot detect any inter turn-fault occurred to the transformer. For that differential relay is used, these relays detect any difference in the transformer accurately and quickly.

For current differential relay, current transformers are used to read the current values on the primary and the secondary side. The readings are sent to the controller, where they are compared and any abnormal condition is detected if the compared value is different from the preset value.

2. Proposed Methodology

Fig. 2 shows that there are two solid-state current sensors i.e. one at the primary side and other at the secondary side of the transformer. The transformer gets 115V as primary voltage from the supply, through a contactor. For this supply, current sensors give the analog current reading to the analog inputs of the Arduino A1 and A2. The current sensors take 5V power supply from the Arduino.

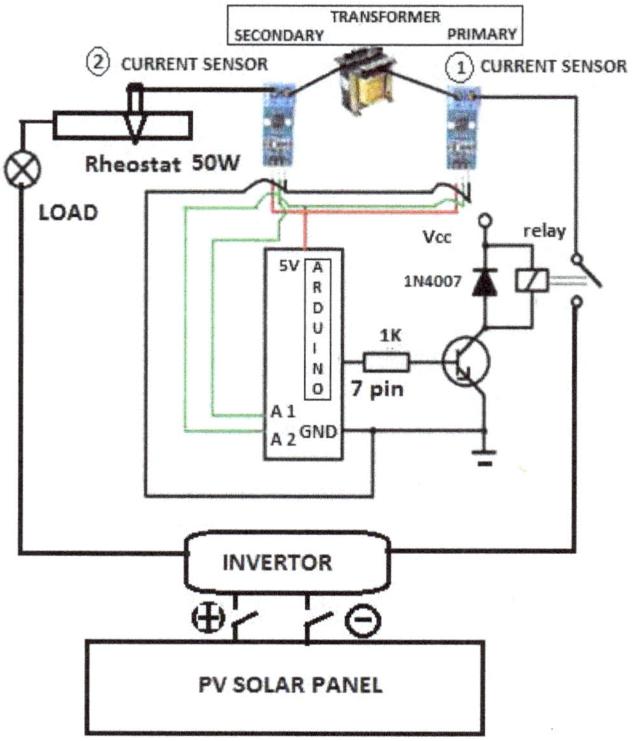

Fig. 2. Proposed protection setup.

Arduino can only handle voltage (between 0V and 5V), so it is necessary to convert this current into an acceptable voltage range. As per [8] a burden is added in resistor circuit as shown in Fig. 3(b) for Fig. 3(a) model (YHDC SCT-013-000) CT.

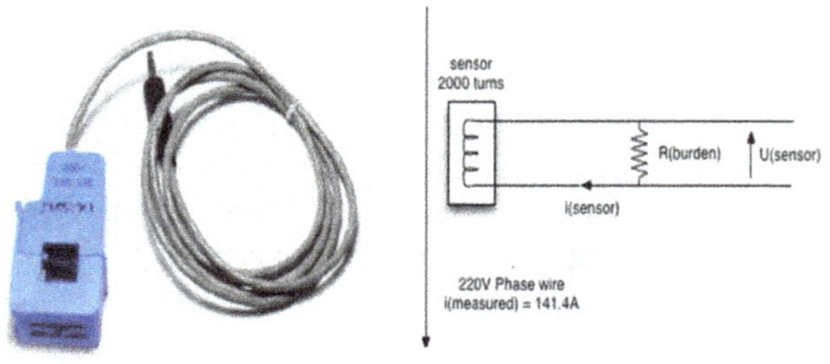

(a) (b)

Fig. 3 (a, b). AC current sensor with I to V circuit.

From Fig 3(b), it can be expressed clearly that the current is alternative around 0 and 5V to maximize measurement resolution and the max voltage at burden resistance is 2.5V. The better Burden resistor value opted using R (burden) = U(sensor)/I(sensor) = 2.5V / 0.0707A = 35.4Ω. The maximum phase error of 4 degree for 18 Ω burden is insignificant (representing a power factor error of less than 0.0025 at unity power factor), but the error of nearly 10 degree with a 120 Ω burden could be troublesome with low current loads having a poor power factor where this input is most likely to be used and such type sensor are used for only AC current measurement. The opted sensor of [9] has applied current flowing through the copper path and it generates a magnetic field based on Hall IC converts into proportional voltage and it also be used for Output voltage proportional to AC or DC currents.

Fig 4 shows the secondary side of the transformer is operating at 15 V(load). A rheostat and light bulb act as load, which indicates the flow of current. The primary side ACS712 current sensor as [10] is found more suitable for the over current protection of the transformer and the secondary side current sensor is used for the differential current value to indicate any internal faults in the transformer. In case of abnormal conditions the Arduino gives a tripping signal to the relay through pin 7, and the relay is connected with the contactor, which disconnects the circuit from the supply.

The sensors were tested to find out the accuracy based on the output of the device has a positive slope [>VI OUT(Q)] when an increasing current flows through the primary copper conduction path (from pins 1 and 2, to pins 3 and 4), which is the path used for current sampling. The internal resistance of this conductive path is 1.2 milliohm, which is typical to provide low power loss.

The reading of the sensor is tabulated in Table 1 and the order variable reading are obtained using rheostat.

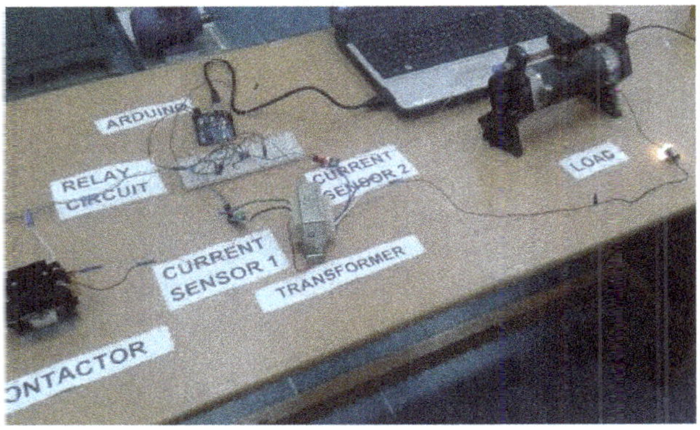

Fig. 4. System under operation status.

Fig. 5. Current reading using a clamp meter.

The sensors take the current reading and it is displayed on a screen connected with the *Arduino* shown in Fig.4 The sensors were used and the accuracy was tested for providing power supply through USB and found that 0.005A is tolerant. The set values were accurately read from the load current and computed as shown in Fig. 5.

Table 1. Current sensor measurements.

Sl. No.	Current (A)
1	0.27
2	0.26
3	0.26
4	0.26
5	0.25
6	0.25
7	0.25
8	0.25
9	0.25
10	0.25

3. Results and Discussions

The load is varied using variable rheostat having small 50W capacity to reduce high current in the primary side. The readings taken by the sensors and then send to the *Arduino*. Usually, a relay will act in an inverse time overcurrent relay mode. The Arduino compares the value with the Pre-set value, if the fault magnitude (over-load) more than pre-set value, then *Arduino* sends the trip signal after a time delay.

If the fault magnitude is suddenly rises to a higher (short circuit) value than the pre-set value, the *Arduino* sends an instant trip signal. If the sensor read more than set value of current, then sensor sends the data to the *Arduino* which compares the value with a preset value and finally protection setup operate after a time delay.

The prototype of transformer is used in the circuit having nearly a current of 1A, for that a preset value is 0.7A and this set value is sufficient to operate the *Arduino*. The load was varied to reach this point and when a value sensed is above 0.7A, The *Arduino* sends the tripping signal to the relay and shows that there is an external fault in the system which caused the tripping. The results for external faults are as tabulated in Table 2.

Table 2. Indicating External Fault.

Sl No.	Primary current (A)	Secondary current (A)	Current ratio	Fault
1	0.68	0.71	0.95	No Fault
2	0.67	0.70	0.96	No fault
3	0.67	0.67	1.00	No Fault
4	0.72	0.71	1.02	Fault Indicated

For the operation of differential relay the transformer internal faults are created and the preset value of the difference in the primary and secondary current was set low. Setting leads to send the trip signal in case of any internal faults.

Table 3. Indicating internal fault.

Sl. No.	Primary current (A)	Secondary current (A)	Current ratio	Fault
1	0.66	0.70	0.94	No Fault
2	0.66	0.70	0.94	No fault
3	0.66	0.68	0.97	No Fault
4	0.67	0.65	1.02	No Fault
5	0.65	0.73	0.89	Fault Indicated

This differential relay finds any possibility of internal faults and later. Differential relay acts as a protection for the transformer from being exposed to current, high temperature and pressure in case of any internal faults, which would further lead to flashover of the transformer. The relay trips the system in the rise in differential current value above the preset value. The preset value was set at 0.9A.Excavation of any change in the difference in current caused by internal faults of the transformer as shown in the Table 3 drafted from result seen on laptop display, which indicate an internal fault. For that type of fault, the currents characteristics of transformer need to be further inspected.

The system setup and the relay are tested in the real time. The result indicates that the relay works as IDMT relay and trips the circuit in case of any abnormal conditions for the overcurrent preset value at 0.7A. The set value is detected and the relay opened the circuit using the contactor. The tabulated fourth reading of Table 2 indicates unacceptable current value, i.e. abnormal current value, which sensed by the relay and disconnects the circuit with help of Arduino. The fifth reading of the Table 2 result shows abnormal condition as an External Fault.

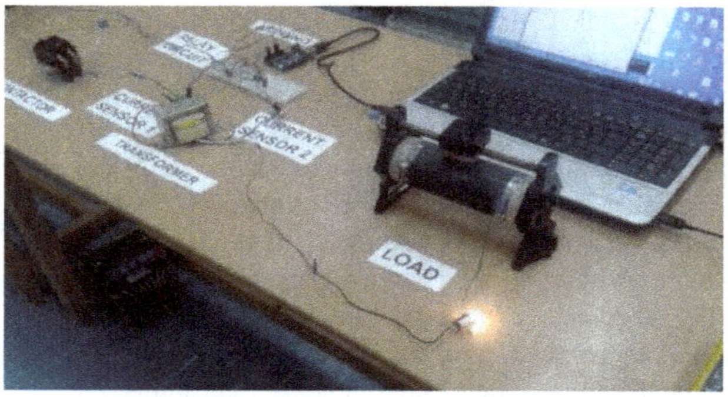

Fig. 6. *Circuit under normal working conditions.*

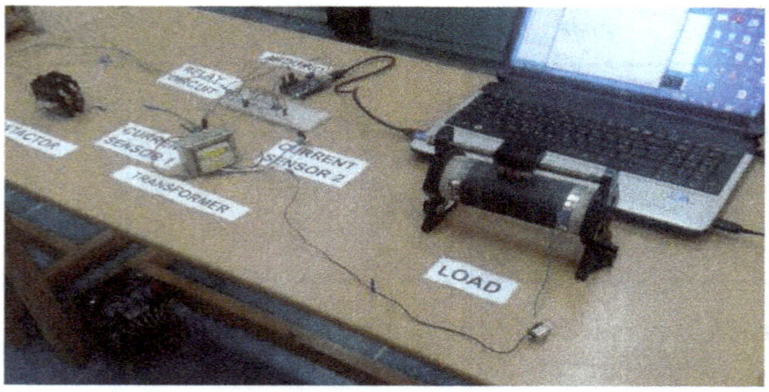

Fig. 7. *Circuit under abnormal conditions.*

Furthermore the connection diagrams for normal conditions and abnormal conditions are presented in Fig 6 and 7 respectively. If there is a change in the load, higher current is drawn from the supply and then sensors read the current value and send the input to the Arduino in the form of voltage signal. This signal is converted into digital form of current value, which is computed by the microcontroller. If the value is higher than the pre-set value the *Arduino* sends the tripping signal to the relay. In the proposed work, the delay time of 500ms was given for current value operated in between 0.7A to 0.75A and otherwise a direct tripping was done for short circuit value.

Considerations for Using High-Impedance or Low-Impedance microprocessor Relays for Bus Differential Protection the transformer can be seen in [10]. There is a wide variety of technologies that are used in today's smart meters for sensing the current submitted [11] delivers the strengths and weaknesses of the most popular sensing technologies. In this case an independent comparison between high and low-impedance bus differential relays is discussed by the authors. Also authors' opinions regarding the superiority of one type of relay over the other for specific considerations discussed in detail [12] for different sensors.

4. Conclusion

The proposed overcurrent and transformer differential protection schemes are tested by creating fault using variable resistance. It is found that, numerical or program based relays have better accuracy and less response time, while clearing the faults and also protection parameters can be varied accordingly. In the proposed *Arduino* the IDMT can be modified or changed using simple programming technique. The actual set values and operational program can be easily viewed in the computer screen or any other preferred monitor. In the developed system, *Hall Effect* based ACS712 current sensors, and *Arduino* does not consume much power, while in operation. This is one of the major advantages of the model developed. The Arduino based program relay set-up has more accurate and flexible in fault detection and clearing, when compared to static as well as electromagnetic relays.

References

[1] Aaditya G. V., Reddy M. S., Rao K. V, Ashish P. C, "project report" Electrical and Electronics Engineering, Gokaraju Rangaraju Institute of Engineering & Technology, Bachupally, Hyderabad, 2010.

[2] Nollette S., 2014, Microprocessor-based relays offer extra value, Control Engineering [Online]: Available from: http://www.controleng.com/single-article/microprocessor-based-relays-offer-extra-value/911ed9a6dde5fa4528a3c087a9ab4c3f.html.

[3] Special report on IEEE Power System Relaying Committee, 2011, 57th IEEE Pulp and Paper Industry Conference. [Online]https://www.eiseverywhere.com/file_uploads/ae8b333bd131b9146e01907d95ec0fcb_Synchronous_Generator_Protection_PPT.pdf.

[4] Ouahdi Dris Farag. M, Elmareimi Rekira Fouad, "Transformer Differential Protection Scheme With Internal Faults Detection Algorithm Using Second Harmonics Restrain and Fifth Harmonics Blocking Logic, 5th International Conf. On Electrical and Electronics Engg., 5-9 December 2007, Bursa, Turkey.

[5] Nader Barsoum, Chai Zen Lee, "Simulation of Power Flow and Protection of a Limited Bus Grid System with Injected Solar Power", Energy and Power Engineering, 2013, 5, 59-69.

[6] Kumar V. L and Garividi S. T. I. C, Microprocessor Relay for protection of Electrical System, Ubiquity, 7(20), 2006, p. 15-23.

[7] Thomas R. M "Electromechanical Relays Versus Solid-State: Each Has Its Place", Sep 2002, [Online] Available from: www.electronicdesign.com/components.

[8] Sairam A., Sandeep P, Vilas B, Manideep K, Kumar V. R, "Project Repot" in Electrical and Electronics Engineering, Gokaraju Rangaraju Institute of Engineering and Technology, Bachupally, Hyderabad, 2014.

[9] Vincent Demay. Current monitoring with non-invasive sensor and arduino, [Online]: Available fom: http://www.homautomation.org/2013/09/17/current-monitoring-with-non-invasive-sensor-and-arduino.

[10] Fully Integrated, Hall Effect-Based Linear Current Sensor IC with 2.1 kVRMS Isolation and a Low-Resistance Current conductor [Online]: Available form: https://maxwell.ict.griffith.edu.au/sok/ees/resources/Current.pdf.

[11] Ken Behrendt, David Costello, Stanley E. Zocholl Considerations for Using High-Impedance or Low-Impedance Relays for Bus Differential Protection 49th Annual Industrial & Commercial Power Systems Technical IEEE Conference, Stone Mountain, Georgia, April 30–May 3, 2013.

[12] Glenn Roemer, FAE, Pulse Engineering Current Sensors in Power Metering Applications, page-1-4 Available form: http://www.pulseelectronics.com/docs/ white_paper.pdf.

An Experimental Study on Heat Transfer Enhancement of Flat Plates Using Dimples

Amjad Khan[1], Mohammed Zakir Bellary[1], Mohammad Ziaullah[1], Abdul Razak Kaladgi[2, *]

[1]Department of Electronics and communication Engineering, P.A College of Engineering, Mangalore, Karnataka, India
[2]Department of Mechanical Engineering, P.A College of Engineering, Mangalore, Karnataka, India

Email address:
amjadece@pace.edu.in (A. Khan), mdzakir87@gmail.com (M. Z. Bellary), mdziya1990@gmail.com (M. Ziaullah), abdulkaladgi@gmail.com (A. R. Kaladgi)

Abstract: Dimples play a very important role in heat transfer enhancement of electronic cooling systems. In the current paper, the fluid flow and heat transfer characteristics of spherical dimples at different angle of orientation from the centre with apex facing the inlet were investigated. The experiment was carried out for laminar Natural convection conditions with air as a working fluid. The overall Nusselt numbers and heat transfer coefficient at different orientation angle of dimples were obtained. From the obtained results, it was observed that the Nusselt numbers and heat transfer coefficient increases with decrease in the orientation angle of dimples.

Keywords: Electronic Cooling, Natural Convection, Dimples, Passive Techniques

1. Introduction

The development of integrated electronic devices with increase level of miniaturization, higher performance and output has increased the cooling requirement of chips considerably. Also, as the chip temperature increases, the stability and efficiency issues increases and the problem of heat dissipation becomes a bottleneck for the development of chips in the electronic industry [1]. Passive heat transfer enhancement techniques are recently used in electronic industries to overcome these issues. In these techniques concavities (dimples), extended surfaces or fins, and protrusions are made on the flat plates to increase the heat transfer rate. Among these, the dimples i.e. concavities are considered important because they not only increase the heat transfer rate but also provide minimum pressure losses (important for pumping power requirements [1]). Dimples increases the heat transfer rate because they produce vortex pairs, increases turbulent mixing, delays growth of the boundary layer, creates reattachment zones necessary for the heat transfer enhancement. And as they do not project into the flow so they contribute less to the foam drag, to produce minimum pressure drop penalties [2]. Another advantage is that in dimple manufacture the removal of material takes place which reduces the cost and weight of the equipments.

Kuethe [3] can be considered as the first person to use dimples on flat surfaces. He observed that the dimples acts as vortex generator, produces turbulent mixing in the flow & increases the heat transfer rate. Afanasyev et al [4] conducted an experiment on friction and heat transfer analysis on flat plates using spherical dimples. And observed an increment of 30-40% in the heat transfer rate. No significant effect on the hydrodynamics of flow was observed in their study. Chyu et al [5] used tear drop type dimples along with hemispherical dimples to study the heat transfer characteristics of air flow in the channel. They observed a considerable increment in the heat transfer rate for the surfaces having dimples (About 2.5 times then their smooth surfaces). Mahmood et al [6] conducted an experiment to investigate the effect of dimples on heat transfer using the flow visualization techniques and concluded that the periodic nature of shedding off of vortices is the main cause of enhancement of heat transfer (much more pronounced at the downstream rims of the dimples). Mahmood et al [7] conducted an experiment to study the effect of Reynolds number, aspect ratio, and temperature ratio in a channel with dimples. They observed through the flow visualization techniques that the secondary vortices that are shed off from the dimples become stronger as the non-dimensional channel height to dimple diameter (H/D) ratio decreases and increases the local Nusselt number in these regions. Rao et al [8] conducted an experiment to investigate the effect of dimple depth on flow and heat transfer characteristics in a pin fin-dimple channel and observed that compared to the pin fin channel, the pin fin dimple channels have improved convective heat transfer performance and especially the pin fin-dimple channel with dimples of higher depth have higher values of Nusselt number.

From the literature review, it is very much clear that dimples have high potential to enhance the heat transfer rate, together with producing minimum pressure drop penalties. The other advantages include low weight, low cost and low fouling rates [9]; however, much of the research either numerical or experimental is on spherical dimples of uniform diameter [4, 10]. And most of the research is confined to flow in the channel or Internal flow (using forced convection), with a very few studies on external flow [10].So the main focus of this study is to investigate the effect of spherical dimples on aluminum/copper test plates under laminar natural flow conditions.

2. Experimental Setup

The fabricated experimental setup used in this study is as shown in the figure below

Figure 1. Experimental setup.

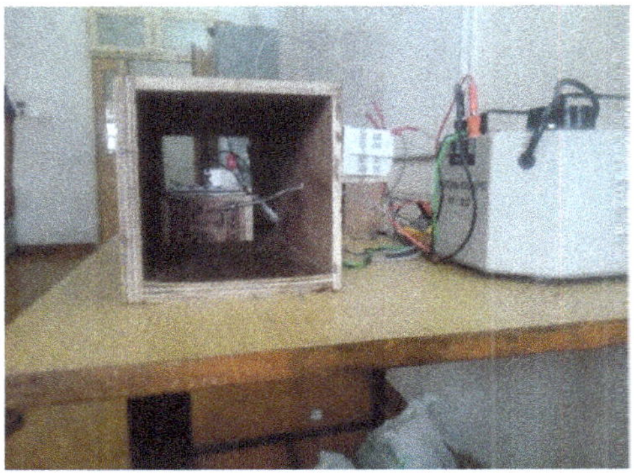

Figure 2. Experimental setup (side view).

It consist of a heater with a capacity of 200 watts, Dimmer stat, Digital temperature, voltmeter and ammeter with J type thermo couple. The test plates were placed on heater and a constant heat flux is supplied through dimmer stat to heater. Air flows parallel to the dimpled test surface. The plate heater is fixed at the bottom of the test plate and was connected to power socket through dimmer stat. Dimmer stat readings were varied to give the required heat input to the test plate. Only top dimpled surface of the test plate was exposed to the air stream from which the convective heat transfer to the air stream would takes place. After reaching a steady state, surface temperature of plate & air temperature were measured with the help of thermocouple.

Table 1. Components and Specifications.

Components	Specification
Heater	100W, 4"x4"
Dimmer stat	6A,230V
Digital Temperature Indicator	6 channel,1200C, 230V
Casing	A wooden casing of size of 8"x8" and 2feet long.
Thermocouple	K-Type, 3000C, 1m long.
Digital Multi-Meter	Voltmeter, Ammeter
Test plate	10x10x2 cm aluminium/copper plates

3. Results and Discussion

Experiments were conducted on aluminum/copper test plates with spherical dimples of non uniform diameters made on flat plates. The different arrangements of dimples used are

Case a. Aluminum plate Dimples with 30^0 orientations with centerline & apex facing the inlet.

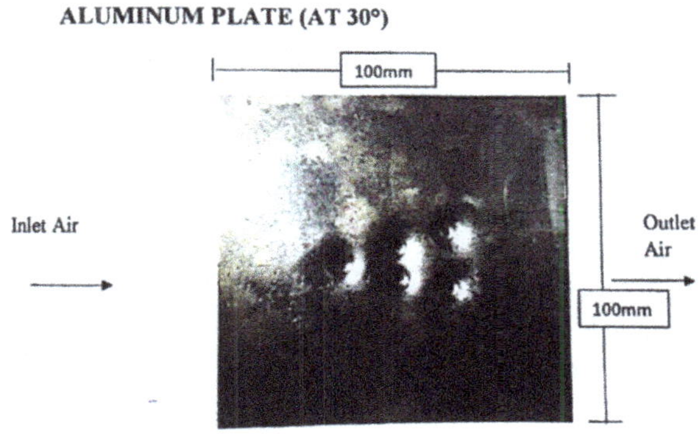

Case b. Dimples with 60^0 orientations with centerline & apex facing the inlet.

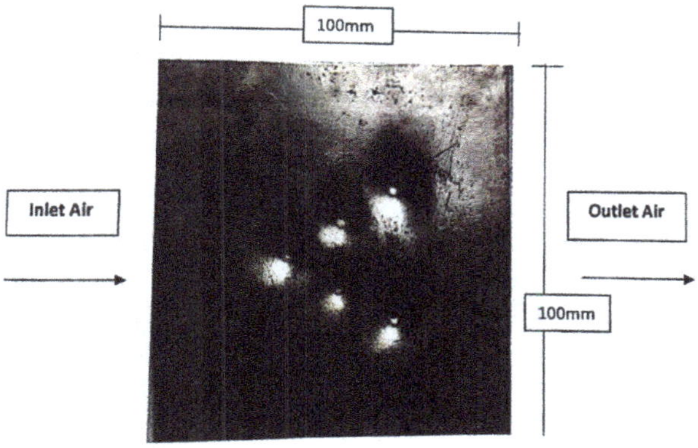

Case c. Dimples with 90^0 orientations with centerline & apex facing the inlet.

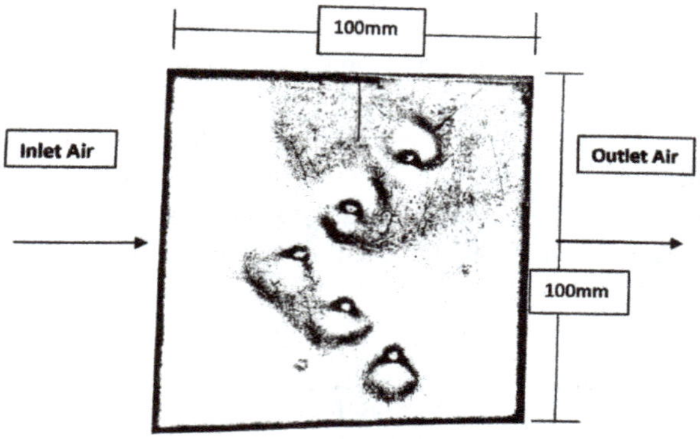

The data obtained were used to find important heat transfer parameters like Nusselt number, heat transfer coefficient, and heat transfer rate. And the experimental findings have been plotted in the form of graphs, mainly

- Nusselt number(Nu) vs. Reynolds number(Re)
- Heat transfer coefficient (h) vs. Reynolds number(Re)
- Heat transfer rate Q vs. Reynolds number(Re)

Figure 3, 4, 5 shows variation of Nusselt number 'Nu', heat transfer coefficient 'h' and heat transfer rate 'Q' with different orientation of dimples. It can be seen that as expected the Nusselt number, heat transfer coefficient 'h' and heat transfer rate 'Q' increases with decrease in orientation angle of dimples. The increase in the Nusselt number as compared to flat plate may due to direct flow impingement on the downstream boundary of the plate and strengthened flow mixing of vortices at the downstream of the plate [1, 11]. The formation of vortex pairs, periodically shedding off from the dimples at low angles, a large up wash regions with some fluids coming out from the central regions of the dimples, are the main causes of enhancement of Nusselt number [6].This happens mostly at low angle orientation of dimples (30^0), as at low angles the dimples are very close to each other so a pronounce effect of dimples is felt. It can also be seen that the variation in the Nusselt number is almost gradual with orientation as expected [12, 13].

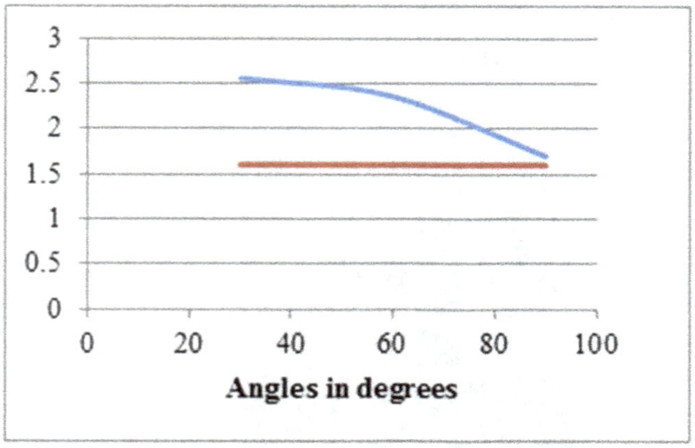

Figure 3. Variation of Nusselt number with orientation of dimples.

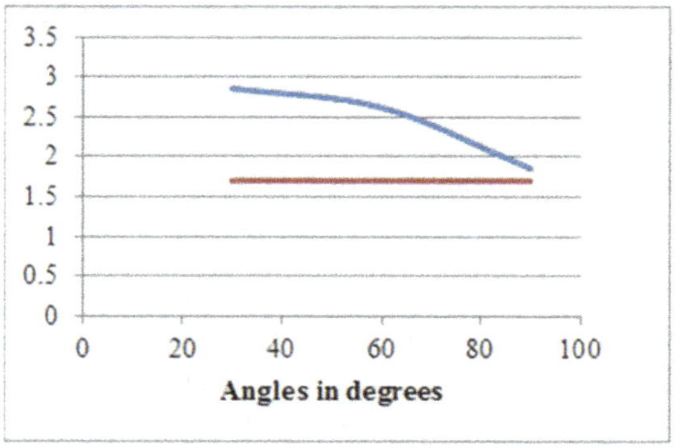

Figure 4. Variation of heat transfer coefficient with orientation of dimples.

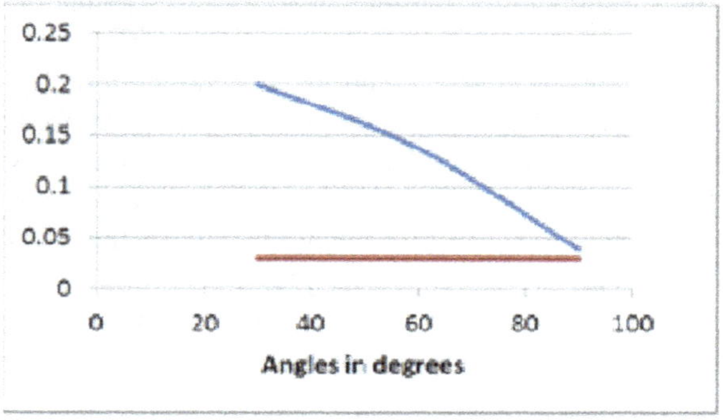

Figure 5. Variation of heat transfer rate with orientation of dimples.

Figure 6, 7, 8 shows variation of Nusselt number 'Nu', heat transfer coefficient 'h' and heat transfer rate 'Q' with different orientation of dimples. It can be seen that as expected the Nusselt number, heat transfer coefficient 'h' and heat transfer rate 'Q' follows the same trend as for the copper plate[16] but with a little lower value because of the low thermal conductivity of aluminum plates as compared to copper plate. It is obvious that 'h' increases as compared to flat plates because the dimple arrangement causes delay in the development of the thermal boundary layer & hence increases the local heat transfer in the reattachment region and increases the heat transfer coefficient [1].It can also be seen that 'Q' increases with decrease in orientation angle of dimples. Because the near-wall turbulent mixing intensity downstream the dimple increases due to the vortex flow shedding from the dimples[6,15].

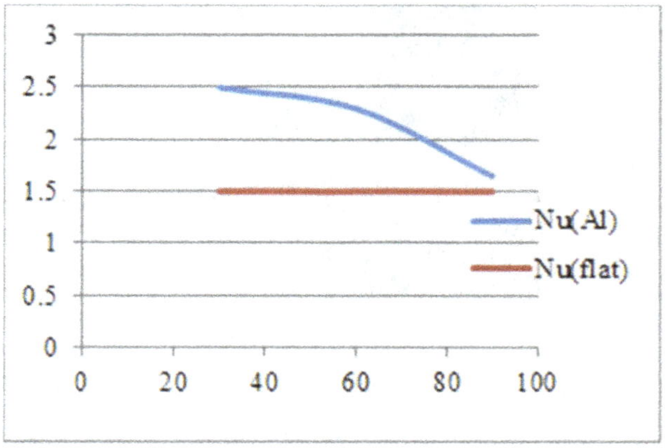

Figure 6. Variation of Nusselt number with orientation of dimples.

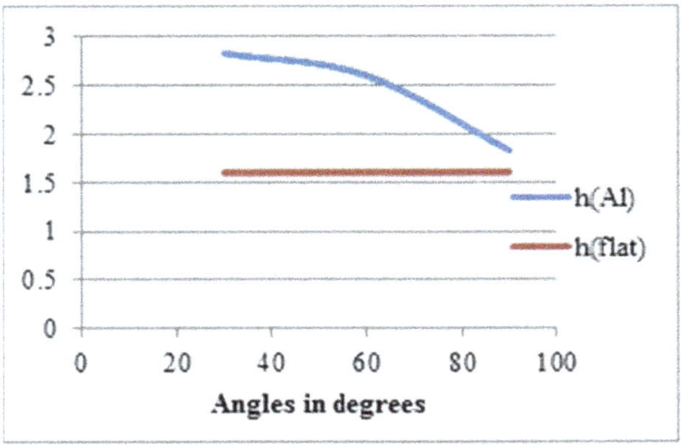

Figure 7. Variation of Heat transfer coefficient with orientation of dimples.

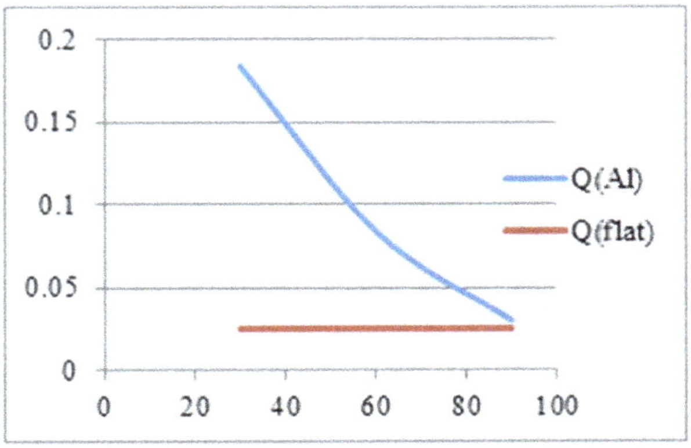

Figure 8. Variation of Heat transfer rate with orientation of dimples.

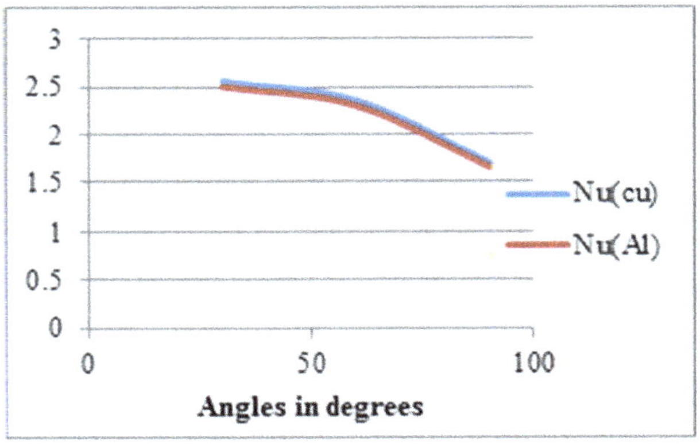

Figure 9. Comparison of Nusselt number.

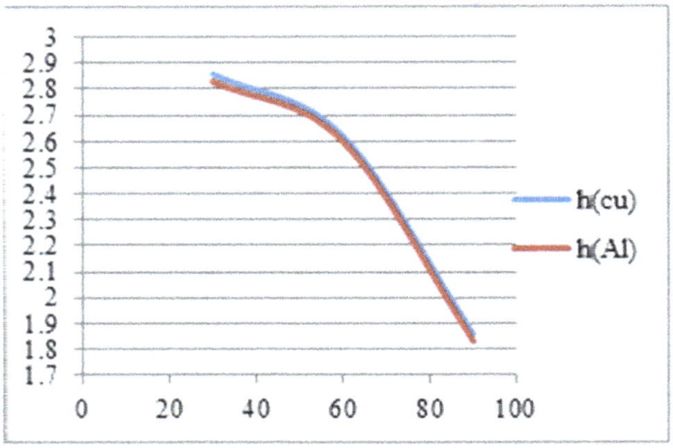

Figure 10. Comparison of Heat transfer coefficient.

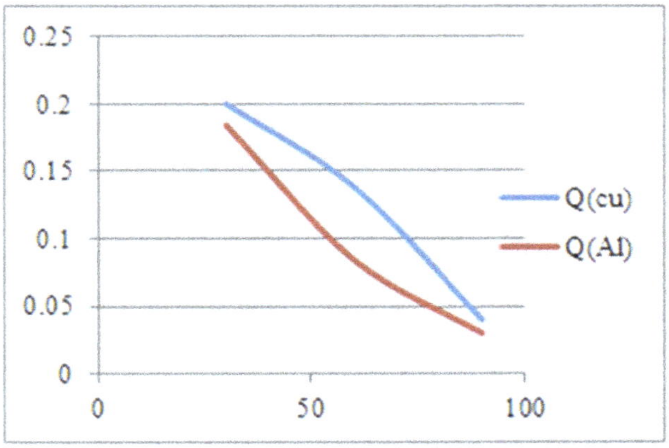

Figure 11. Comparison of Heat transfer rate.

Figure 9, 10, 11 shows Comparison of Nusselt number 'Nu', heat transfer coefficient 'h' and heat transfer rate 'Q' with different orientation of dimples. It can be seen that both the plates follows the same trend i.e. with decrease in orientation angle the value increase because at low angles (30^0) the dimples are close as compared at higher angles (90^0).So a pronounce effect of dimples is felt.

4. Conclusion

In this experimental work an investigation of the effect of air flow over a flat plate with different orientation of dimples on the flat plate is carried out. The main conclusions of the work are:

- Nusselt number 'Nu', and heat transfer coefficient 'h' increases with different orientation of dimples as compared to the flat plate due to direct flow impingement on the downstream boundary and strengthened flow mixing by the vortices at the downstream.
- It can be seen that as expected the Nusselt number, and heat transfer coefficient 'h' curve for aluminum plate follows the same trend as for the copper plate[16] but with a little lower value because of the low thermal conductivity of aluminum plates.
- Nusselt number 'Nu' and heat transfer coefficient 'h' are higher for low angle orientation of dimples because at low angles the dimples are close to each other hence a pronounce effect of dimples is felt which is obvious from the figure.

References

[1] Zhang, D., Zheng, L., Xie, G., and Xie, Y.,An Experimental Study on Heat Transfer enhancement of Non-Newtonian Fluid in a Rectangular Channel with Dimples/Protrusions, Transactions of the ASME, Vol. 136, pp.021005-10, 2014.

[2] Beves, C.C., Barber, T.J., and Leonardi,E., An Investigation of Flow over Two-Dimensional Circular Cavity. In 15th Australasian Fluid Mechanics Conference, the University of Sydney, Australia, pp.13-17, 2004.

[3] Kuethe A. M., Boundary Layer Control of Flow Separation and Heat Exchange. US Patent No. 1191, 1970.

[4] Afanasyev,V.N.,Chudnovsky,Y.P.,Leontiev,A.I.,andRoganov,P. S.,Turbulent flow friction and heat transfer characteristics for spherical cavities on a flat plate. Experimental Thermal Fluid Science, Vol. 7, Issue 1, pp. 1–8, 1993.

[5] Chyu, M.K., Yu, Y., Ding, H., Downs, J.P., and Soechting, F.O., Concavity enhanced heat transfer in an internal cooling passage. In Orlando international Gs Turbine & Aero engine Congress & Exhibition, Proceedings of the 1997(ASME paper 97-GT-437), 1997.

[6] Mahmood,G.I.,Hill,M.L.,Nelson,D.L.,Ligrani,P.M.,Moon,H.K ., and Glezer,B., Local heat transfer and flow structure on and above a dimpled surface in a channel. J Turbomach, Vol.123, Issue 1, pp: 115–23, 2001.

[7] Mahmood, G. I., andLigrani, P. M., Heat Transfer in a Dimpled Channel: Combined Influences of Aspect Ratio, Temperature Ratio, Reynolds Number, and Flow Structure. Int. J. Heat Mass Transfer, Vol.45, pp.2011–2020, 2002.

[8] Xie, G. N., Sunden, B., and Zhang, W. H., Comparisons of Pins/Dimples Protrusions Cooling Concepts for an Internal Blade Tip-Wall at High Reynolds Numbers. ASME J. Heat Transfer, Vol.133, Issue 6, pp. 0619021-0619029, 2011.

[9] Gadhave,G., and Kumar.P., Enhancement of forced Convection Heat Transfer over Dimple Surface-Review.International Multidisciplinary e - Journal .Vol-1, Issue-2, pp.51-57, 2012

[10] Katkhaw, N., Vorayos, N., Kiatsiriroat, T., Khunatorn, Y., Bunturat, D., and Nuntaphan., A. Heat transfer behavior of flat plate having 450 ellipsoidal dimpled surfaces. Case Studies in Thermal Engineering, vol.2, pp. 67–74, 2014

[11] Patel,I.H .,andBorse ,S.H. Experimental investigation of heat transfer enhancement over the dimpled surface. International Journal of Engineering Science and Technology,Vol.4, Issue 6,pp.3666–3672, 2012.

[12] Faheem Akhtar, Abdul Razak R Kaladgi and Mohammed Samee, Heat transfer augmentation using dimples in forced convection -an experimental approach. Int. J. Mech. Eng. & Rob. Res. Vol4,Issue 1 ,pp 150-153,2015.

[13] Faheem Akhtar, Abdul Razak R Kaladgi and Mohammed Samee, Heat transfer enhancement using dimple surfaces under natural convection—an experimental study, Int. J. Mech. Eng. & Rob. Res. Vol 4, Issue1, pp 173-175, 2015.

[14] Yu Rao ,Yamin Xu,Chaoyi Wan,A Numerical Study of the Flow and Heat Transfer in the Pin Fin-Dimple Channels With Various Dimple Depths, Journal of Heat Transfer ,Transactions of the ASME,Vol. 134, pp-071902-1-9, 2012.

[15] P.M. Ligrani, J.L. Harrison, G.I. Mahmood, M.L. Hill, Flow structure due to dimple depression on a channel surface, Phys. Fluids 13 (2001) 3442–3451.

[16] Hasibur Rahman Sardar and Abdul Razak Kaladgi ,Forced Convection Heat Transfer Analysis through Dimpled Surfaces with Different Arrangements, American Journal of Energy Engineering Vol 3 ,Issue 3,pp 37-45,2015.

Detection and Identification of PQ Disturbances Using S-Transform and Artificial Intelligent Technique

Ahmed Hussain Elmetwaly[1], Abdelazeem Abdallah Abdelsalam[2, *], Azza Ahmed Eldessouky[3], Abdelhay Ahmed Sallam[3]

[1]Dept. of Electrical Engineering, El Shorouk Academy, Cairo, Egypt
[2]Dept.of Electrical Engineering, Suez Canal University, Ismailia, Egypt
[3]Dept. of Electrical Engineering, Port-Said University, Port-Said, Egypt

Email address:
eng.ahmedhussain7@gmail.com (A. H. Elmetwaly), aaabdelsalam@eng.suez.edu.eg (A. A. Abdelsalam), azzaeldessouky@yahoo.com (A. A. Eldessouky), aasallam@ucalgary.ca (A. A. Sallam)

Abstract: This paper proposes a new technique based on S-transform time-frequency analysis and Fuzzy expert system for classifying power quality (PQ) disturbances. The S-transform is a new time frequency analysis method. It has the features of both continuous wavelet transform (CWT) and short time Fourier transform (STFT). Through S-transform time-frequency analysis, a set of feature components are extracted for identifying PQ disturbances such as; the amplitude of the S-transform matrix and the total harmonic distortion (THD). The two parameters are the inputs to Fuzzy-expert system that uses some rules on these inputs to characterize the PQ events in the captured waveform (e.g. sag, swell, interruption, surge, sag with harmonic and swell with harmonic). Several simulation using Matlab environment and practical data are used to validate the proposed technique. The results depict that the proposed technique has the ability to accurately identify and characterize PQ disturbances.

Keywords: Power Quality, S-Transform, Fuzzy Expert System

1. Introduction

In a power system; switching, faults, dynamics, or nonlinear loads cause various types of power quality (PQ) disturbances such as surges, harmonics, interruptions, sags, swells, etc. [1]. In order to improve the quality of power supply, it is necessary to make clear the sources and causes of PQ disturbances before appropriate mitigating actions can be taken. To analyze PQ disturbances, short time discrete Fourier transform (STFT) [2, 3] is mostly often used. This transform has been successfully used for stationary signals where properties of signals do not evolve in time. For non-stationary signals, the STFT does not track the signal dynamics properly due to the limitations of a fixed window width chosen in advance. Thus, STFT cannot be successfully used to analyze transient signals comprising both high and low-frequency components.

On the other hand, wavelet transform [4- 6] is a notable tool for detection, localization and classification of the disturbances. However, the noises will lower down the performance of wavelet as the spectrum of noises overlaps with that of power quality disturbances.

Kalman filter can be employed to detect and to analyze voltage event [7, 8]. The results of Kalman filter depend on the model of the system used as well as the appropriate selection of the filter parameters is not guaranteed, the rate of convergence of the results will be slow or the results will diverge.

The S-transform (ST) [9-11] is an extension of continuous wavelet transforms (CWT) and STFT. Because of its good time-frequency characteristic, it is very adequate for PQ disturbances analysis and classification. The classification of PQ disturbances can be done by applying artificial intelligent techniques like artificial neural network [12], Fuzzy logic [13], and support vector machine (SVM) [14].

This paper proposes a Fuzzy expert system for making a decision based on the features extracted from S-transform. These features are the amplitude of the captured waveform and the total harmonic distortion. The tested waveforms are generated using Matlab environment software and also an IEEE practical data.

2. The Proposed Classification Methodology

The block diagram of the proposed system as shown in Fig. 1 comprises two stages to: (i) evaluate the value of amplitude and total harmonic distortion (THD) of the captured wave using S-transform equations and (ii) classify the disturbance using Fuzzy-expert system according to the evaluated values.

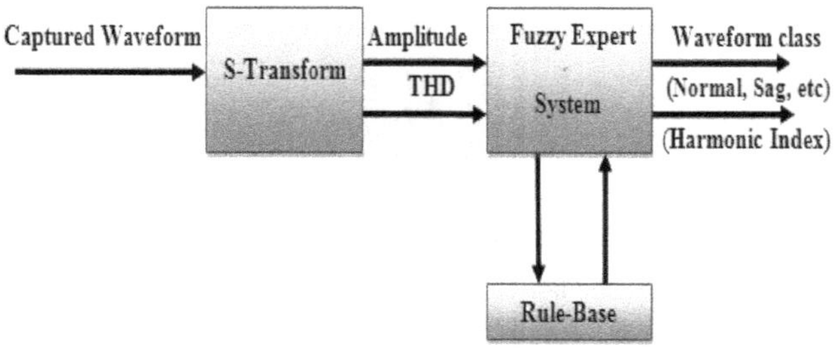

Figure 1. *The block diagram of the proposed system*

2.1. The S-Transform

Let $P[kT]$, $k = 0, 1, \ldots, N-1$ denote a discrete time $P(t)$, with a time sampling interval of T. The discrete Fourier transform of the signal can be obtained as follows:

$$P\left[\frac{n}{NT}\right] = \frac{1}{N}\sum_{k=0}^{N-1} P(kT)e^{-\left(\frac{i2\pi nk}{N}\right)} \tag{1}$$

where, $n = 0, 1, \ldots, N-1$ and the inverse discrete Fourier transform is:

$$P[kT] = \frac{1}{N}\sum_{n=0}^{N-1} P(n/nT)e^{\left(\frac{i2\pi nk}{N}\right)} \tag{2}$$

In the discrete case, the S-Transform is the projection of the vector defined by the time series $P(KT)$, onto a spanning set of vectors, the spanning vectors are not orthogonal and the elements of the S-Transform are not independent. Each basis vector (of the Fourier transform) is divided into localized vectors by an element- by- element product with the shifted Gaussians, such that the sum of these localized vectors is original basis vector. The S-Transform of a discrete time series $P(KT)$ is given by [15]:

$$S\left[\frac{n}{NT}, jT\right] = \frac{1}{N}\sum_{m=0}^{N-1} P(m + n/nT)\, G(n,m)e^{\left(\frac{i2\pi mj}{N}\right)} \tag{3}$$

where $G(n,m) = e^{-(2\pi^2 m^2/n^2)}$ = Gaussian function and $j, m, n = 0, 1, \ldots, N-1$.

The following steps are adapted for computing the discrete S-Transform:

1 Perform the discrete Fourier transform of the original time $P(KT)$ (with N points and sampling interval T to get $P(m/nT)$ using the FFT routine. This is only done once.

2 Calculate the localizing Gaussian $G(n,m)$ for the required frequency $(\frac{n}{nT})$.

3 Shift the spectrum $P(m/nT)$ to $P(m + n/nT)$ for the frequency n/nT (one point addition)

4 Multiply $P(m + n/nT)$ by $G(n,m)$ to get $B(\frac{n}{nT}, \frac{m}{nT})$ (N multiplication).

5 Inverse Fourier transforms of $B(\frac{n}{nT}, \frac{m}{nT})$ m/NT to j to give the row of $S[\frac{n}{nT}, jT]$ corresponding to frequency n/nT.

6 Repeat steps 3, 4, and 5 until all the rows of $S[\frac{n}{nT}, jT]$ corresponding to all discrete frequencies n/nT have been defined.

From Equation (3), it is seen that the output from the S-transform is a $(N \times M)$ matrix called the S-matrix whose rows pertain to frequency and columns to time. Each element of the S-matrix is complex valued. The choice of windowing function is not limited to the Gaussian function; other windowing functions are also implemented successfully.

2.2. The Fuzzy Expert System

Fuzzy logic [16, 17] refers to a logic system which represents knowledge and reasons in an imprecise or Fuzzy manner for reasoning under uncertainty. Unlike the classical logic systems, it aims at modeling the imprecise modes of reasoning that play an essential role in the human ability to infer an approximate answer to a question based on a store of knowledge that is inexact, incomplete, or not totally reliable. It is usually appropriate to use Fuzzy logic when a mathematical model of a process does not exist or does exist but is too difficult to encode and too complex to be evaluated fast enough for real time operation. The accuracy of the Fuzzy logic systems is based on the knowledge of human experts; hence, it is only as good as the validity of the rules. Fig. 2 shows the construction of Fuzzy expert system.

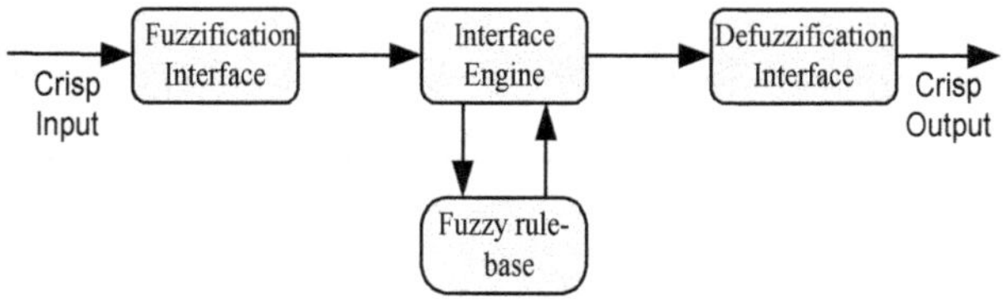

Figure 2. *The construction of a Fuzzy expert system*

2.3. Implementation of the Proposed Methodology

Many analysis results of PQ disturbances could be obtained from ST matrix, $S[\frac{n}{NT}, jT]$. The first extracted parameter is the amplitude 'A' which is the locus of maximum value of elements present in the column of the S-matrix corresponding to the time. The total harmonic distortion (THD) is the second extracted parameter. It is calculated using the FFT routine, Equation (1) where,

$$THD = \frac{\sqrt{v_2^2 + v_3^2 + v_4^2 + \cdots + v_n}}{v_1} \tag{4}$$

Where, v_1 is the fundamental wave amplitude, $v_2, v_3, v_4, \ldots, v_n$ are the other frequencies (harmonic) order amplitudes.

For classifying the disturbance waveforms, five Fuzzy sets

are chosen for the amplitude A, the first input of Fuzzy-expert, designated as VSA (very small amplitude), SA (small amplitude), NA (normal amplitude), LA (large amplitude), and VLA (very large amplitude). The total harmonic distortion (THD), the second input of Fuzzy-expert, has two Fuzzy sets that are designed as (Small value of THD) and (Large value of THD).

The input variables membership functions of the Fuzzy expert system are shown in Figs. 3 and 4.

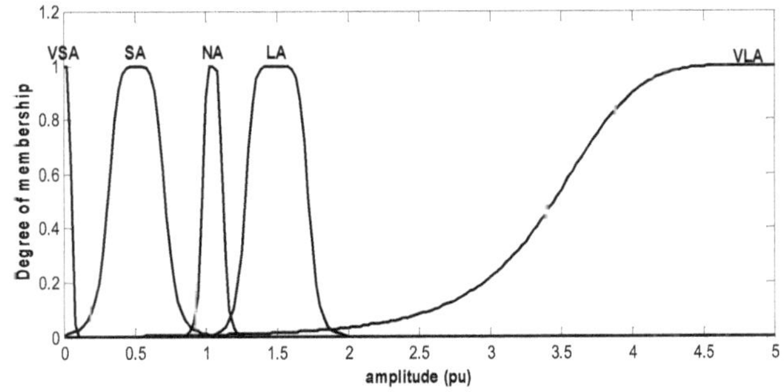

Figure 3. *The First input membership function*

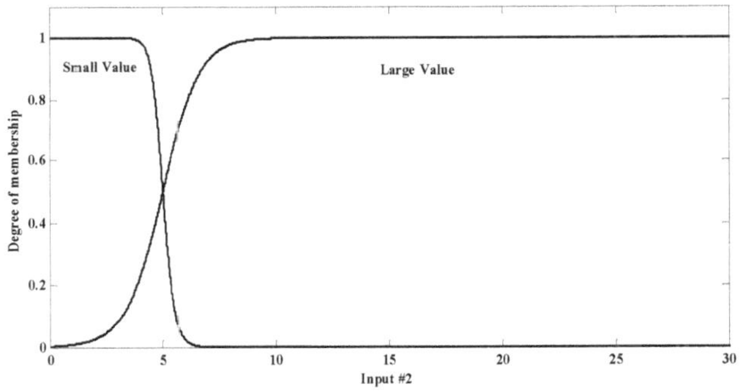

Figure 4. *The second input membership function*

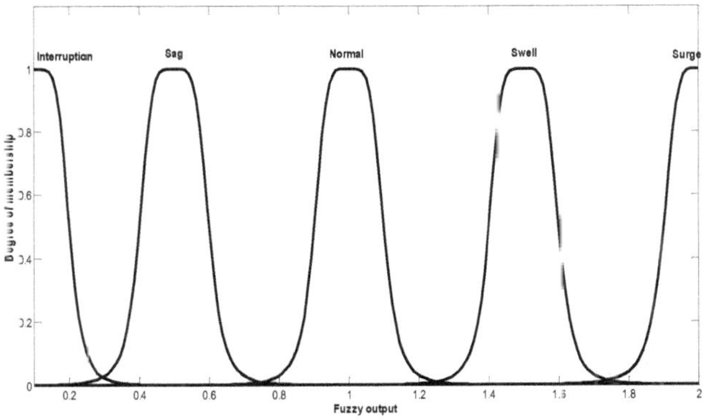

Figure 5. *The first Fuzzy output membership function*

The output membership function is defined by five sets. These sets are designated as interruption, sag, normal, swell, and surge.

The first output of the Mamdani Fuzzy system, Fig. 5, can assume values between 0 and 2 for the output, where,

Interruption = 0, Sag= 0.5, Normal = 1, Swell = 1.5, Surge = 2.

The second output of the Mamdani Fuzzy system, Fig. 6, can assume values between 0 and 1 for the output where,

Pure wave = 0, Distorted wave =1.

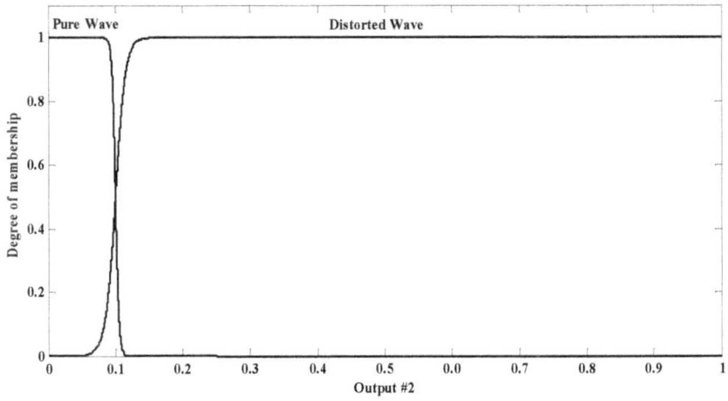

Figure 6. *The second Fuzzy output membership function*

The parameters of amplitude membership function are determined according to the definition of each PQ event.

The brief rule sets of Fuzzy expert system are below:

1 If Input #1 is SA and the Input #2 is Small value, Then the output #1 is Sag, and the output #2 is (pure wave)
2 If Input #1 is VSA and the Input #2 is Small value, Then output #1 is Interruption, and output #2 is (pure wave)
3 If Input #1 is LA and Input #2 is Small value, Then output #1 is Swell, and output #2 is (pure wave)
4 If Input #1 is VLA and Input #2 is Small value, Then output #1 is Surge, and output #2 is (pure wave)
5 If Input #1 is NA and Input #2 is Small value, Then output #1 is Normal, and output #2 is (pure wave)
6 If Input #1 is NA and Input #2 is Large value, Then output #2 is (Distorted wave)
7 If Input #1 is SA and Input #2 is Large value, Then output #1 is Sag, and output #2 is (Distorted wave)
8 If Input #1 is LA and Input #2 is Large value, Then output #1 is Swell, and output #2 is (Distorted wave)

3. Simulation Results

In this study, power quality disturbances signals are seven signal disturbances including voltage sag, swell, interruption, surge, harmonic distortion, sag with harmonic and swell with harmonic.

The generated waveform has a frequency of 50 Hz and the voltage waveform sampled at a rate of 1.6 kHz, i.e. 32 samples per cycle. The general equation of generated wave form is:

$$S(t) = A_1 \sin(wt) + A_2 \sin(3wt) + A_3 \sin(5wt) \quad (5)$$

where, A_1 is the amplitude of the fundamental wave and equal to 1 p.u, and A_2 & A_3 are the amplitudes of the third and fifth harmonic order imposed on the fundamental sine wave. The following case studies are presented to illustrate the aptness of the proposed system.

3.1. Normal Voltage

The normal voltage is the rated operating voltage at rated frequency, also there is not any PQ disturbance on the fundamental wave. The general equation form of the generated wave is $\mathrm{S(t)} = \mathrm{A_1}\sin(\mathrm{wt})$, where, $\mathrm{A_1} = 1$ p.u and putting $\mathrm{A_2}$ & $\mathrm{A_3}$ equal to zero as shown in Fig. 7-a. Fig 7-b shows the output of the S transform by searching the maximum of each column of the S matrix. Fig. 7-c shows the frequency contour of the captured wave using FFT analysis. The period of disturbance and the total output of the Fuzzy expert system are shown in the Figs. 8-a, 8-b and 8-c, respectively. From Fig 8-a, the magnitude equals one all over the period which is Normal. The second Fuzzy output equals zero, this means that the waveform does not contain harmonic distortion.

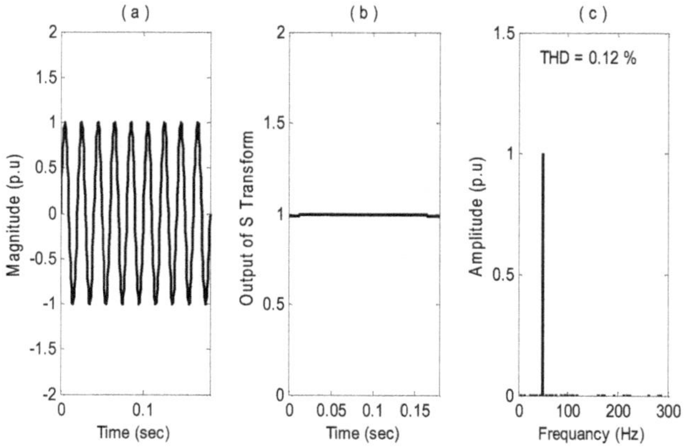

Figure 7. Normal voltage: (a) waveform, (b) magnitude time spectrum and (c) frequency contour

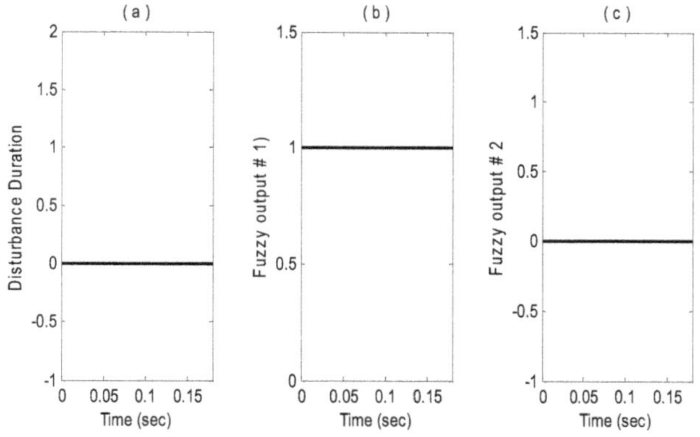

Figure 8. Normal voltage: (a) disturbance duration, (b) Fuzzy output #1 (c) Fuzzy output #2

3.2. Voltage Sag

Voltage sag is a decrease of 10–90% of the rated system voltage for duration of 0.5 cycles to 1 min. The generated wave equation is $S(t) = A_1\sin(wt)$, where, $A_1 = 1$ p.u. and A_2 & A_3 are equal to zero. By applying a decrease to the voltage magnitude to be 0.5 p.u for 3 cycles as shown in Fig. 9-a. Figs. 9-b and 9-c show the output of S transform and the frequency contour of the captured wave in order to calculate the total harmonic distortion. Fig. 10-a shows the period of disturbance while Fig. 10-b gives the classification of disturbance according to the amplitude. In this case the magnitude drop from one to 0.5 which means that the waveform contains sag. The second Fuzzy output equals zero, this means that the waveform does not contain harmonic distortion as shown from Fig. 10-c.

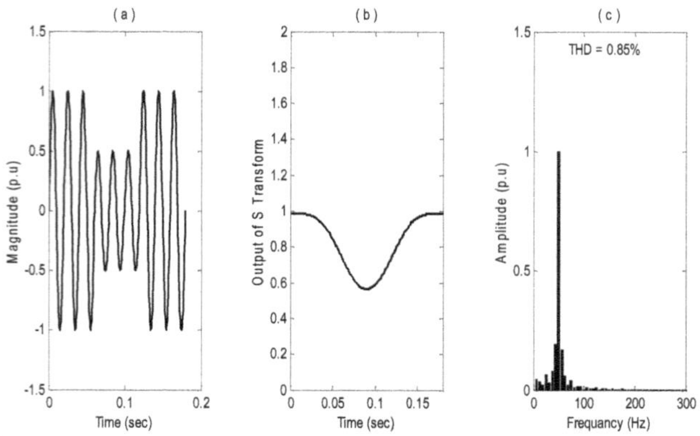

Figure 9. Voltage sag: (a) waveform, (b) magnitude time spectrum, (c) frequency spectrum

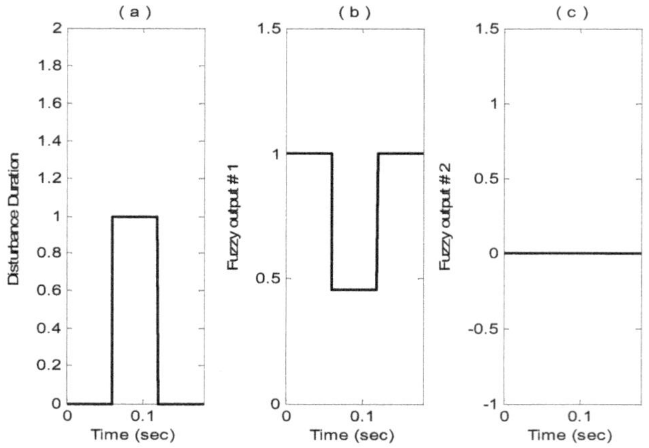

Figure 10. Voltage sag: (a) disturbance duration, (b) Fuzzy output #1 (c) Fuzzy output #2

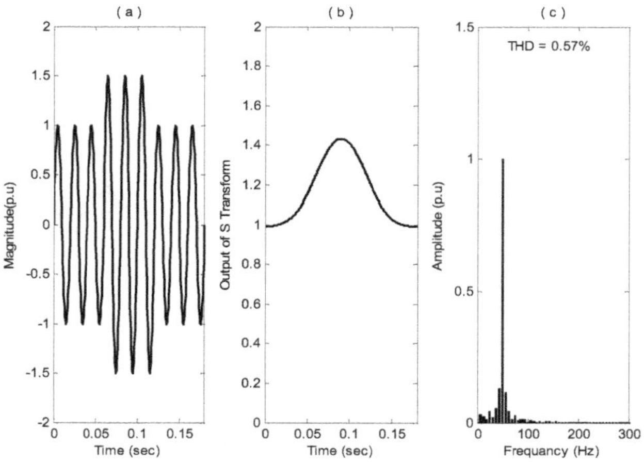

Figure 11. Voltage swell: (a) waveform, (b) magnitude time spectrum, (c) frequency spectrum

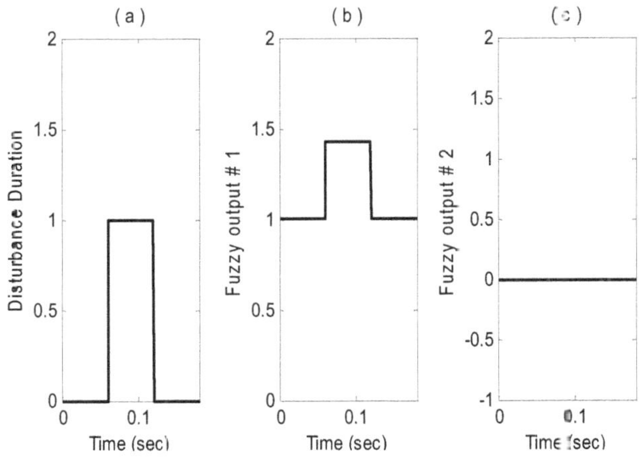

Figure 12. Voltage swell: (a) disturbance duration, (b) Fuzzy output #1 (c) Fuzzy output #2

3.3. Voltage Swell

In the case of voltage swell, there is a rise of 10 to 90% in the voltage magnitude for 0.5 cycles to 1 min. the equation of the generated wave is $S(t) = 1\ \sin(wt)$ and $A_2 \,\&\, A_3$ are equal to zero. Increasing of the voltage to 1.5 p.u for three cycles, the magnitude rises from one to 1.5 p.u as shown in Figs. 11-b and 12-b, this means that the waveform contains swell. The second Fuzzy output equals zero, this means that the waveform does not contain harmonic distortion.

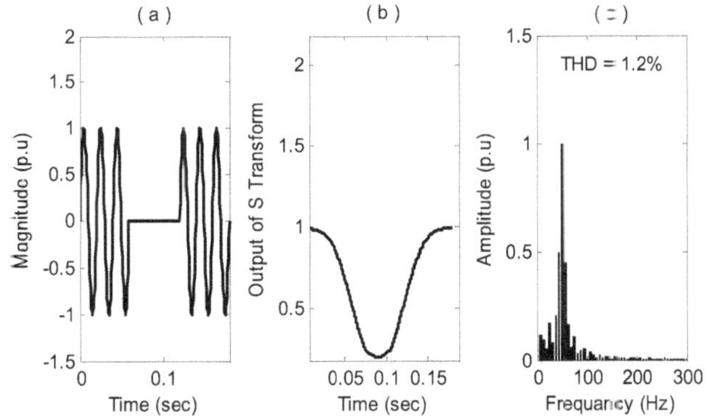

Figure 13. Voltage interruption: (a) waveform, (b) magnitude time spectrum and (c) frequency spectrum

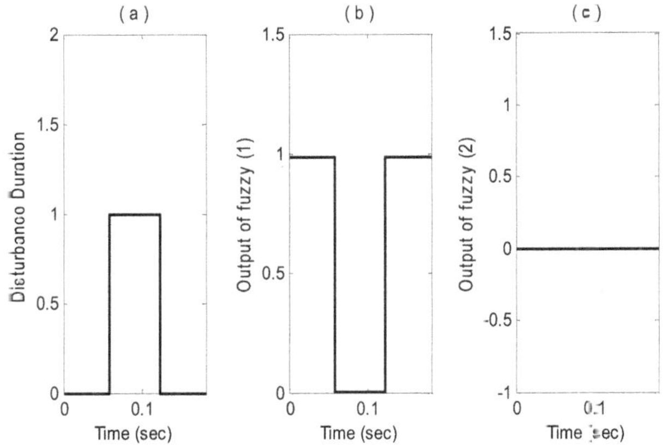

Figure 14. Voltage interruption: (a) Disturbance duration, (b) Fuzzy output #1 and (c) Fuzzy output #2

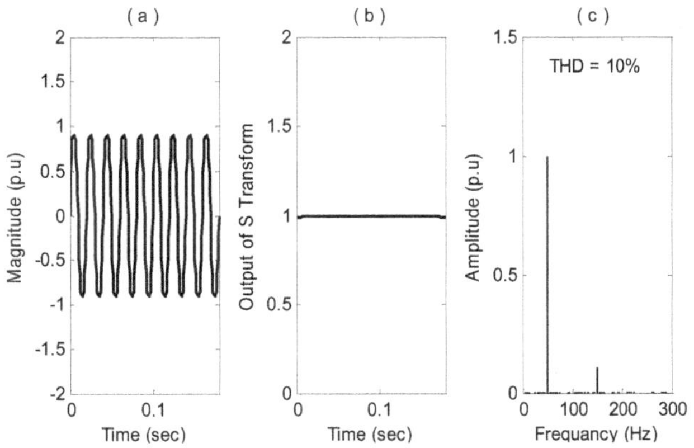

Figure 15. Voltage distortion: (a) waveform, (b) magnitude time spectrum and (c) frequency spectrum

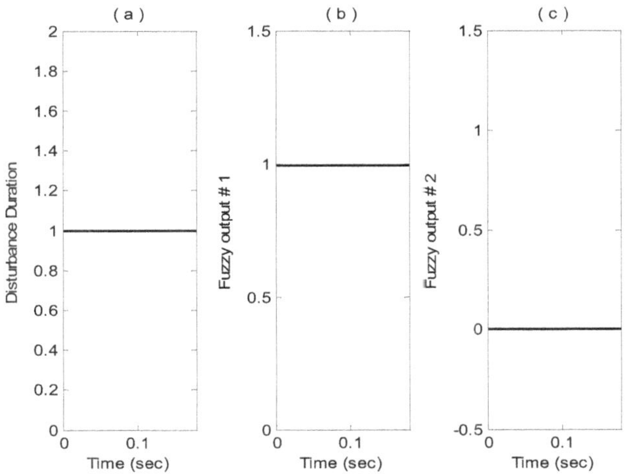

Figure 16. Voltage distortion: (a) disturbance duration, (b) Fuzzy output #1 and (c) Fuzzy output #2

3.4. Voltage Interruption

An interruption may be seen as a loss of voltage on a power system. Such disturbance describes a drop of 90-100% of the rated system voltage for duration of 0.5 cycles to 1 min. The generated wave equation is $S(t) = 1\ \sin(wt)$ with $A_2 \,\&\, A_3$ are equal to zero. By applying an interruption for three cycles as shown in Fig. 13-a, the magnitude drop from one to zero is shown in Figs. 13-b and 14-b. Fig. 14-c illustrates the second Fuzzy output, in this case the harmonic index of the captured wave is equal to zero. So, the wave is pure.

3.5. Voltage Distortion

Distortion of the voltage waveform occurs when the harmonic is generated. This is done by adding the third harmonic to the original sine wave so that the generated wave is represented by $S(t) = 1\ \sin(wt) + A_2 \sin(3wt)$ with $A_2 = 0.105$ p.u and the total harmonic distortion is 10%. Fig. 15-a shows the waveform with a third harmonic for nine cycles while Fig. 16-c shows the second Fuzzy output, in this case the harmonic index of the captured wave is equal to one. Hence, the wave is distorted.

3.6. Sag with Harmonics

The sag with harmonic disturbance is done by adding the third harmonic to a sag waveform so that it can be represented by the equation $S(t) = 1\sin(wt) + A_2\sin(3wt)$ with $A_2 = 0.105$ p.u. Fig 17-a shows the generated waveform with a sag of 0.5 p.u and a third harmonic for the three cycles. Hence, the total harmonic distortion is 10%. Fig. 18-b shows the first Fuzzy output, in this case, the amplitude is decreased from 1 to 0.5 p.u which means sag. The second Fuzzy output is shown in Fig. 18-c with harmonic index of the captured wave equals one so that the wave is distorted.

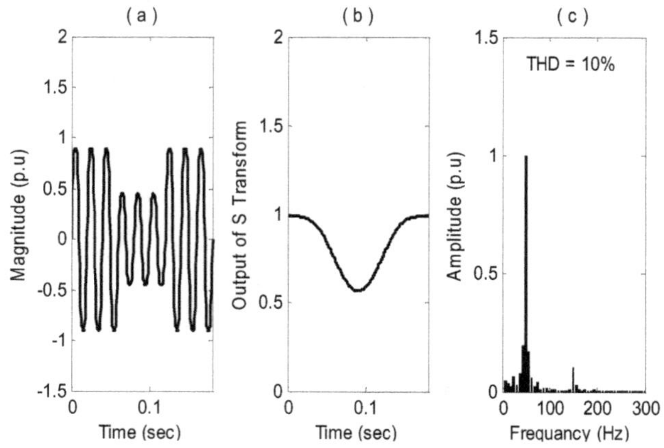

Figure 17. *Sag with harmonic: (a) waveform, (b) magnitude time spectrum and (c) frequency spectrum*

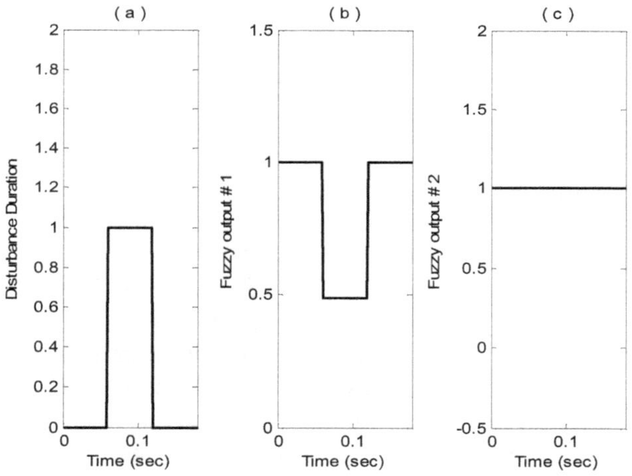

Figure 18. *Sag with harmonic: (a) disturbance duration, (b) Fuzzy output #1 and (c) Fuzzy output #2*

3.7. Swell with Harmonic

The swell with harmonic is done by adding the third harmonic to a sag wave and is represented by the equation $S(t) = A_1\sin(wt) + A_2\sin(3wt)$ with $A_1 = 1$ p.u and $A_2 = 0.105$ p.u. Applying swell of 1.5 p.u and a third harmonic for three cycles so that the total harmonic distortion is 10% as shown in Fig. 19-a. The first Fuzzy output shown in Fig. 19-b illustrates an increase of the amplitude of the waveform from 1 to 1.5 p.u which means swell. The second Fuzzy output equals one, this means that the waveform contains harmonic distortion.

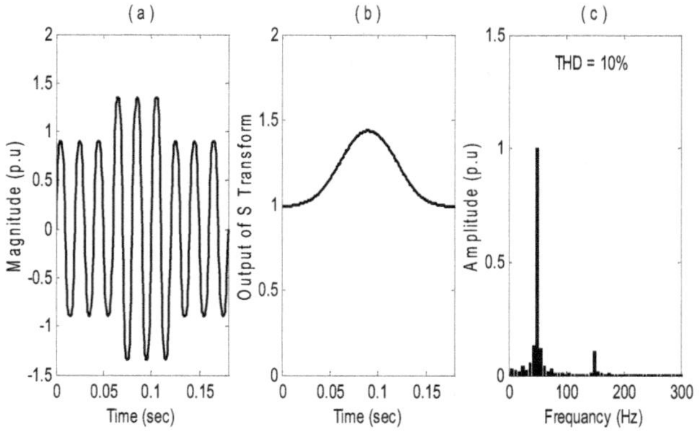

Figure 19. *Swell with harmonic: (a) waveform, (b) magnitude time spectrum and (c) frequency spectrum*

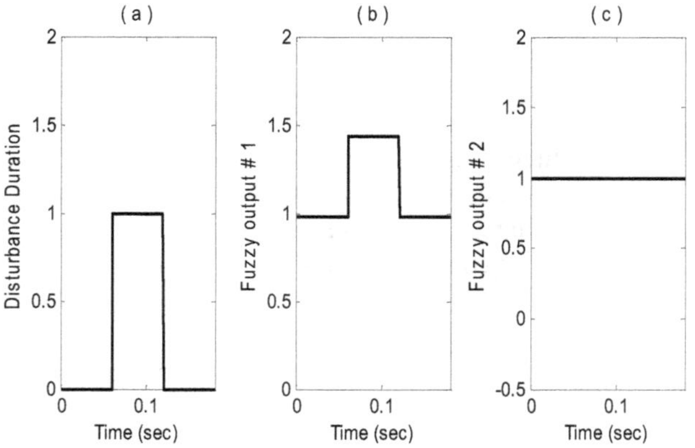

Figure 20. *Swell with harmonic: (a) disturbance duration, (b) Fuzzy output #1 and (c) Fuzzy output #2*

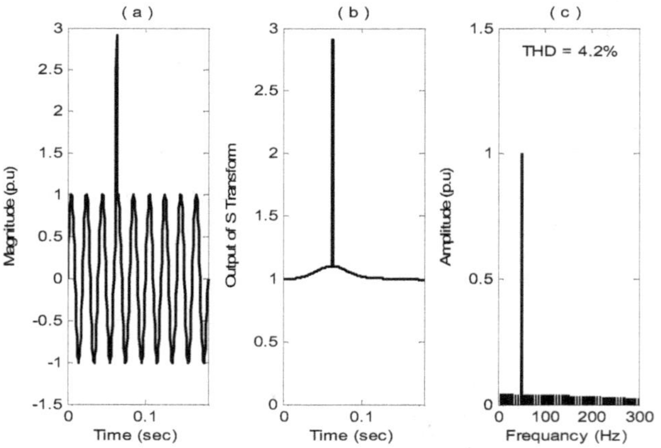

Figure 21. *Voltage surge: (a) waveform, (b) magnitude time spectrum and (c) frequency spectrum*

3.8. Voltage Surge

The surge occurs when the amplitude is suddenly increased from 1 to 3 p.u .for one-quarter cycle. In this case, the waveform is represented by $S(t) = A_1\sin(wt) + A_2\sin(3wt)$ with $A_1 = 1$, $A_2 = 0$ p.u. The amplitude is suddenly increased from 1 to 3 p.u as shown in Fig. 21-b.

The first Fuzzy output shown in Fig. 22-b demonstrates a rises in amplitude to be (two) which indicate a case of surge. The second Fuzzy output equals zero, this means that the waveform does not contain harmonic distortion and the results are shown in Figs. 21 and 22.

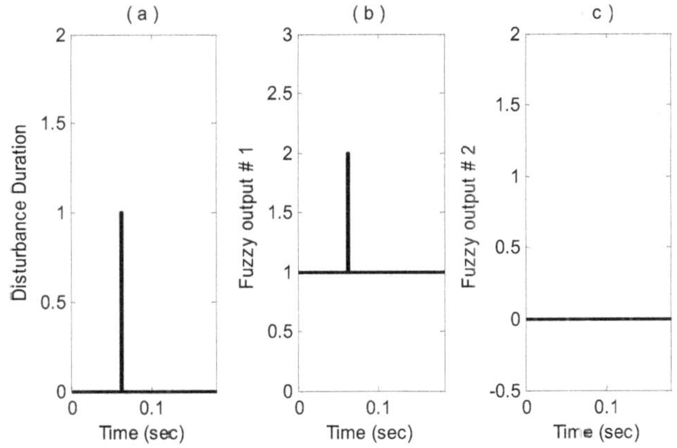

Figure 22. Voltage surge: (a) disturbance duration, (b) Fuzzy output #1 and (c) Fuzzy output #2

4. Practical Data Results

This section presents some of results obtained by applying the new approach on practical data. The practical data are obtained from the IEEE Project Group 1159.2 [19]. The sample frequency used is $F_s = 15360$ Hz, or 256 samples per 60 Hz cycle.

4.1. Case Study #1

Considering the captured waveform doesn't have any disturbance as shown in Fig 23-a. The Fuzzy first output is one which is normal and the second Fuzzy output equals zero, as shown in Fig. 24. This means that the waveform does not contain harmonic so that the wave is pure.

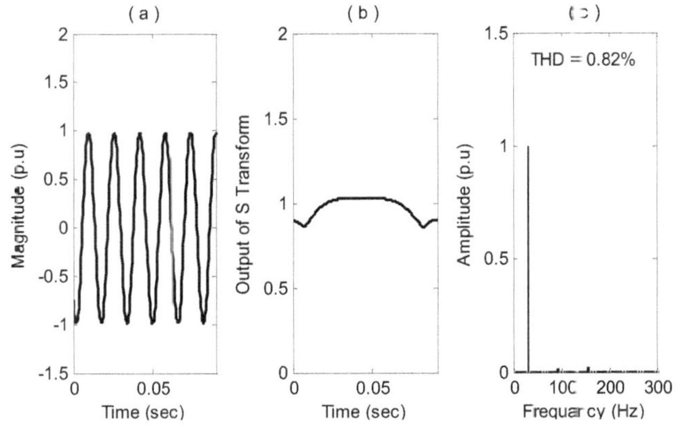

Figure 23. Case #1: (a) waveform, (b) magnitude time spectrum and (c) frequency spectrum

4.2. Case Study #2

In this case study, the sag in voltage waveform is detected using the S transform and is characterized using the results of Fuzzy expert system. Fig. 25-a shows the voltage waveform with voltage sag. Fig. 26-b shows the output of the Fuzzy system which equals 0.5. This means that the waveform contains sag. The output #2 of Fuzzy expert system equals zero, this means that this waveform does not contain a harmonic distortion.

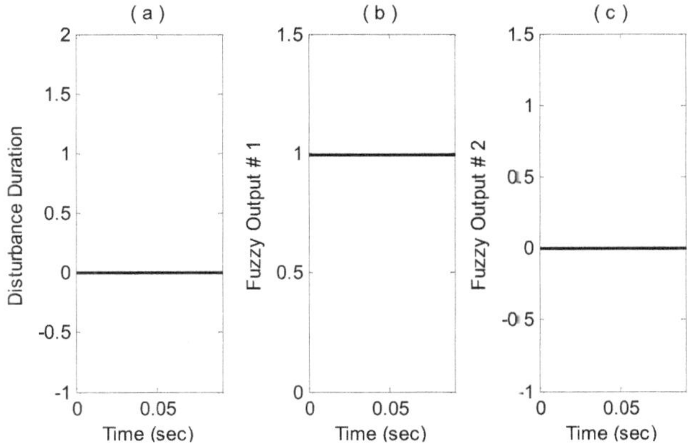

Figure 24. Case #1: (a) disturbance duration, (b) Fuzzy output #1 and (c) Fuzzy output #2

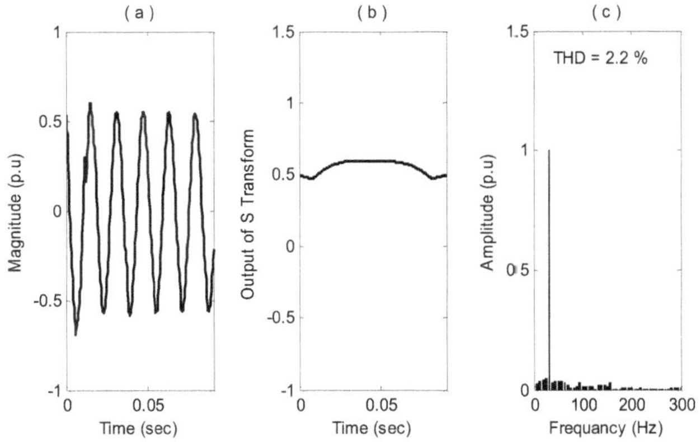

Figure 25. Case #2: (a) waveform, (b) magnitude time spectrum and (c) frequency spectrum

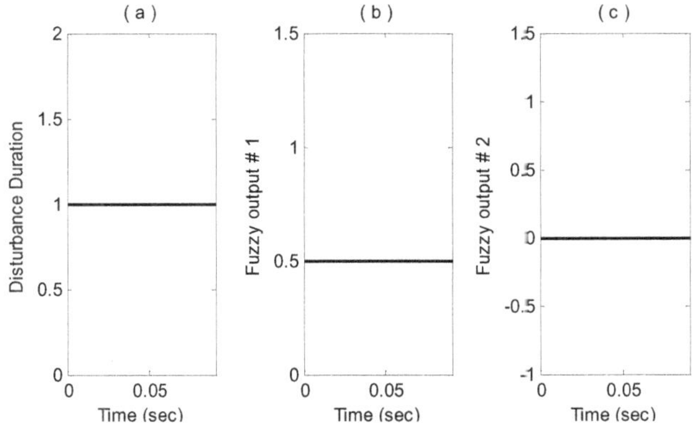

Figure 26. Case #2: (a) disturbance duration, (b) Fuzzy output #1 and (c) Fuzzy output #2

4.3. Case Study #3

The voltage waveform of this case is shown in Fig. 27-a. This waveform contains the sag power quality event with harmonics. As can be seen in Fig 28, the Fuzzy output clearly points the sag PQ event in the waveform, Fuzzy output #1is equal to 0.5 which refers to sag. The output #2 of Fuzzy expert system equals one, this means that the waveform contains harmonic.

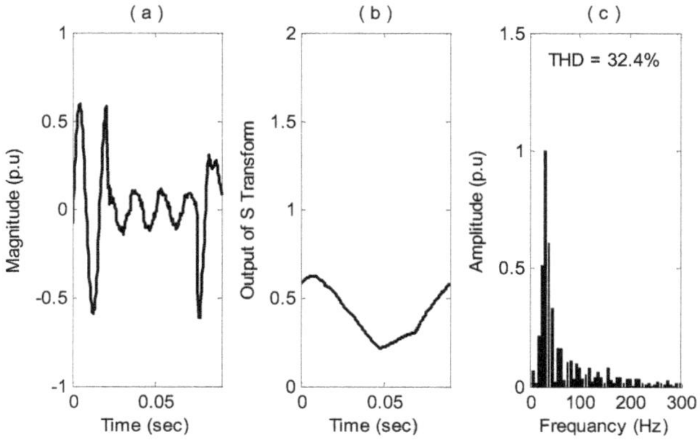

Figure 27. *Case #3: (a) waveform, (b) magnitude time spectrum and (c) frequency spectrum*

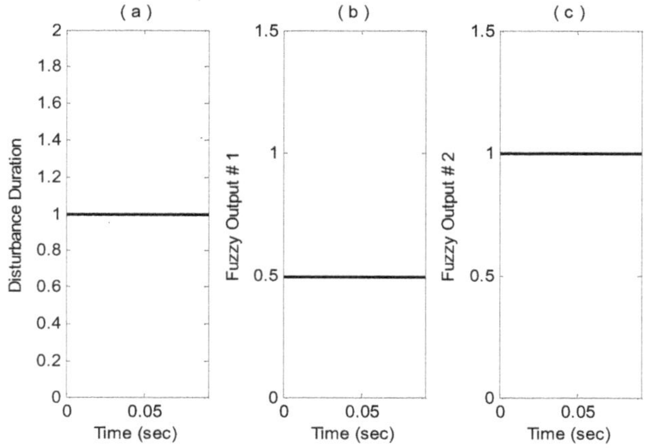

Figure 28. *Case #3: (a) disturbance duration, (b) Fuzzy output #1 and (c) Fuzzy output #2*

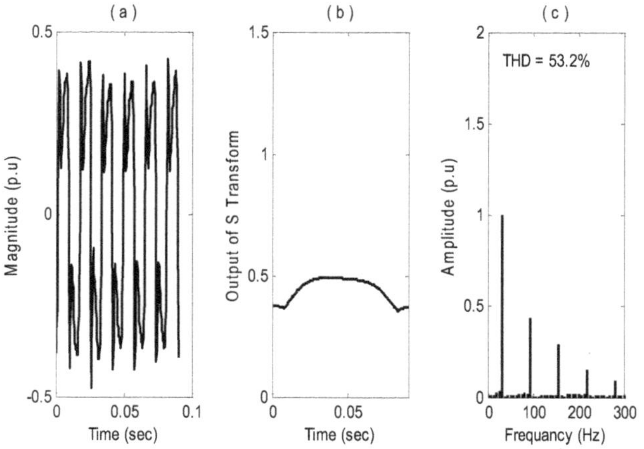

Figure 29. *Case #4: (a) waveform, (b) magnitude time spectrum and (c) frequency spectrum*

4.4. Case Study #4

In this case, the test waveform contains two power quality events; sag and harmonic distortion, with different harmonic order as shown in Fig 29-a. The output of S-transform is shown in Fig. 29-b. The duration of the disturbance is shown in Fig. 30-a. Fig. 30-b shows the first Fuzzy output which equals 0.5. This means that the waveform contains sag. The second Fuzzy output equals one, i.e. the wave form is distorted.

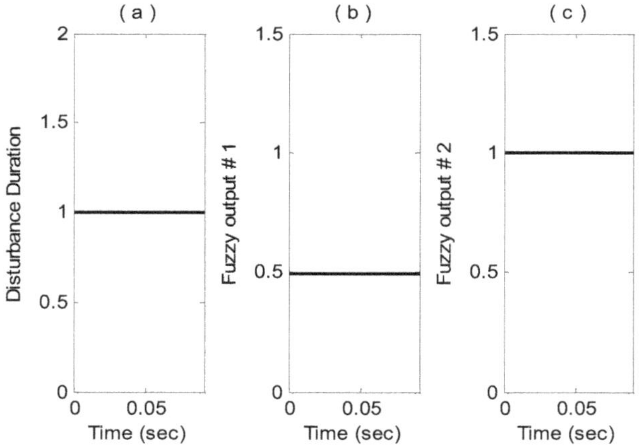

Figure 30. *Case #4: (a) disturbance duration, (b) Fuzzy output #1 and (c) Fuzzy output #2*

5. Comparison Between the Proposed Technique and Previous Published Works

By comparing the performance of the proposed technique with those of other methods used for classification and identification of PQ disturbance such as the WT-based ANN method [6], WT and rule-based methodology [18], it is found that, there are a lot of parameters (more than 50 parameters) had to be determined to classify power disturbances in the previous mentioned methods. In this paper, using the ST-based method, five types of single disturbance and two complex disturbances (sag with harmonic and swell with harmonics) can be classified using less number of calculated parameters which make the calculation period too short. On the other hand the wavelet transform coefficients at every scale belong to scope frequencies, not to a single frequency. Hence, it could not count the frequency amplitude accurately. ST has better frequency distinguish ability than WT. It is easily to identify the fundamental frequency of the of the disturbance signal of voltage sag, swell, and interruption and the other frequencies in case of harmonics providing a better visual disturbance degree comparing to DWT.

6. Conclusions

This paper presents a hybrid technique for characterizing PQ disturbances. The hybrid technique is based on S-transform for extracting two parameters, amplitude and harmonic indication from the captured distorted waveform. The two parameters are the inputs to Fuzzy-expert system that uses some rules on these inputs to characterize the PQ events in the captured waveform. The results show that the proposed hybrid technique has the ability to identify and classify the power system disturbances with high accuracy and small computation time comparing with other methods.

References

[1] IEEE Std 1159-1995, IEEE Recommended Practice for Monitoring Electric Power Quality, IEEE Inc., New York, pp.1-59, 1995.

[2] G. T. Heydt, P. S. Field, C. C. Liu, D. Pierce, L. Tu, G. Hensley, "Applications of the windowed FFT to electric power quality assessment", IEEE Trans Power Deliv., vol. 14, no. 4, pp. 1411–14166. 1999.

[3] Y. H. Gu, M. H. J. Bollen,"Time-frequency and time-scale domain analysis of voltage disturbances", IEEE Trans. Power Deliv., vol. 15, no. 4, pp. 1279–84, 2000.

[4] O. Poisson, P. Rioual, M. Meunier, "Detection and measurement of power quality disturbances using wavelet transform," IEEE Trans Power Deliv., vol. 15, no. 3, pp. 1039–1044, 2000.

[5] A. M. Gaouda, S. H. Kanoun, M. M A. Salama, A. Y. Chikhani, "Pattern recognition applications for power system disturbance classification", IEEE Trans. Power Deliv., vol. 17, no. 3, pp. 677–683, 2002.

[6] Z. L. Gaing, "Wavelet-based neural network for power disturbance recognition and classification," IEEE Trans. Power Deliv., vol. 19, no. 4, pp. 1560–1568, Oct. 2004.

[7] Abdelazeem A. Abdelsalama, Azza A .Eldesouky, and Abdelhay A. Sallam, "Classification of power system disturbances using linear Kalman filter and Fuzzy-expert system," Electr. Power Energy Syst., vol 43, no. 1 pp. 688–695, 2012.

[8] C. I. Chen, G. W. Chang, R. C. Hong, H. M. Li, "Extended real model of Kalman filter for time-varying harmonics estimation," IEEE Trans Power Deliv., vol. 25, no. 1, pp. 17–26, 2010.

[9] X. Xiao, F. Xu, H. Yang, "Short duration disturbance classifying based on S transform maximum similarity", Int J Electr Power Energy Syst., vol. 31, no. 7, pp. 374–78, 2009.

[10] S. Suja, Suja Jovitha, "Pattern recognition of power signal disturbances using S Transform and TT Transform," Int J Electr Power Energy Syst., vol. 32, no. 1, pp. 37–53, 2010.

[11] P. K. Dash, B. K. Panigrahi, G. Panda. "Power quality analysis using s-transform," IEEE Transactions on Power Delivery, vol. 18, no. 2, pp. 406–411, 2003

[12] S. Santoso, E. J. Powers, W.M. Grady, and A. C.Parsons, "Power quality disturbance waveform recognition using wavelet-based neural classifier -part 2: application," IEEE Trans. Power Delivery, vol. 15, pp 222-228, Jan. 2000.

[13] Y. Liao, J-B. Lee, "A Fuzzy expert system for classifying power quality disturbances," International Journal of Electrical Power and Energy Systems, vol. 26, no. 3, pp. 199–205, 2004.

[14] W-M. Lin, C. Wu, C-H. Lin, F. S. Cheng, "Classification of multiple power quality disturbances using support vector machine and one-versus-one approach," International Conference on Power System Technology, vol. 2, pp. 1–8, 2006.

[15] R. G. Stockwell, L. Mansinha, and R. P. Lowe, "Localization of the complex spectrum: The S-transform," IEEE Trans. Signal Process., vol. 44, no. 4, pp. 998–1001, Apr. 1996.

[16] K.Passino, S.Yurkovich, Fuzzy Control, Longman: Addison Wesley, 1998.

[17] S. Guo, L. Peter, "A reconfigurable analog Fuzzy logic controller," Proceeding of the Third IEEE conference on IEEE World congress on computational intelligence, vol. 1, pp. 124-128, June 1994.

[18] M. Kezunovic and L. Yuan, "A novel software implementation concept for power quality study," IEEE Trans. Power Del., vol. 17, no. 2, pp. 544–549, Apr. 2002.

[19] IEEE project group P1159.2. <http://www.standards.ieee.org>.

12

Dynamic Harmonic Domain Modelling of Space Vector Based UPFC

Devendra Manikrao Holey, Vinod Kumar Chandrakar

Department of Electrical Engineering, G. H. Raisoni College of Engineering, Nagpur, India

Email address:
dev_mh@yahoo.com (D. M. Holey), vc_vkc@yahoo.co.in (V. K. Chandrakar)

Abstract: This paper presents analytical frequency domain method for harmonic modeling and evaluation of Space Vector Pulse Width Modulation (SVPWM) based unified power flow controller (UPFC). SVPWM is the best among all the PWM techniques. It gives a degree of freedom of space vector placement in a switching cycle. Dynamic modeling technique use for space vector modulation (SVM) based Unified power flow controller (UPFC) for harmonic analysis using dynamic harmonic domain. Performance of the device is evaluated in Dynamic harmonic domain simulation studies using MATLAB environment. The switching function spectra are necessary for harmonic transfer Matrix is calculated using Fourier series.

Keywords: Harmonic Domain, Voltage Source Converter (VSC), Space Vector Modulation (SVM), UPFC, STATCOM, SSSC

1. Introduction

SVM technique is a digital modulation, to generate PWM load line voltages which is equal to a given load line voltage. By properly selecting the switching states of the inverter and the calculation of the appropriate time period for each state. Low switching frequency PWM techniques, such as phase shifted carrier technique, have been proposed to produce a controllable output voltage satisfying harmonic requirements for utility system [4]. Among other PWM techniques, the Space Vector Modulation (SVM), switching patterns are generated on basis of three-phases, leading to lower switching frequencies than carrier-based techniques shown in [5]. Also, SVM utilizes the DC bus voltage better, and therefore can extend output voltage to closer the square wave operation. These advantages make the SVM a better candidate for especially high power applications. A low switching frequency Delayed Sampling Technique in conjunction with the SVM appropriate for GTO-based high power converter application was proposed in [6]. High quality output voltages is obtained by Low switching frequency SVM strategies [13].

First Harmonic in power system caused by highly non-linear devices degrades its performance. Forced Commutated VSCs are the main building block for low and medium power application. Due to recent development in the semiconductor technology and availability of high power switches e.g. Insulated Gate Bipolar Transistor (IGBT) and Gate Turn Transistor (GTO) have widespread acceptance in for high power VSC's, which are used for FACTS controllers. FACTs devices are solution of some power quality problems and also create some power quality problems such as harmonics. The power system harmonic analysis is the calculation of the magnitude, phase of fundamental and higher order harmonic of system signals. The generation of harmonics in present power system is due to large size of power converter. To reduce the harmonics in the system filter and modern switching pattern are used. The increasing prevalence of flexible AC transmission system (FACTS) devices makes having accurate model these devices essential. One attracting method for modeling the steady state performance of these devices is frequency domain analysis [1]. Harmonic phasor contain both positive and negative frequency terms for phase dependence. FACTS devices are characterized by their switching nature.

UPFC is the most important FACTs controller for regulating voltage and power flow in the transmission line. It is made up of two VSCs, one is shunt connected and other is in series. DC capacitor is connected between two VSCs. Series converter provides both active and reactive support whereas shunt converter provide necessary power to series converter and also provide reactive power support to the transmission line.

Previous work conducted in Dynamic harmonic domain has been primarily focus on modeling PWM, multi-module based UPFC. The power Quality index can be assess directly from the DHD modeling. Active, reactive, apparent powers as well as power factor are some important power Quality indices. For the linear circuit the indices are defined in term of fundamental frequency whereas for nonlinear circuit, when nonlinear element are present in a circuit such as electronics devices, given on basic of Fourier coefficient or given in terms of Total Harmonic Distortion (THD). For assessment of accurate power Quality indices precise calculation of harmonic component is needed during transient period. Other authors have already made their contribution regarding modeling of FACTS devices and power system element using DHD method [7-8].

There are two methods of DHD modelling first one is that direct mathematical mapping of all system equations and input in frequency domain other method is that does not mapped all equations and input in frequency domain gives more accurate result during transients. In second method transient due to circuit parameters changes are used. Second method is more suitable for calculation of instantaneous power quality indices, protection analysis and real time application all through it is violating causality and spurious dynamics result provided enough harmonic consider [3]. The time-variant system is converted to time-invariant system and then it is expressed in the dynamic symmetric components. For high power applications, the space vector modulation technique is preferred.

An analytical model of Unified Power Flow Controller (UPFC) for unbalanced operation is given in [12]. It is assumed that the PWM control eliminates low harmonic distortion. But in real systems this is not always true. The harmonic domain model of a VSC is proposed for HVDC-VSC back-to-back and HVDC-VSC transmission shows that VSC HVDC system generate harmonic and interact with on the power system [14]. The transient model and control system of UPFC is presented for the fast dynamic response in [16]

Mathematical model of unified power flow controller (UPFC) for steady state, transient stability a eigen value studies is given [17] The application of a space vector modulation (SVM) strategy for a multimodule back-to-back HVDC converters for low switching frequency that minimizes ac-side low-order voltage harmonics. SVM strategy based on sequential sampling technique by appropriately introducing phase shift for the corresponding voltage harmonics of the converter modules and maintaining the fundamental voltage components of VSC in-phase to obtain maximum ac-side voltage is presented in [18]. It eliminates the requirement of transformer arrangement for harmonic reduction, only simple arrangement of transformer can be used [18].

Dynamic harmonic domain model for UPFC is already developed by other authors for PWM converters. Now a day advance switching techniques such SVPWM are used for the high power application due to low switching frequency and easy to implement digitally. Dynamic harmonic domain model for SVPWM based UPFC does not found in literature. To extend the results obtained by other authors in the modeling of UPFC FACTS device, the proposed SVPWM VSC based UPFC model is developed in order to obtained the evolution in the of harmonic interference of the UPFC. Dynamic Harmonic analysis of PWM based UPFC is shown in [7]. This paper present the Dynamic analysis of SVPWM based UPFC during steady state and dynamic condition is given. This gives the information about harmonic indices during transient condition also. This can be used for control system design for Space vector based UPFC as space vector modulation scheme which is suitable for digital implementation. The paper is organized as follows. The second section provides fundamentals of DHD method. Third section is fundamental of Space Vector based pulse width modulation based voltage source converter; Fourth section is deal with DHD modeling of SVPWM VSC used for UPFC; this arrangement is used in this paper

2. Dynamic Harmonic Domain Basic

The system given by Ordinary Differential Equations (ODE) can be transform to an alternative arrangement called the Dynamic Harmonic Domain, based on the approximation of the system by Fourier series over a period of the fundamental frequency. Linear periodic system (LTP) is converted into linear time domain (LTI) system using dynamic harmonic domain (DHD).

Consider a LTP system is given by

$$\begin{matrix} \dot{x}\ t & A\ t\ x\ t & B\ t\ u\ t \\ y(t) & C\ t\ x\ t & D\ t\ u\ t \end{matrix} \tag{1}$$

Where A (t) is given as

$$A\ t \quad a_{h} e^{jh\ _0 t} \quad \cdots \quad a_{1} e^{j\ _0 t} \quad a_0 \quad a_1 e^{j\ _0 t} \quad \cdots \quad a_h e^{jh\ _0 t} \tag{2}$$

Where h is the highest harmonics of interest and ω0 is the fundamental frequency of the system. Equation (1) can be represented in the Fourier series, the ordinary differential equation to an alternative DHD representation.

Therefore

$$\begin{matrix} \dot{X} & A\ \ S\ X & BU \\ Y & CX & DU \end{matrix} \tag{3}$$

Variable given in the equation (1) can transform into the vectors in the equation (3) with their coefficient related to the harmonic components of their instantaneous signals as,

$$X=\left[X_{-h}(t)\cdots X_{-1}(t)\,X_0(t)\,X_{1(t)}\cdots X_h(t)\right]$$

S is the matrix of differential and given by

$$S=diag\left[-jh\omega_0 t\cdots-j\omega_0\;\;0\;\;j\omega_0 t\cdots jh\omega_0\right]$$

The matrices A, B, C and D has Toeplitz structure and their time domain counterpart such that

$$A=\begin{bmatrix} A_0 & A_{-1} & \cdots & A_{-h} & & & \\ A_1 & \ddots & \ddots & \ddots & \ddots & & \\ \vdots & \ddots & A_0 & A_{-1} & \ddots & \ddots & \\ A_h & \ddots & A_1 & A_0 & A_{-1} & \ddots & A_{-h} \\ & \ddots & \ddots & A_1 & A_0 & \ddots & \vdots \\ & & \ddots & \ddots & \ddots & \ddots & A_{-1} \\ & & & A_h & \cdots & A_1 & A_0 \end{bmatrix}$$

The steady state solution is obtained directly from equation (3), considering ($\dot{X}\quad 0$) gives,

$$\begin{matrix} X \quad (S \quad A)^{\;1}BU \\ Y \quad CX \quad DU \end{matrix} \tag{4}$$

The matrices A, B, C and D are constant and input U is also constant. The solution of X and Y is obtained from equation (4). The solution of equation (4) can be used to initialize the DHD simulation in time domain.

3. Space Vector Modulation

3.1. Principles of SVPWM

Signal processing view point space vector modulation is a digital modulation technique. SVPWM is based in such a way that there are only two independent variables in a three-phase system. Orthogonal coordinates are used to represent the 3-phase voltage in the phasor diagram. A three-phase voltage vector represented by complex space vector as in equation 1 neglecting zero sequence components [10-11]

$$\begin{matrix} V \\ V \end{matrix} \quad \frac{2}{3}\begin{matrix} 1 & \frac{1}{2} & \frac{1}{2} \\ 0 & {}^{3}\!/_{2} & {}^{3}\!/_{2} \end{matrix} \quad \begin{matrix} Van \\ Vbn \\ Vcn \end{matrix} \tag{5}$$

SVM use the combinations of switching states to approximate the locus of Vref. In α-β plane a hexagon centered at origin of αβ plane, identifies the space vectors shown in fig. 1. Which is divide into six sectors. Each sector covers the space corresponding to 600.

The distinct possible switching states of the 2-level VSC are represented as eight voltage vectors, out of which six are active states (V1-V6) and two are null states (V0, V7). The active states contribute output line voltage as +Vdc or –Vdc, where as null states do not contribute any output voltage for VSC. The eight voltage vectors are shown in Table 1. In table 1 denotes ON state of the switch and 0 denotes OFF state of the switch.

The reference vector is synthesized by the three adjacent switching vectors. For example, when Vref falls into sector I as shown in Fig. 2, it can be synthesized by V1, V2 and V0. The optimum PWM modulation is expected if the sampling rate is as high as possible, only two non-zero switching states adjacent to the reference and one zero switching state are used to synthesize the reference vector, and the cycle wherein the average voltage vector becomes equal to the reference vector consists of three successive switching states only. This gives

$$V_{ref}T_s \quad V_1T_1 \quad V_2T_2\; V_0T_0 \tag{6}$$

Where Ts is the period of the switching cycle, T1 and T2 are the switching times of the vectors V1 and V2.

T1 and T2 are calculated as

$$T1 \quad \frac{Vref}{\frac{2}{3}}Ts\frac{\sin(60 \qquad)}{\sin 60} \tag{7}$$

$$T2 \quad \frac{Vref}{\frac{2}{3}}Ts?\frac{\sin}{\sin 60} \tag{8}$$

$$T_0 \quad T_s \quad T_1 \quad T_2 \tag{9}$$

Similar calculation is applied to sector II to VI Vector V8 can be used in place of V7. The choice is based on the requirement to minimize average number of switching per cycle.

The Maximum value of V_{ref} is obtain when θ=30^0 and V_{ref} is given by

$$\max V_{ref}=\cos 30^{\circ}\sqrt{\frac{2}{3}}V_{dc} \tag{10}$$

This is the maximum value of line to line voltage injected by the converter. The maximum magnitude of Vref is also the radius of circle inscribed in the hexagon shown in fig. 1 The Square wave converter generates a space vector of magnitude $\sqrt{6}/\pi$ Vdc the maximum value of the modulation index as

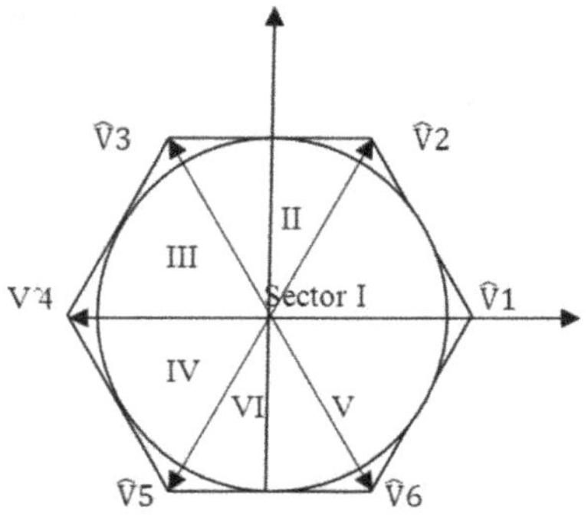

Fig. 1. Switching Vector of 2-level converter in αβ plane.

$$m_{\max} = \frac{\pi}{\sqrt{6}\sqrt{2}} = 0.907 \quad (11)$$

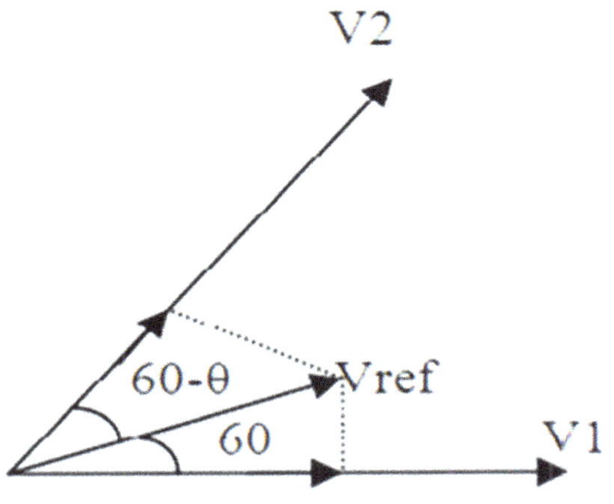

Fig. 2. Representation of reference vector.

Table 1. 2-Level Inverter Voltage Vectors Voltage Vectors.

S. No.	S_a	S_b	S_c	Line to Neutral voltage		
				Van	Vbn	Vcn
1	1	0	0	Vdc	0	0
2	1	1	0	Vdc	Vdc	0
3	0	1	0	0	Vdc	0
4	0	1	1	0	Vdc	Vdc
5	0	0	1	0	0	Vdc
6	1	0	1	Vdc	0	Vdc
7	1	1	1	Vdc	Vdc	Vdc
8	0	0	0	0	0	0

3.2. Switching Vector Model of VSC

The General switching function is obtained in time domain for space vector modulation. The harmonic content in switching function is given by Fourier series.

$$Sa \quad t \quad \overset{h}{\underset{n \quad h}{}} \quad Sa\, e^{jn \quad 0t} \quad (12)$$

$$S_{b(\ t)} \quad \overset{h}{\underset{n \quad h}{}} \quad S_b e^{jn \quad 0t} \quad (13)$$

$$S_{c(\ t)} \quad \overset{h}{\underset{n \quad h}{}} \quad S_c e^{jn \quad 0t} \quad (14)$$

Where Sa, Sb, Sc are switching function obtained by using SVPWM algorithm. The line switching vector is defined as

$$\begin{matrix} S_{ab} & S_a & S_b \\ S_{bc} & S_b & S_c \\ S_{ca} & S_c & S_a \end{matrix} \quad (15)$$

The switching vector for harmonic domain is defined as

$$S_1 \quad \begin{matrix} S_{ab} \\ S_{bc} \\ S_{ca} \end{matrix} \quad , S_2 \quad S_{ab} \quad S_{bc} \quad S_{ca} \quad (16)$$

4. Dynamic Harmonic Domain Modelling of SVPWM Based UPFC

The Unified Power Flow Controller (UPFC) is the important tool for real-time control of AC transmission system. It used to control the transmitted real and reactive-power flows through a transmission line, improving the transient stability margins, damping power oscillations and providing voltage support. DHD model of UPFC is presented in [7] considering selective harmonic elimination method. In preceding discussion we extent model proposed in [7] considering space vector modulation techniques. Figure 3 shows the equivalent circuit of UPFC connected the transmission lines. It consists of two VSCs connected to common DC capacitor. One VSC connected in shunt act as a STATCOM and other connected in series act as SSSC. Re+jXe shows the resistance and impedance of coupling transformer. The three-phase voltages and currents on the AC side of the SSSC are $V_{ABC1}(t)$ and, $i_{ABC}(t)$, The three-phase voltages and currents on the AC side of the STATCOM are $V_{abc1}(t)$ and, $i_{abc}(t)$ respectively and the DC side voltage vdc(t), DC side current $i_1(t)$, $i_2(t)$. The voltage and current on AC side in terms of switching function is given as

$$\begin{matrix} V_{abc1} \quad t & p_{S1} \quad t \ .v_{dc}(t) \\ V_{ABC1} \quad t & p_{S2} \quad t \ .v_{dc}(t) \end{matrix} \quad (17)$$

$$\begin{matrix} i_1 \quad t & q_{S1} \quad t \ .i_{abc}(t) \\ i_2 \quad t & q_{S2} \quad t \ .i_{ABC}(t) \end{matrix} \quad (18)$$

$V_{ABC1}(t)$ and $i_{ABC}(t)$ are three phase voltage and current vectors given by:

$$V_{ABC1} \quad t \quad v_{A1}(t) \quad v_{B1}(t) \quad v_{C1}(t) \quad (19)$$

$$i_{ABC} \quad t \quad i_A(t) \quad i_B(t) \quad i_C(t)^T \quad (20)$$

Where ps1(t), ps2(t) and qs1(t), qs2(t) are transformation vectors [12], [13], which are given by:

$$P_{s1} = \begin{bmatrix} S_{ab1}(t) \\ S_{bc1}(t) \\ S_{ca1}(t) \end{bmatrix} \quad P_{s2} = \begin{bmatrix} S_{ab2}(t) \\ S_{bc2}(t) \\ S_{ca2}(t) \end{bmatrix} \quad (21)$$

$$q_{s1}(t) = \left[S_{ab1}(t) \quad S_{bc1}(t) \quad S_{ca1(t)} \right]$$
$$q_{s2}(t) = [S_{ab2}(t) \quad S_{bc2}(t) \quad S_{ca2}(t)] \tag{22}$$

The time domain state equation of UPFC is:

$$\frac{dV_{dc}\ t}{dt} \quad \frac{1}{C}(i_1\ t \quad i_2\ t) \tag{23}$$

Substitute the values of i1(t), i2(t) gives:

$$\frac{dV_{dc}\ t}{dt} \quad \frac{1}{C}(q_{s1}\ t\ i_{abc}\ t \quad q_{s2}\ t\ i_{ABC}\ t) \tag{24}$$

$$\frac{di_{abc}\ t}{dt} \quad \frac{R_e}{L_e} i_{abc}\ t \quad V_{abc}\ t \quad V_{abc1}\ t \tag{25}$$

$$\frac{di_{ABC}\ t}{dt} \quad \frac{R_e}{L_e} i_{ABC}\ t \quad V_{ABC}\ t \quad V_{ABC1}\ t \tag{26}$$

Current in AC side of STATCOM and SSSC in terms of switching function is

$$\frac{di_{abc}\ t}{dt} \quad \frac{R_e}{L_e} i_{abc}\ t \quad \frac{1}{L_e}(V_{abc}\ t \quad p_{s1}\ t\ V_{dc}\ t) \tag{27}$$

$$\frac{di_{ABC}\ t}{dt} \quad \frac{R_e}{L_e} i_{ABC}\ t \quad \frac{1}{L_e}(V_{ABC}\ t \quad p_{s2}\ t\ V_{dc}\ t) \tag{28}$$

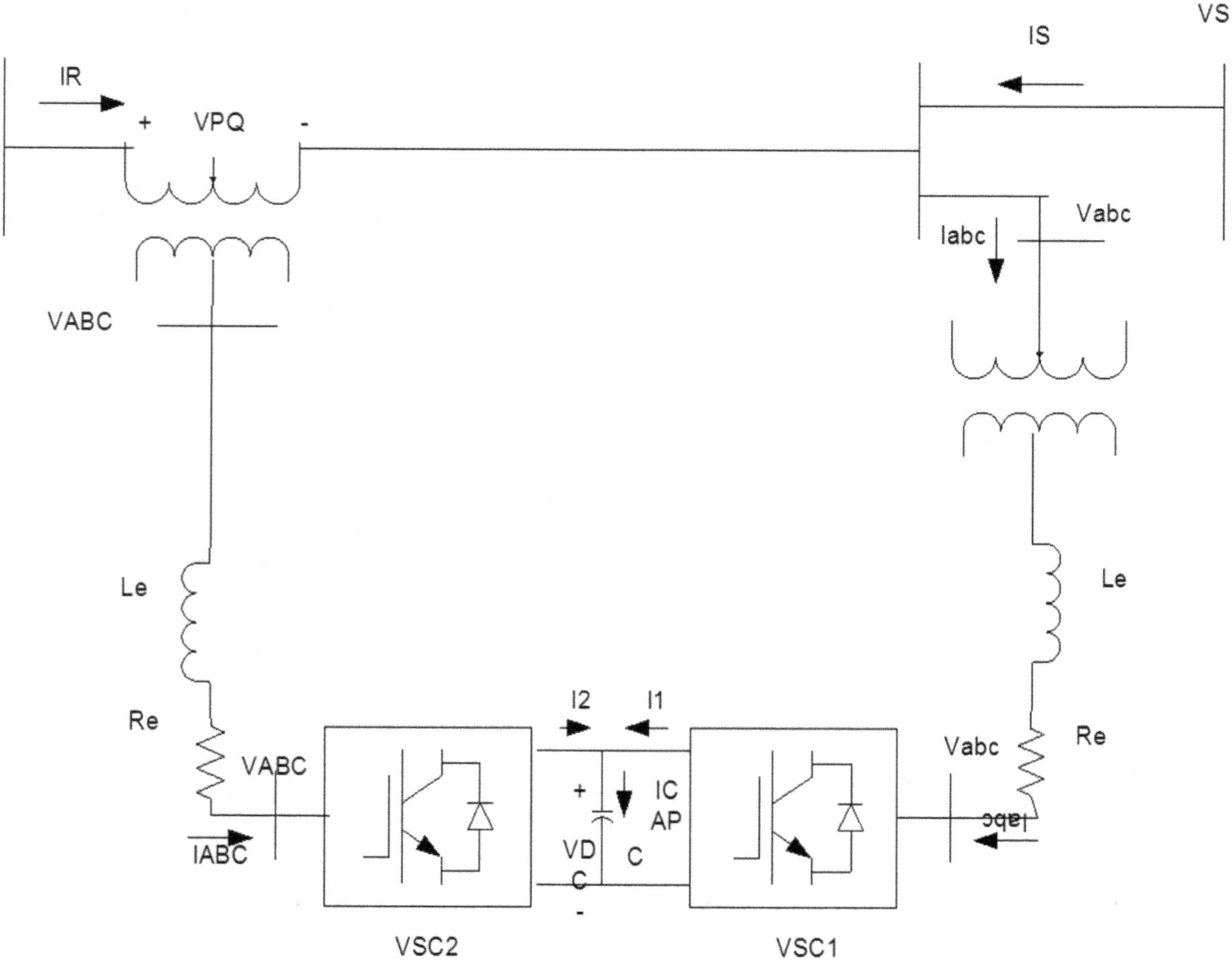

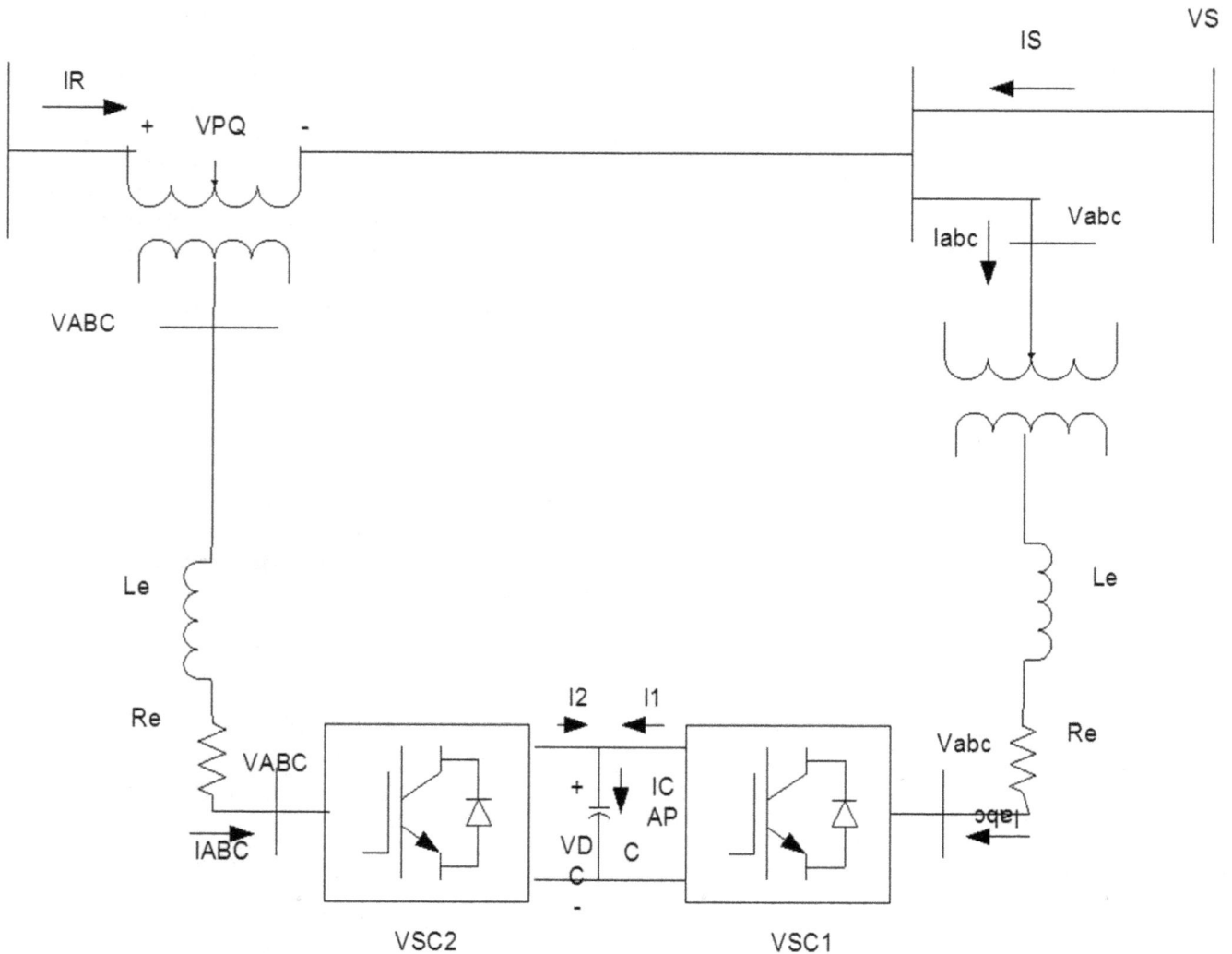

Fig. 3. *Unified Power flow controller (UPFC).*

The linear time periodic equations (24, 27 & 28) in matrix form will be written as:

$$\begin{bmatrix} \dfrac{di_S\ t}{dt} \\ \dfrac{di_R\ t}{dt} \\ \dfrac{dV_{dc}\ t}{dt} \end{bmatrix} = \begin{bmatrix} \dfrac{R_e}{L_e} & 0 & \dfrac{1}{L_e}(P_{s1} \quad P_{s2}) \\ 0 & \dfrac{R_e}{L_e} & \dfrac{1}{L_e}P_{s2} \\ \dfrac{1}{C}q_{s1}\ t & \dfrac{1}{C}(q_{s1}\ t \quad q_{s2}\ t) & 0 \end{bmatrix} \begin{bmatrix} i_S\ t \\ i_R\ t \\ V_{dc}\ t \end{bmatrix} \quad \frac{1}{L_e} \begin{bmatrix} 2 & 1 & 0 \\ 1 & 1 & 0 \\ 0 & 0 & 0 \end{bmatrix} \begin{bmatrix} V_S\ t \\ V_R\ t \\ 0 \end{bmatrix} \tag{29}$$

The equation (29) can be transfer to linear time invariant equation considering theory of DHD analysis as:

$$\begin{bmatrix} \dot{I}_S\ t \\ \dot{i}_R\ t \\ \dot{V}_{dc}\ t \end{bmatrix} = \begin{bmatrix} \dfrac{R_e}{L_e}U_1 \quad D(jh\ _0) & O4 & \dfrac{1}{L_e}(P_{s1} \quad P_{s2}) \\ O4 & \dfrac{R_e}{L_e}U_1 \quad D(jh\ _0) & \dfrac{1}{L_e}P_{s2} \\ \dfrac{1}{C}Q_{s1}\ t & \dfrac{1}{C}(Q_{s1}\ t \quad Q_{s2}\ t) & D(jh\ _0) \end{bmatrix} \begin{bmatrix} I_S\ t \\ I_R\ t \\ V_{dc}\ t \end{bmatrix} \quad \frac{1}{L_e} \begin{bmatrix} 2U1 & U1 & O2 \\ U1 & U1 & O2 \\ O1 & O1 & O3 \end{bmatrix} \begin{bmatrix} V_S\ t \\ V_R\ t \\ 0 \end{bmatrix} \tag{30}$$

The initial condition is obtained by considering derivatives of state variables as zero gives:

$$\begin{bmatrix} I_S\ t \\ I_R\ t \\ V_{dc}\ t \end{bmatrix} = \begin{bmatrix} \dfrac{R_e}{L_e}U_1 \quad D(jh\ _0) & O4 & \dfrac{1}{L_e}(P_{s1} \quad P_{s2}) \\ O4 & \dfrac{R_e}{L_e}U_1 \quad D(jh\ _0) & \dfrac{1}{L_e}P_{s2} \\ \dfrac{1}{C}Q_{s1}\ t & \dfrac{1}{C}(Q_{s1}\ t \quad Q_{s2}\ t) & D(jh\ _0) \end{bmatrix}^{1} \frac{1}{L_e} \begin{bmatrix} 2U1 & U1 & O2 \\ U1 & U1 & O2 \\ O1 & O1 & O3 \end{bmatrix} \begin{bmatrix} V_S \\ V_R \\ 0 \end{bmatrix} \tag{31}$$

5. Simulation

In order to access dynamic harmonics response including power quality indices, The per-phase inductive reactance and resistance of the coupling transformer and the capacitance of the dc capacitor are R, = 0.04Ω, L = 0.2 mH and C = 5000μF, respectively.

Under steady state conditions the bus per phase voltages V_R and V_S in volts at 50 Hz are

$$V_{Ra}(t) = \sin \omega_0 t,\ V_{Sa}(t) = \sin \omega_0 t$$
$$V_{Rb}(t) = \sin\left(\omega_0 t - 120^0\right),\ V_{Sb}(t) = \sin\left(\omega_0 t - 120^0\right)$$
$$V_{Rc}(t) = \sin\left(\omega_0 t + 120^0\right),\ V_{Sc}(t) = \sin\left(\omega_0 t + 120^0\right)$$

Assume disturbances in the voltages starting at 0.04 seconds and lasting for 0.005 seconds. During disturbances voltages on the VS bus is 150% of the original value. The simulation was started at to = 0 seconds with final time tf = 0.1 seconds and an integration time step is 0.001s. 50 harmonics are considered. System is simulated using MATLAB software.

Fig. 4.a. Shows current at terminal of VSC1 which is shunt connected. Its RMS value is shown in fig. 4.c. Shows that during disturbances fundamental component of STATCOM current increases. Fig. 4.b. shows harmonic component of SSSC current. Fig. 4.d. shows the RMS value of SSSC current. It is observed that during disturbances RMS value of fundamental current increases. Fig. 4.e. shows the harmonic component of voltage at the terminal of VSC1. It shows those harmonic components are well within the range. Fig. 4.f. shows the harmonic component of the voltage on the capacitor. Fundamental component of capacitor voltage is constant by using SVPWM techniques without using any control to maintain capacitor voltages constant. It shows that during disturbances exhibits the dynamic behavior of only the 1st, 3rd, 5th and 7th harmonic components with time for the phase-a is shown in fig. 4.

UPFC's dynamic power quantities of all the three phases are shown in Fig. 5. fig. 5.a. shows the active power absorb by STATCOM. It shows that STATCOM absorb the more active power during disturbances. Fig. 5.b. shows the apparent power on for the VSC1 and for VSC2. It shows that during disturbances SSSC supplied the apparent power to the system. Fig. 5.c. shows the active power supplied by SSSC. During steady state SSSC is not supplying power to the system but STATCOM is absorbing the active power to compensate the looses occurs in the SSSC and STATCOM circuits. Fig. 5.d. and Fig. 5.f. shows the reactive power supplied by SSSC and STATCOM whereas Fig. 5.c. shows the distorted power at STATCOM terminals.

The voltage and the current THD of the SSSC and STATCOM in all three phases are shown in Fig. 6. THD are within the range as per standards.

6. Conclusion

This paper presents the Space Vector based switching strategy for a UPFC that utilize the voltage source converter to minimize the harmonic at the point of common contact. The linear time periodic equation is converted into linear time invariant system which is done using dynamic harmonic domain for calculation of harmonic interference in the system during transient and evaluated based on dynamic harmonic domain algorithms using MATLAB code.

The proposed model is used to calculate harmonic interference produced by space vector based UPFC. It gives accurate result and information about harmonics indices during transient operation of system as compared to time domain simulation. Harmonics indices are important for designing control system for the system.

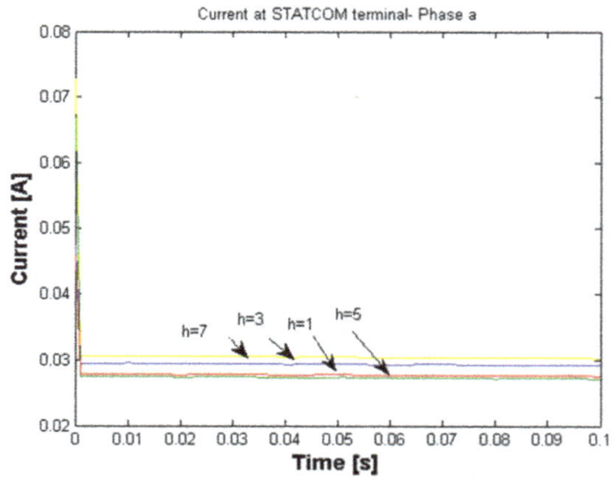

4.a. Current at STATCOM terminal Phase A

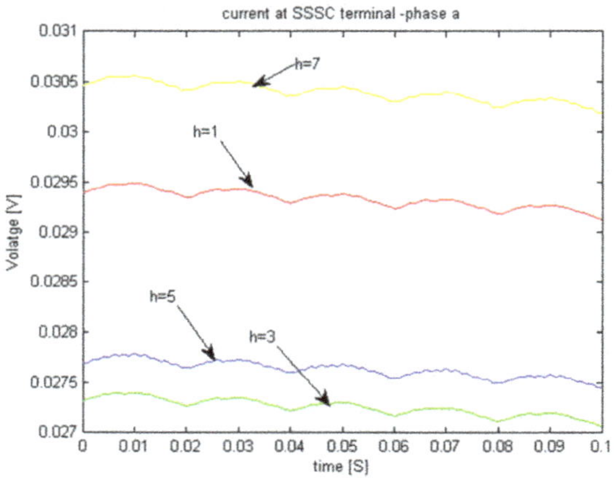

4.b. Current at SSSC terminal Phase A

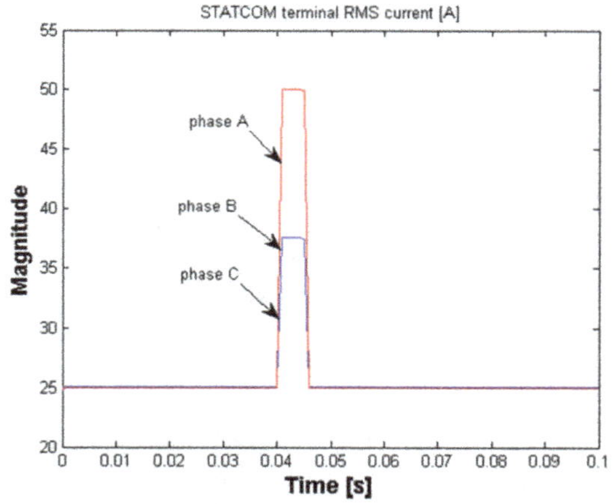

4.c. RMS Value of STATCOM Current

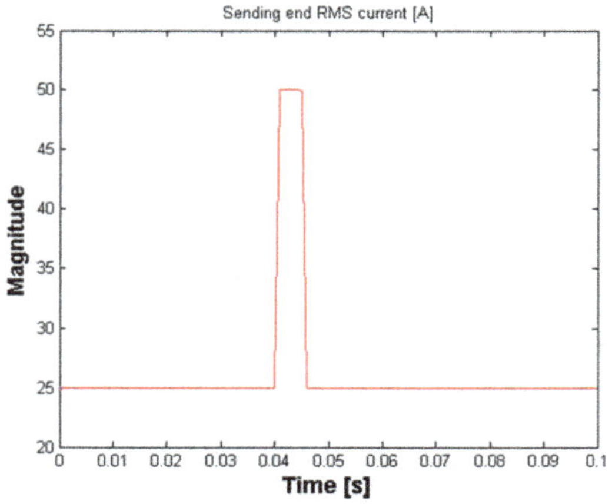

4.d. RMS value of SSSC Current

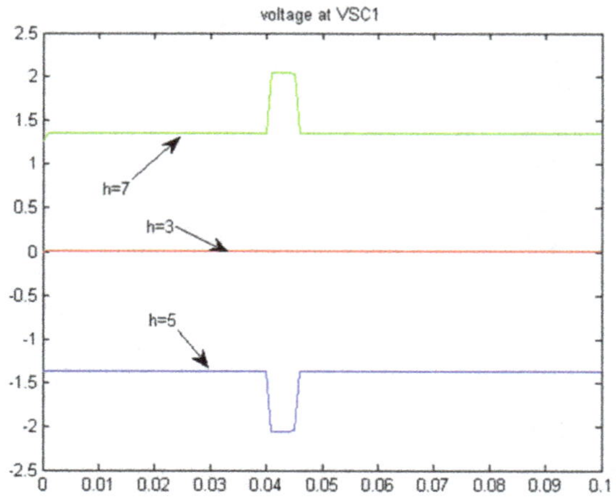

4.e. Voltage of at terminal of VSC1

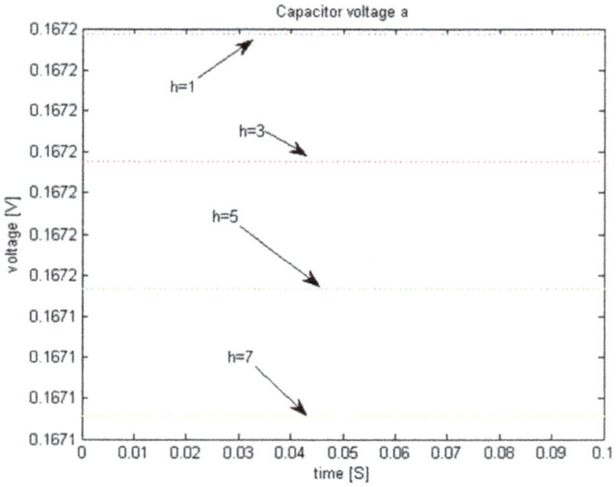

4.f. Voltage on DC capacitor

Fig. 4. *Voltage and current at Converter terminal.*

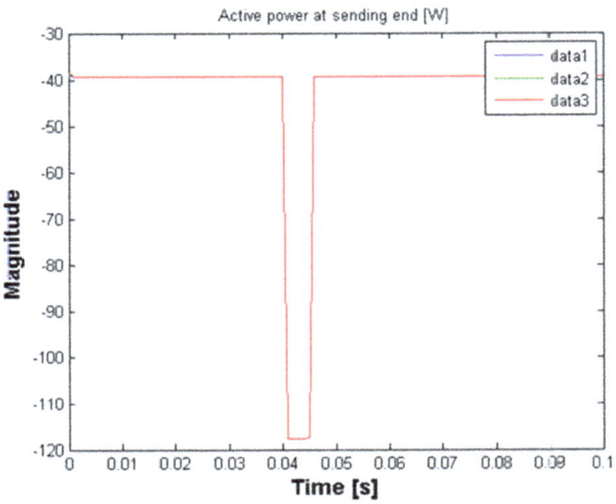

5.a. Active power at STATCOM Terminal

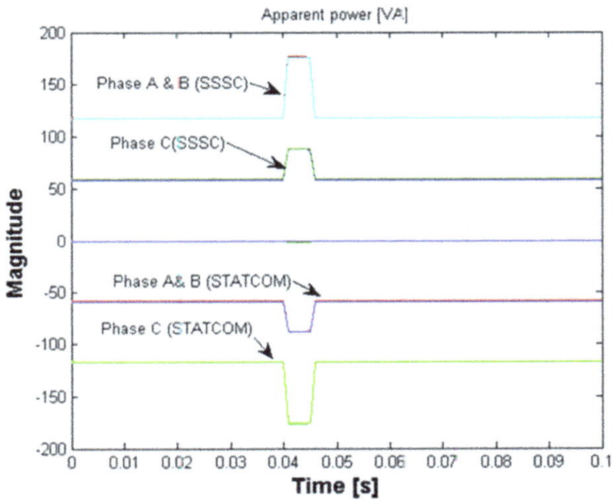

5.b. Apparent power at STATCOM and SSSC terminals

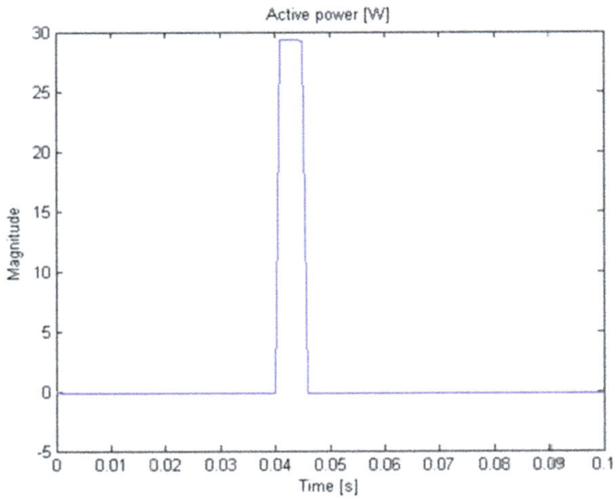

5.c Active power at SSSC Terminal

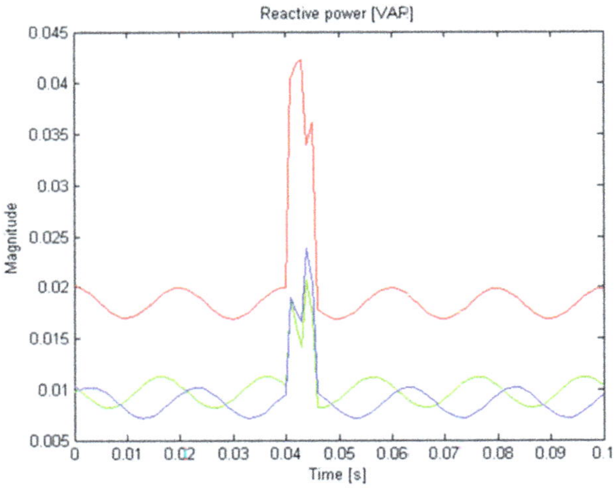

5.d. Reactive power at SSSC Terminal

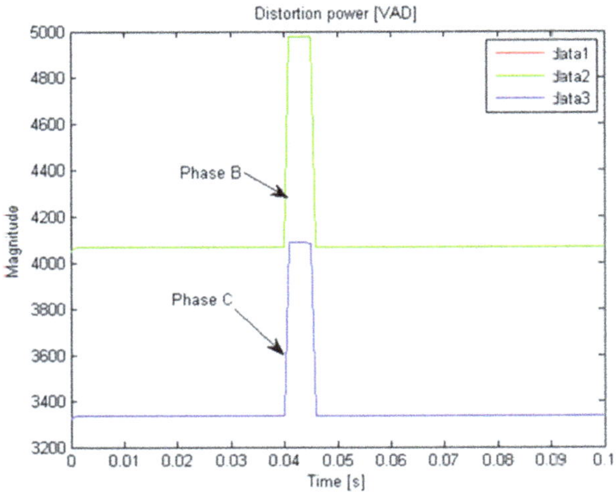

5.e. Distorted Power at STATCOM Terminal

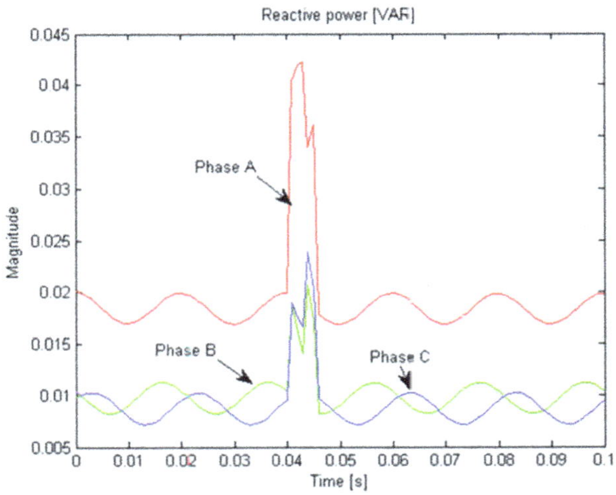

5.f. Reactive power at STATCOM Terminal

Fig. 5. *UPFC terminal electric quantities.*

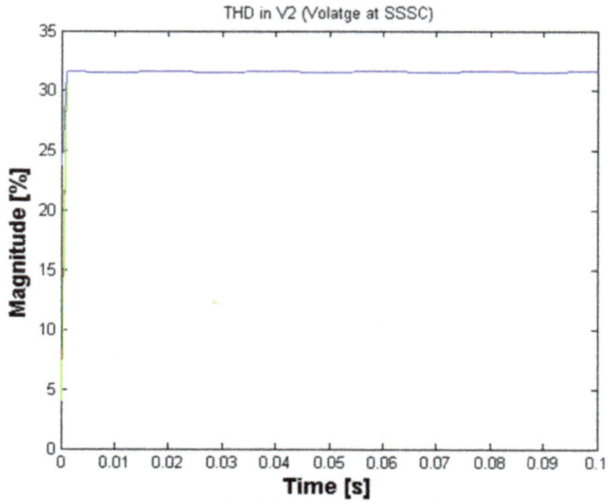

6.a. THD in Voltage at SSSC

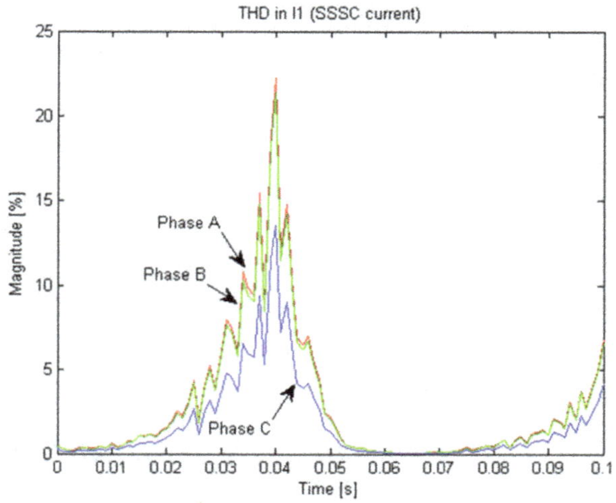

6.b. THD in Current at SSSC

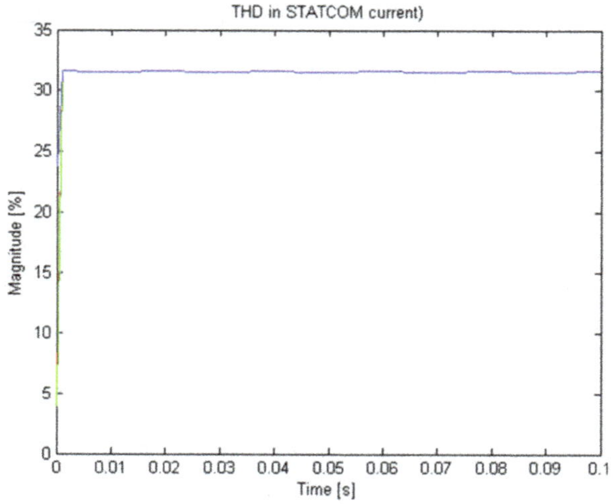

6.c. THD in STATCOM current

Fig. 6. *Total harmonic distortion in voltage and current.*

References

[1] C. D. Collins, G. N. Bathurst, N. R. Watson and A. R. Wood," Harmonic domain Approach of STATCOM modelling", IEE proceeding Generation Transmission, Distribution, Vol 152 no.2, pp. 194-200 March 2005.

[2] Juan Segundo – Rmirez and Aurelio Medina," Modelling of FACTS deives based on SPWM VSC", IEEE Transaction on power Delivery Vol.24 No.4, October 2001.

[3] Farhad Yahyaie, and Peter W. Lehn, "On Dynamic Evaluation of Harmonics Using Generalized Averaging Techniques IEEE Transactions On Power Systems", Vol. 30, No. 5, September 2015.

[4] K. L. Lian and P. W. Lehn," Steady-State Solution of a Voltage-Source Converter with Full Closed-Loop Control" IEEE Transactions On Power Delivery, Vol. 21, No. 4, Pp. 2071-2080 October 2006.

[5] Abner Ramirez," The Modified Harmonic Domain: Interharmonics," IEEE Transactions On Power Delivery, Vol. 26, No. 1, Pp. 235-241 January 2011.

[6] J. Jesus Rico, Manuel Madrigal and Enrique Acha," Dynamic Harmonic Evolution Using the Extended Harmonic Domain," IEEE Transactions On Power Delivery, Vol. 18, No. 2, Pp 587-594april 2003.

[7] Bharat Vyakaranam, Manuel Madrigal, F. Eugenio Villaseca and Rick Rarick," Dynamic Harmonic Evolution in FACTS via the Extended Harmonic Domain Method," Power and Energy Conference at Illinois (PECI), 2010. Year: 2010 Pages: 29–38.

[8] Pável Zúñiga-Haro," Harmonic Modeling of Multi-pulse SSSC," IEEE Bucharest Power Tech Conference, June 28th - July 2nd, Bucharest, Romania.

[9] Abner Ramirez and J. Jesus Rico," Harmonic/State Model Order Reduction of Nonlinear Networks." IEEE Transactions on Power Delivery, Year 2015.

[10] P. Tripura1, Y. S. Kishore Babu 2, Y. R. Tagore," Space Vector Pulse Width Modulation Schemes for Two-Level Voltage Source Inverter," ACEEE Int. J. on Control System and Instrumentation, Vol. 02, No. 03, October 2011.

[11] Babita Nanda," Total Harmonic Distortion of Dodecagonal Space Vector Modulation." International Journal of Power Electronics and Drive System (IJPEDS) Vol. 4, No. 3, September 2014, pp. 308-313.

[12] P. C. Stefanov A. M. Stankovic," Dynamic Phasors in Modeling of UPFC Under Unbalanced Conditions." International Conference on Power System Technology, 2000. Proceedings. PowerCon 2000. Pages: 547-552 vol. 1.

[13] M. Saeedifard, A. R. Bakhshai, G. Joos, P. Jain," Modified Low Switching Frequency Space Vector Modulators for High Power Multi-Module Converters," Applied Power Electronics Conference and Exposition, 2003. APEC '03. Eighteenth Annual IEEE Pages: 555–561.

[14] M Madrigal and E Acha," Harmonic Modelling Of Voltage Source Converters For Hvdc Stations," Seventh International Conference on AC-DC Power Transmission, 2001 Pp: 125–131.

[15] Yao Shu-jun, Song Xiao-yan, Wang Yan, Yan Yu-xin, Yan Zhi," Research on dynamic characteristics of Unified Power Flow Controller (UPFC)," 4th International Conference on Electric Utility Deregulation and Restructuring and Power Technologies (DRPT), 2011pp: 490–493.

[16] Ali Ajami, S. H. Hosseini and G. B. Gharehpetian," Modelling and Controlling of UPFC for Power System Transient Studies," Transactions On Electrical Eng., Electronics, And Communications Vol.5, No.2, pp: 29-35 August 2007.

[17] A. Nabavi-Niaki M. R. Iravani," Steady-State And Dynamic Models Of Unified Power Flow Controller (Upfc) For Power System Studies." IEEE Transactions on Power Systems, Vol. 11, No. 4, pp: 1937-1943 November 1996.

[18] Maryam Saeedifard, Hassan Nikkhajoei, Reza Iravani, and Alireza Bakhshai," A Space Vector Modulation Approach for a Multimodule HVDC Converter System," IEEE Transactions On Power Delivery, Vol. 22, NO. 3, pp 1643-1654. JULY 2007.

Influence of Power Quality Problem on the Performance of an Induction Motor

Amaize Aigboviosa Peter[1], Ignatius Kema Okakwu[2], Emmanuel Seun Oluwasogo[3], Akintunde Samson Alayande[4], Abel Ehimen Airoboman[2]

[1]Department of Electrical and Information Engineering, College of Engineering, Covenant University, Ota, Nigeria
[2]Department of Electrical/Electronics Engineering, University of Benin, Benin City, Nigeria
[3]Department of Electrical and Computer Engineering, Kwara State University, Malete, Nigeria
[4]Department of Electrical Engineering, Faculty of Engineering and the Built Environment, Tshwane University of Technology, Pretoria, South Africa

Email address:
amaizepeter@yahoo.com (P. A. Amaize), igokakwu@yahoo.com (I. K. Okakwu), emmanueloluwasogo@yahoo.com (E. S. Oluwasogo), alayandeakintundesamson@gmail.com (A. S. Alayande), abelarrow@yahoo.com (A. E. Airoboman)

Abstract: This paper presents the application of MATLAB® Simulink as a useful tool for predicting the performance of an induction motor. The influence of power quality problem on the performance of an induction motor is critically investigated. Mathematical modelling of an induction motor subjected to an unsymmetrical voltage conditions are presented. The results obtained from the simulation reveal the presence of rotor noise and vibration during operation of induction motor under voltage unbalance.

Keywords: MATLAB® Simulink, Induction Motor, Power Quality, Unsymmetrical Voltage

1. Introduction

The widening gap between the power supplied and power demanded is as a result of the increase in the number of domestic, commercial and industrial loads. As such, there has been an increasing stress towards energy management in the industrial sector as they are the major consumers. The continuously varying load demand by domestic consumers has therefore led to a power quality problem. This problem is a major concern to the power system engineers in recent years [1]. Power quality problem or disturbanceis mainly concerned with the deviations of voltage and/or current from the ideal values [2]. Voltage variations and unbalance seem to be the most commonly occurring power quality problems within as a result of unequal distribution loads across the power network [3]. Three-phase induction motors are widely used in industrial, commercial and residential systems whose working performance could be greatly affected when driven with an unsymmetrical three-phase voltages [4], [5]. Industrial utilities make significant amount of investment in order to achieve higher energy efficiency. However, the lowered performance variations are mainly due to the quality of the incoming supply [6]. Hence, the knowledge of possible variation in performance due to the impact of voltage variation and unbalance is a necessity.

The contributions of this paper are in three folds: To determine the percentage of unbalance voltage that is tolerable for effective operational performance of an induction motor, to determine the impact of the rated voltage, under-voltage and over-voltage unbalance on induction motor operational performance and to simulate the performance of a three-phase induction motor.

This paper investigates the level of difference in the operating performance of induction motors working under balanced and unbalanced voltage conditions. In section 2, mathematical details of the voltage unbalance in the model of induction motors is presented. Section 3 gives the details of the motor parameters used for the study. Section 4 presents the results and discussion obtained from the simulation while the conclusion is given in section 5.

2. Voltage Unbalance Modelling in Induction Motors

For the sake of simplicity, steady-state performance of three-phase induction motors is usually carried out by neglecting the core loss and friction and windage loss components[7]. However, in industrial situations, the utility energy bill is dependent on components such as the power factor of the plant, total active power usage and overall efficiency of operation. Therefore, accurate estimation of losses is extremely important to avoid significant errors in the efficiency estimation [8]. The core loss depends on the applied voltage while friction and windage loss depends on the operating speed. The power input on no-load is only to account for the no-load losses in the form of stator copper loss, core lossand windage and friction loss.

The modified steady-state per phase equivalent circuit is that takes into account the core loss and friction and windage loss under running conditions is shown in figure 1 [9].

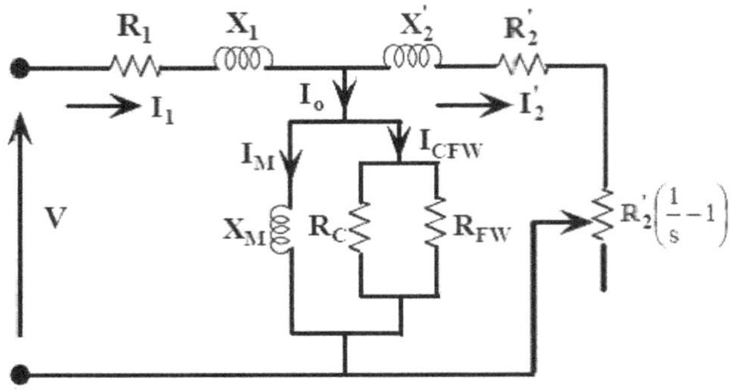

Figure 1. Per-phase equivalent circuit of an induction motor.

where V is the applied voltage, R_1 and $X_1$are stator resistance and reactance respectively, R'_2 and X'_2 are equivalent rotor resistance and reactance as referred to the stator. $R_C$is the core loss resistance, R_{FW}represents the resistance of the friction and windage loss, X_M is the magnetizing reactance, s is the operating slip, I_1 is the stator current, I_o is the no-load current component and I'_2 is the rotor current referred to stator side.

The equivalent circuit parameters of X_1, X'_2, X_M, R_C and R_{FW} can be obtained from the no-load and blocked rotor tests data as presented in reference [9].

By applying symmetrical component technique, under the condition of asymmetry, the per phase induction motor equivalent model can easily be resolved into positive sequence and negative sequence equivalent circuits. Let V_{RY}, V_{YB} and V_{BR} be the measured line-to-line voltage with V_{RY} being selected as the reference phasor.

For the positive sequence equivalent circuit,

$$V_p\angle\theta_p = \frac{V_{RY}\angle 0 + aV_{YB}\angle\theta_{YB} + a^2V_{BR}\angle\theta_{BR}}{3} \quad (1)$$

$$I_{1P}\angle\theta_{CP} = \frac{V_P\angle\theta_{VN}}{Z_P\emptyset_P} \quad (2)$$

For negative sequence equivalent circuit.

$$V_N < \theta_N = \frac{V_{RY}<0 + aV_{YB}<\theta_{YB} + a^2V_{BR}<\theta_{BR}}{3} \quad (3)$$

$$I_{1N}\angle\theta_{CN} = \frac{V_N\angle\theta_{VN}}{Z_N\angle\emptyset_N} \quad (4)$$

where, $V_P\angle\theta_{VP}$ and $V_N\angle\theta_{VN}$are the positive sequence and negative sequence voltages, $I_{1P}\angle\theta_{CP}$ and $I_{1N}\angle\theta_{CN}$are the positive sequence and negative sequence stator currents,$Z_P\angle\theta_P$ and $Z_N\angle\theta_N$are the positive sequence and negative sequence input impedances while the operatora is given by $1\angle 120°$.

Thus, under voltage unbalance conditions, the induction motor can be thought of as two separate motors in operation, one operating with a positive sequence voltage V_P and slip 's', and other operating with a negative sequence voltage V_N and slip ' (2 - s) ' [7].

The current in each phase can therefore be expressed as

$$I_R\angle\theta_R = I_P\angle\theta_{CP} + I_N\angle\theta_{CN} \quad (5)$$

$$I_Y\angle\theta_Y = a^2I_P\angle\theta_{CP} + aI_N\angle\theta_{CN} \quad (6)$$

$$I_B\angle\theta_B = aI_P\angle\theta_{CP} + a^2I_N\angle\theta_{CN} \quad (7)$$

The actual power output is the sum of the positive and negative power output components given as

$$P_O = P_P + P_N \quad (8)$$

where

Positive sequence power output,

$$P_P = \frac{3(I'_{2N})^2R'_2\,(s-1)}{(2-s)} \quad (9)$$

Negative sequence power output,

$$P_N = \frac{3(I'_{2N})^2R'_2\,(1-s)}{(2-s)} \quad (10)$$

where, I'_{2P} and I'_{2N} are positive and negative sequence rotor current components.

For steady state operation, the torque developed by motor, T_M equals the load torque, T_L

$$Tm = T_L \quad (11)$$

Under conditions of voltage unbalance, we have

$$Tm = Tp + Tn \quad (12)$$

where T_P and T_N denote the positive and negative sequence torque components respectively.

The total power input can be expressed as

$$P_{IN} = Real[3(V_PI^*_P + V_NI^*_N)] \quad (13)$$

where equation (13) indicates the conjugate value.

Motor efficiency is given by

$$\%\eta = \frac{P_P+P_N}{P_{IN}} \times 100\% \quad (14)$$

For a qualitative analysis of power quality problem, under-voltage and over-voltage unbalance conditions, at the distribution end and the point of utilization, in three-phase power systems need to be investigated. Some causes of voltage unbalance are the uneven distribution of single-phase loads in three-phase power systems, asymmetrical transformer winding impedances, open-Y, open-Δ transformer banks, incomplete transposition of transmission lines, blown fuses on three-phase capacitor banks, etc.[10-17].

Based on the foregoing, performance analysis of equipment in power systems under voltage unbalance condition is very important. Three-phase induction motor is one of the most widely used equipment in industrial, commercial and residential applications for energy conversion purposes. Because of various techno-economic benefits, the three phase induction motors are used more than ever before. However, most of them are connected directly to the electric power distribution system and they are exposed to unbalanced voltages. In theoretical point of view, the unbalanced voltages induce negative sequence current which produces a backward rotating field in addition to the forward rotating field produced by the positive sequence one[18]. The interaction of these fields produces pulsating electromagnetic torque and ripple in speed [19], [20]. Such condition has severe negative effects on the performance of an induction motor.

The effect of voltage unbalance is more pronounced in three-phase induction motors. When a three-phase induction motor is supplied by an unbalanced system, the resulting line currents show a degree of unbalance that is several times the voltage unbalance.

3. Data Analysis and Simulink

The technical data of a simple three-phase induction motor investigated is presented in table 1.The induction motor is totally enclosed fan cooled (TEFC) with a cast aluminium squirrel cage. The models for the motor under balance and unbalanced conditions are created using MATLAB®simulink workspace shown in figure 2. The measured voltage for each line-to-neutral voltage are as follows:

R-N (190V), Y-N (190V) and B-N (194V)

Table 1. *Induction motor data.*

Parameter	Value	Parameter	Value
Rated Voltage (V)	$415_{(L\text{-}L)}$	Stator Resistance	1.115Ω
Power (kW)	1492	Stator Inductance	0.005974Ω
Frequency (Hz)	50	Rotor Resistance	1.083Ω
Speed(rpm)	1500	Rotor Inductance	0.005974Ω
Mutual Inductance	0.2037		
Number of Poles	4		

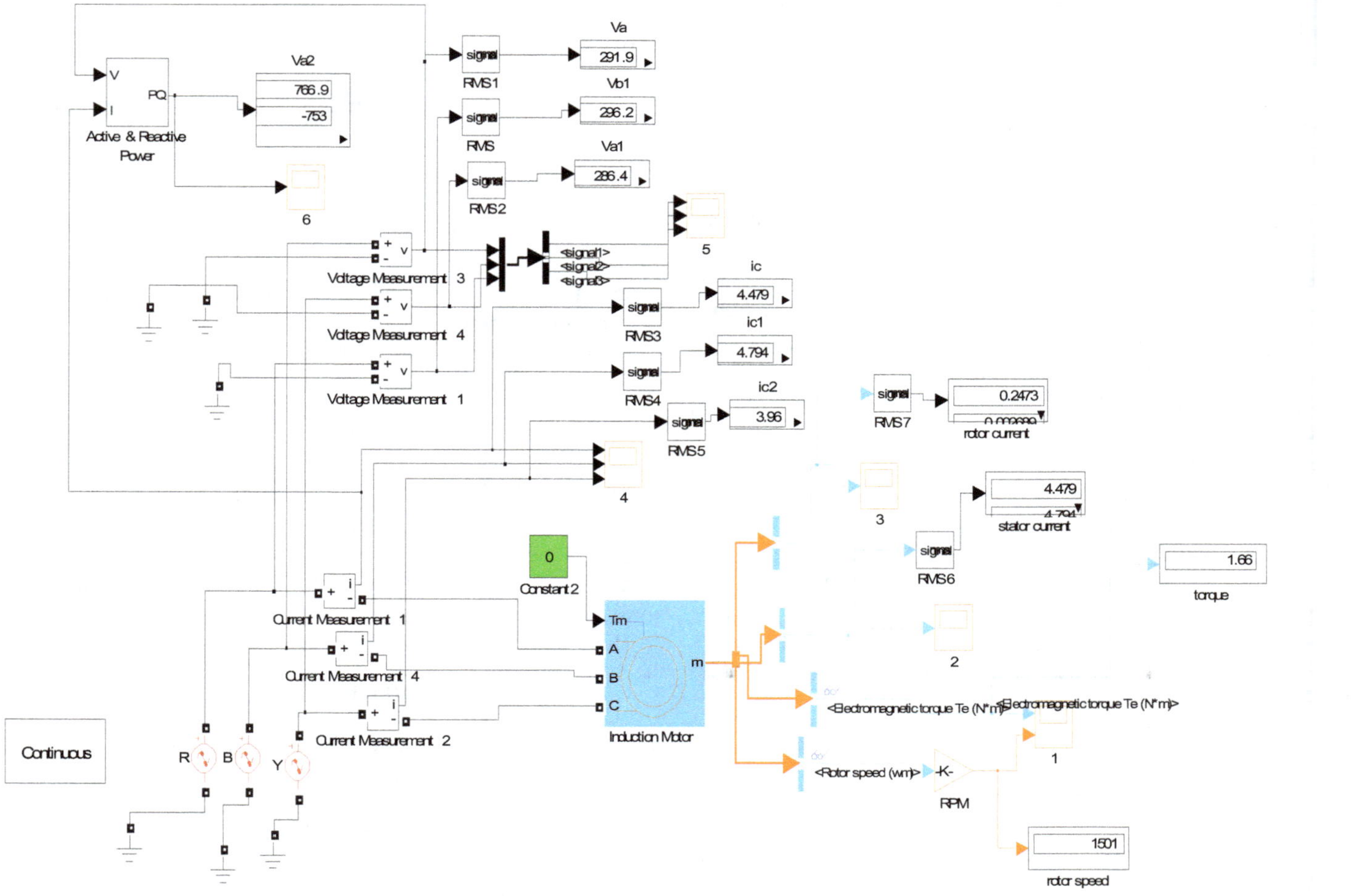

Figure 2. *Simulink model for the balanced and unbalanced conditions.*

The influence of unbalanced voltage on its performance is evaluated under rated conditions with balanced voltage at no-load. The motor is tested with three types of three-phase voltage unbalance factors which are the rated, under-voltage and over-voltage factors as shown in table 2.

Table 2. Voltage unbalance factors.

1.05	Over-Voltage Unbalance
1.00	Rated Voltage Unbalance
0.95	Under-Voltage Unbalance

For the analysis of the under-voltage unbalanced condition, the positive sequence voltage is fixed at 95% of the rated voltage and the simulation performed for different values of VUFs (Voltage unbalance factors) between 2% and 10%. For the case of the ratedvoltage unbalanced condition, the positive sequence voltage was fixed at the rated voltage and simulation conducted for different grades of VUF from 2% to 10%. Finally, the over-voltage unbalanced condition is studied with the positive sequence voltage fixed at 105% of the rated voltage and simulation performed for different values of VUFs.

4. Results and Discussion

The simulation results obtained from the analysis of the induction motor under balanced and unbalanced conditions are presented graphically in time domain.

4.1. Balanced Voltage

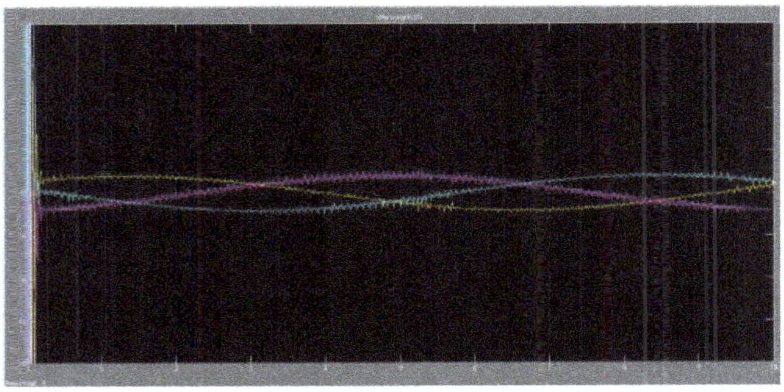

Figure 3. *Rotor currents on no load balanced voltage.*

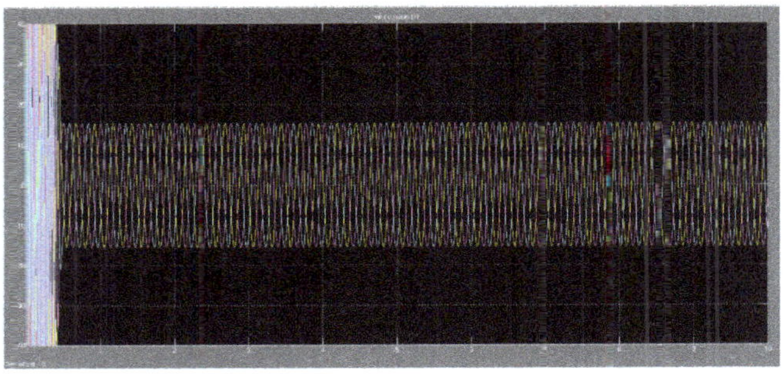

Figure 4. *Rotor currents on full load balanced voltage.*

The three-phase stator currents waveforms shown in figures 3 and 4 are steady and the induced rotor current waveforms are uniform and linear as seen in figures 5 and 6. The Induced rotor currents on no-load and on full load are found to be 0.5A and 15A respectively while stator currents are 3.5A and 17A on no-load and full load respectively.

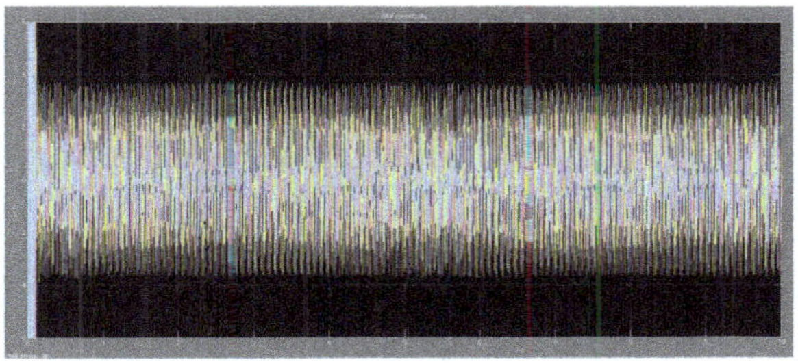

Figure 5. *Stator currents on no load balanced voltage.*

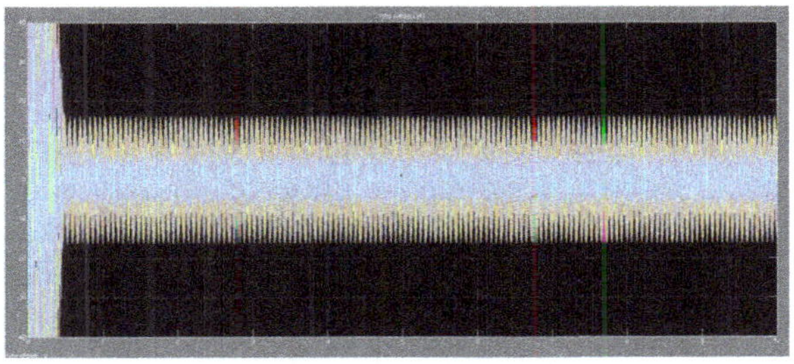

Figure 6. *Stator currents on full load balanced voltage.*

4.2. Unbalanced Voltage

4.2.1. Voltage Unbalance – Under-Voltage

Figures 7, 8 and 9 show the waveforms produced under unbalanced voltage conditions.The waveforms indicate rippled and cloudy stator and rotor currents on full load at under voltage unbalanced conditions. Rotor current are above indeterminate with average value of 23A and 18A on 75%, and 50% of load respectively. While the stator currents are indeterminate with average value of 25A and 20A on 75% and 50% of full load respectively.

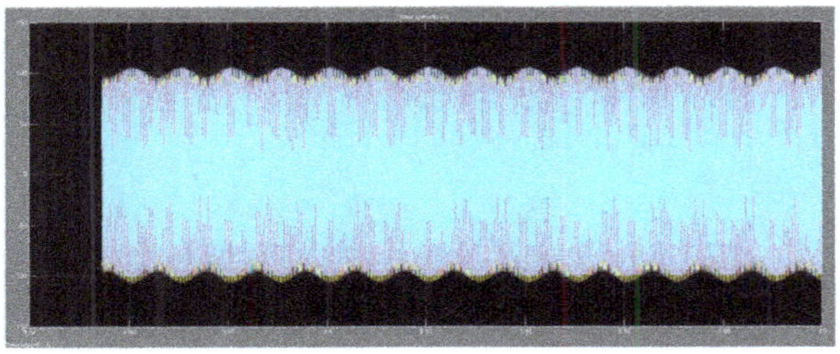

Figure 7. *Rotor currents on full load at under-voltage unbalance.*

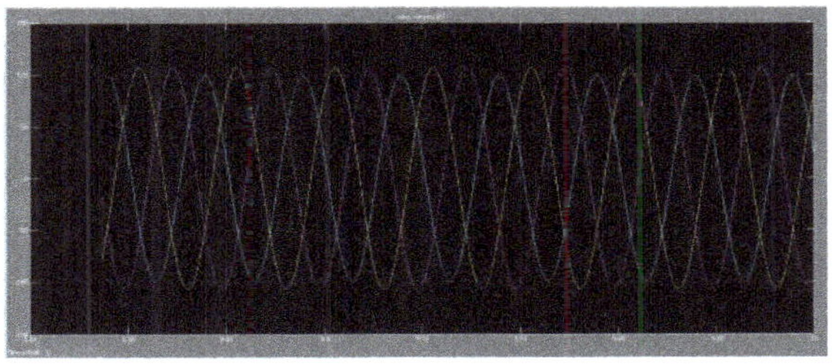

Figure 8. *Stator currents at full load of under-voltage unbalance.*

At under-voltage unbalance condition, for all values of VUF, the motor's torque and speed were undefined for full load operation as shown in figure 9. However, at reduced loads, the outputs were determinate with indications of growing gross ripples as the value of VUF increases.

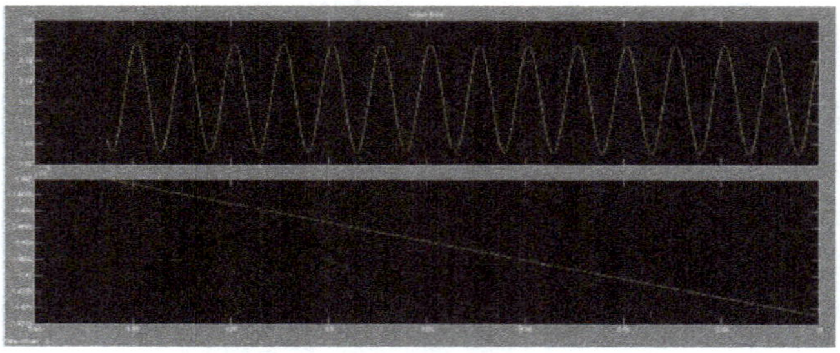

Figure 9. Electromagnetic torque and rotor speed at full load torque at under voltage unbalance (VUF of 2-10% with positive sequence voltage fixed at 0.95 of rated voltage).

4.2.2. Voltage Unbalance – Over-Voltage

The waveforms of the rotor and stator currents at full load over-voltage unbalance condition are presented in figures 10 and 11 respectively. The estimated values of rotor current are 25A, 18A and 12A on full load, 75% and 50% of full load respectively. Stator currents are also estimated to be 28A, 20A and 14A on 75% and 50% of full load respectively.

Figure 10. Rotor current at full load over-voltage unbalance.

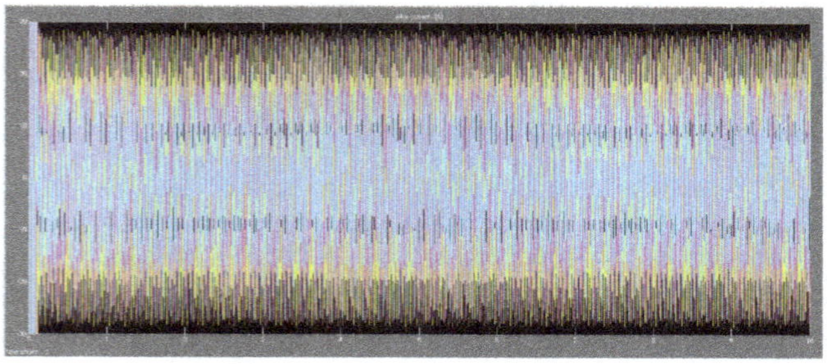

Figure 11. Stator current at full load over voltage unbalance.

4.2.3. Voltage Unbalance – Rated Voltage

The waveforms indicated rippled and cloudy stator and rotor currents on full load at rated voltage unbalanced conditions as shown in figures 12 and 13 respectively. Rotor current are indeterminate, 20A, 13A on full load, 75%, and 50% of load respectively. Stator currents are indeterminate, 22A, and 15A on full load, 75%, and 50% of load respectively.

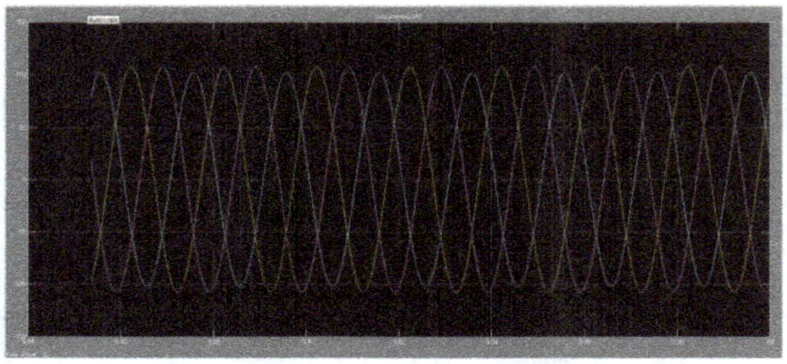

Figure 12. Stator currents at full load rated voltage unbalance.

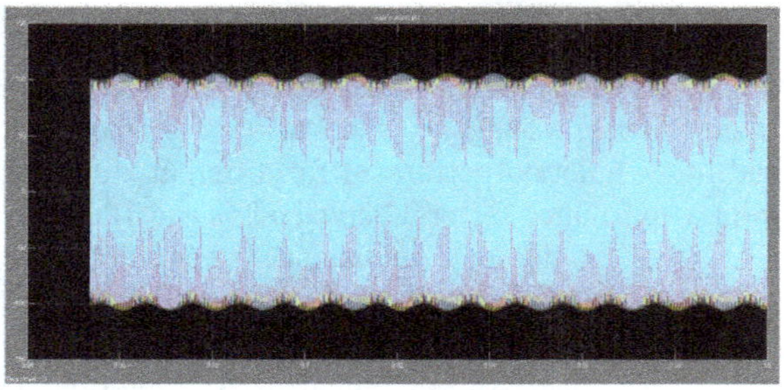

Figure 13. Rotor currents at full load rated voltage unbalance.

5. Conclusion

This paper has shown that the presence of ripples in both stator and rotor currents' waveforms, in all cases of unbalance voltages when motor is on load, indicate the presence of harmonics. The simulation results also showthat more current is drawn by the stator and more is induced in the rotor as the load increases in all cases of unbalance. This implies that increase in copper losses accompany voltage unbalance, which may lead to increase heating, horsepower load and thus a reduced rated output power.

At over-voltage unbalance however, motor indicated fair performance at VUF of 2-4% with a load reduction of 50%. At rated voltage unbalance with VUF of 2-4%, good performance was observed on load reduction of 50%. Above VUF of 4% for all types of unbalance, motor operation became grossly inefficient and load reduction did improve operational performance of the induction motor.

From the findings of this paper, it has shown that there is a noteworthy difference in the performance of an induction motor under unbalanced source voltages compared to balanced source voltages. The results proved that the operational performance of an induction motor can be studied using simulated result from MATLAB® Simulink without going through the arduous analytical method. Since unbalanced conditions cannot be completely eradicated, it is therefore essential that motors be protected against all types of unbalances with NEMA, IEC and IEEE specifications and appropriately derated for effective and efficient performance.

References

[1] Ezer, D., Hanna, R. A. and Penny, J. (2002). "Active Voltage Correction for Industrial Plants", IEEE Trans. Industry Applications. 38(6), pp1641-1646.

[2] Bollen, M. H. J. (2000). "Understanding Power Quality Problems," IEEE Press, New York.

[3] Jouanne, A. and Banerjee, B. (2001). "Assessment of Voltage Unbalance", IEEE Trans. Power Delivery 16, pp 782-790.

[4] Jouanne .A. and Banerjee, B. October 2001 "Assessment of Voltage Unbalance," IEEE Transactions on Power Delivery, Vol. 16, No. 4, pp. 782-790.

[5] Kersting, W. H. (2001) "Causes and Effects of Unbalanced Voltages Serving an Induction Motor," IEEE Transactions on Industry Applications, Vol. 37, No. 1, pp. 165-170.

[6] De Almeida, A.T., Ferreira, F.J.T.E., Both, D., (2005). "Technical and Economical Considerations in the Application of Variable-Speed Drives with Electric Motor Systems." IEEE Transactions on Industry Applications 41, pp. 188–199.

[7] Johan, A. (2011). "Investigation of Issues Related to Electrical Efficiency Improvements of Pump and Fan Drives in Buildings." Thesis for the Degree of Doctor of Philosophy, Department of Energy and Environment, Chalmers University of Technology Goteborg, Sweden

[8] Wang, Y. J. (2001). "Analysis of Effects of Three-Phase Voltage Unbalance on Induction Motor with Emphasis on the Angle of the Complex Voltage Unbalance Factor." IEEE Trans. Energy Conversion 16(3), pp 270–275.

[9] Ibiary, Y. (2003). "An Accurate Low-Cost Method for Determining Electric Motors Efficiency for the Purpose of Plant Energy Management. IEEE Trans. Industry Applications 39(4), pp 1205-1210.

[10] Kothari, D. P. and Nagrath, I. J. (2004). Electric Machines, 3rd Edition, Tata McGraw Hill, New Delhi, India.

[11] Annette, J. and Banerjee, B. B, May 2000 "Voltage Unbalance: Power Quality Issues, Related Standards and Mitigation Techniques," Electric Power Research Institute, Palo Alto, CA, EPRI Final Rep..

[12] Woll, B.J, January/February 1975 "Effect Of Unbalanced Voltage on The Operation of Polyphase Induction Motors," IEEE Trans. Industry Applications, vol. IA-11, No. 1, pp. 38

[13] Schmitz, N. L and Berndt, M. M, February 1963. "Derating Polyphase Induction Motors Operated with Unbalanced Line Voltages," IEEE Trans. Power App. Syst., pp. 680–686

[14] Williams, J. W, April 1954. "Operation of 3-phase induction motors on unbalanced voltages," AIEE Trans. Power App. Syst., Vol. PAS-73, pp. 125–133.

[15] Seematter, S. C and Richards, E. F, Sept./Oct. 1976. "Computer Analysis of 3-Phase Induction Motor Operation Of Rural Open Delta Distribution Systems," IEEE Trans. Ind. Appl., Vol. IA-12, pp. 479–486.

[16] Muljadi, .E, Schiferl, R. and Lipo, T. A, May/June 1985. "Induction Machine Phase Balancing by Unsymmetrical Thyristor Voltage Control," IEEE Trans. Industry Applications, Vol. IA-21, No. 4, pp. 669–678.

[17] Lee, C. Y, June 1999 "Effects of Unbalanced Voltage on the Operation Performance of A Three-Phase Induction Motor," IEEE Trans. Energy Conversion, Vol. 14, No. 2, pp. 202–208.

[18] Smith, D. R, Braunstein, H. R, and Borst, J.D, April 1988. "Voltage Unbalance In 3 and 4-Wire Delta Secondary Systems," IEEE Trans. Power. Delivery, Vol. 3, No. 2, pp. 733–741.

[19] Krause P.C, 1986. Analysis of Electric Machinery, McGraw-Hill, 1986, New York.

[20] Alwash, J. H. H., Ikhwan, S.H., March 1995. "Generalised Approach to the Analysis of Asymmetrical Three-Phase Induction Motors", IEE Proceedings Electric Power Applications, Volume: 142, Issue: 2, Page(s): pp87-96.

Analysis of Partial Discharge Patterns for Generator Stator Windings

Tae-Sik Kong, Hee-Dong Kim, Tae-Sung Park, Kyeong-Yeol Kim, Ho-Yol Kim

Korea Electric Power Corporation (KEPCO) Research Institute, Daejeon, South Korea

Email address:
kongts@kepco.co.kr (Tae-Sik Kong), hdkim@kepco.co.kr (Hee-Dong Kim), parkts@kepco.co.kr (Tae-Sung Park), k2yeol@kepco.co.kr (Kyeong-Yeol Kim), hoyolkim@kepco.co.kr (Ho-Yol Kim)

Abstract: Forced outage of generators due to stator winding insulation failure can result in significant financial loss because of the high cost repair and loss of production. In recent years, the demand for insulation diagnosis is increasing to prevent unexpected failures, as the capacity of generators has increased. Insulation diagnosis is composed of the insulation resistance measurement, polarization index measurement, dissipation factor (DF) tip-up test, AC current increasing ratio measurement, and the partial discharge (PD) measurement. In this paper, the results of the PD measurement and PD pulse pattern analysis performed on a healthy generator and two generators that experienced dielectric breakdown failure during operation is presented.

Keywords: Generator, Stator Winding, Insulation Failure, PD Pattern, AC Current, Dissipation Factor

1. Introduction

Defects in the insulation system of generator stator windings can be produced during manufacturing or due to thermal, mechanical, electrical, or chemical deterioration when the generator is operated for a long time. With insulation degradation due to a combination of such operating stresses, voids can be formed inside the insulation material, and dielectric breakdown may eventually result from partial discharge activity [1-3].

A forced outage of a generator during operation due to dielectric breakdown of the stator windings requires long repair time and hence results in enormous economic loss. Therefore, the importance of insulation diagnosis that evaluates the soundness of generator stator winding insulators is increasing. There exists a dielectric strength test method for verifying whether dielectric strength is sufficient by applying 2E + 1kV (E : rated voltage) for 1 minute and checking whether the insulation withstands the voltage without failing. However, such a test method is only used in special cases such as a shop test or an acceptance test for quality assurance of newly manufactured windings and not for maintenance testing of generators in the field. One of the methods for testing the dielectric strength for these generators is a high-potential test, which applies a voltage of 1.25 ~ 1.5 times the rated voltage for 1 min, and in South Korea, this test method is only used in limited numbers of cases where the dielectric strength at failure or restoration is evaluated in the field [4-6]. Conventional dielectric tests for generator stator windings use a test voltage lower than the rated voltage and mainly use the insulation resistance measurement, polarization index (PI) measurement, AC current increasing ratio measurement, dissipation factor (DF) tip-up measurement, and partial discharge (PD) measurement [7-9].

Insulation diagnosis tests for generators stator winding with dielectric breakdown due to overheated copper conductors and abraded stator windings with induced external discharge noise of generators is carried out and presented in this paper. The PD patterns measured for normally deteriorated generators are analyzed, and the correlation between the causes of the defects and the PD patterns are shown.

2. Experimental Procedure

The polarization index (PI) was measured using a commercially available automatic insulation tester (Megger, S1-5010) at DC 5 kV in individual phases before applying AC voltage to the stator windings. Commercially available equipment, namely, Schering bridge (Tettex Instruments), coupling capacitor, and PD detector (Tettex Instruments, TE 571), were used to measure AC current, dissipation factor, and PD magnitude, respectively. The Schering bridge consists of a high voltage (HV) supply (Type 5283), a bridge (Type 2818), and a resonating inductor (Type 5285). A HV supply and control system (Tettex Instruments, Type 5284), Schering bridge (Tettex Instruments, Type 2816), resonating inductor (Tettex Instruments, Type 5288), coupling capacitor, coupling unit, and PD detector were used to measure AC current, dissipation factor, and PD magnitude in 15 kV generator stator windings. The HV supply and control system, Schering bridge, and resonating inductor were used to obtain the AC current and dissipation factor measurements. For PD measurements, AC voltage was applied to the generator stator winding through a connected HV supply and control system. The coupling capacitor (Tettex Instruments, 4,000 pF) amplified signals from the winding, which were sent to the coupling unit (Tettex Instruments, AKV 572) and then to the PD detector (Tettex Instruments, TE 571) that measures the magnitude and pattern of PD. The frequency band of the PD detector ranged from 40 to 400 kHz.

3. Test results and Discussion

3.1. Failure Due to Copper Conductor Overheating

This is a case where a large-scale water cooling steam turbine generator was tripped because of the destruction of the main insulation of the stator winding caused by cooling water supply discontinuance. The generator, which was manufactured with thermal class B insulation material with a maximum allowable temperature of 130 °C, experienced a discontinuance of the cooling water supply. The discontinuance was caused by an error in the coolant supply system for approximately 10 min during normal operation, and the temperature of the stator windings increased rapidly to approximately 200 °C. This caused the main insulation of the stator winding to fail, and a ground fault eventually tripped the generator. The melted compound of the connecting joint between the cooling water box and the stator winding caused by overheating is shown in Figure 1, and the failed insulation of the stator winding is shown in Figure 2.

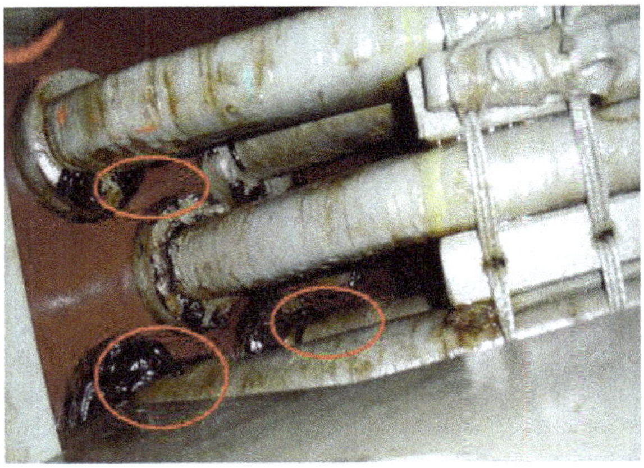

Figure 1. *Compound melting due to overheating.*

Figure 2. *Stator winding insulation failure.*

To check the insulation condition of the stator windings except for the bar with dielectric breakdown, the insulation resistance measurement, polarization index measurement, AC current increasing ratio test, dissipation factor tip-up measurement, and partial discharge measurement were performed. The AC current increasing ratio test consists of measurement of the increment in the current as the test voltage is increased, as shown in Figure 3. The dissipation factor tip-up test consists of measurement of the increment in the dissipation factor (tan δ) with respect to its value at the initial test voltage, as shown in Figure 4. The AC current increasing ratio (ΔI) and dissipation increment factor (Δtan δ) are closely related to the partial discharge. There is no partial discharge at low voltage, but partial discharge begins to occur at the void within the insulation system as the test voltage is increased, eventually increasing the values of the AC current and dissipation factor. Because these test measure the current and dissipation factor of the whole insulation system, it is used to find the average deterioration condition of the insulation material.

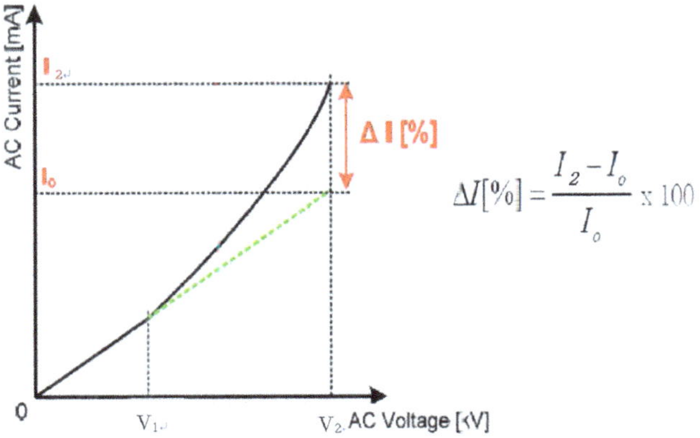

Figure 3. *Voltage vs. AC current.*

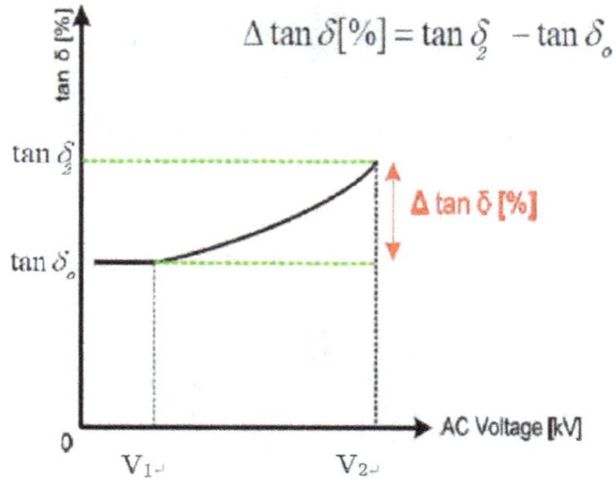

Figure 4. *Voltage vs. Dissipation factor.*

The line to line voltage of the generator was 26 kV, and test voltage was increased to 32.5 kV (125% of its line-line voltage) for AC dielectric withstand voltage test, while holding this voltage for one minute. The AC current and dissipation factor were measured in 1 kV increments as the test voltage was increased.

The AC current and DF measurement results at 15kV (phase voltage) were measured in two years prior to the failure. The same voltage of 15 kV was applied to measure the values of ΔI and Δtan δ in order to compare the test results. The results show that both ΔI and Δtan δ increased after the failure, as shown in Table 1. This implies that PD started to occur more than two years prior to the failure, and also implies that the number of voids, which cause discharging within the insulation system, increased because of stator winding thermal overheating.

Table 1. *Results of AC current increasing ratio and dissipation factor Tip-Up measurements.*

Phases	AC current (ΔI) [%]		Dissipation factor (Δtanδ) [%]	
	Before failure	After failure	Before failure	After failure
A	1.79	3.08	1.05	1.27
B	2.14	2.08	0.97	1.09
C	2.06	2.08	1.08	1.26

Table 2. *PD measurement results.*

Phases	Before failure		After failure	
	Discharge [pC]	Predominance	Discharge [pC]	Predominance
A	7,100		30,680	
B	6,900	+PD ≒ -PD	21,030	+PD < -PD
C	7,300		29,770	

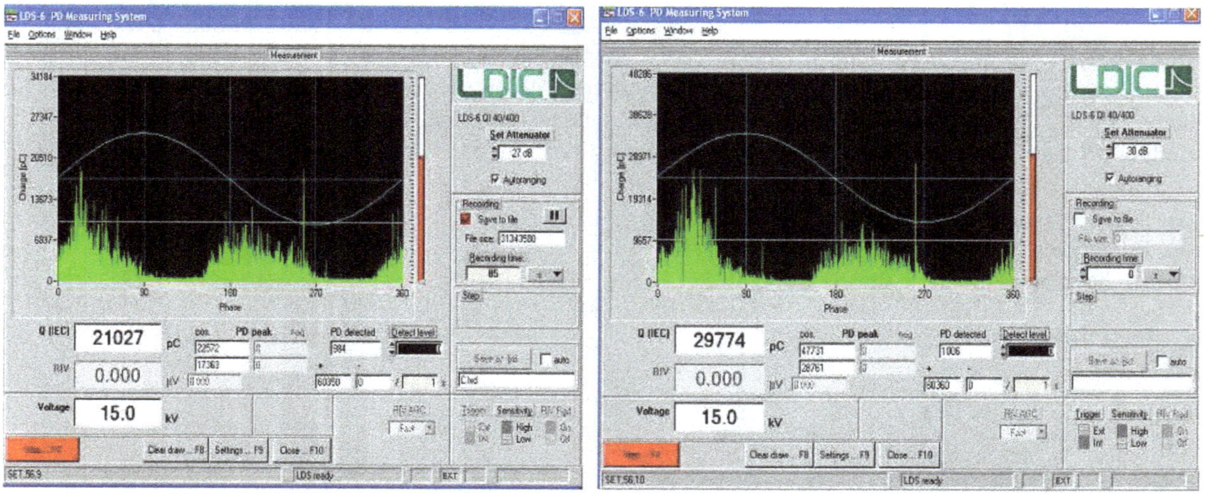

Figure 5. *PD pattern for copper conductor overheating.*

In the PD measurement, the maximum value of the multiple discharge pulses generated in the insulation system is measured. The largest discharge pulse is assumed to occur at the largest defect point, and the PD measurement is used to measure the level of PD activity in the largest defect in the insulation system. In the measurement, the voltage was 15 kV (phase voltage), and the results are shown in Table 2. The results show that the level of PD increased significantly after the failure and that there was also a change in the discharge pattern. Before the failure, the size and number of the negative discharge pulses, which appeared when the test voltage is positive, were similar to those of the positive discharge pulses, which appeared when the test voltage is negative. However, negative pulses were larger than the positive pulses after the failure due to thermal overheating, as shown in Figure 5. This pattern appears mainly in the gap between the copper conductor and strand insulation [10]. Furthermore, it was found that the insulation was not in good condition because the PD size was close to 30,000 pC or greater [11].

This pattern is produced because of delamination between the copper conductor and the insulation because of the differential thermal expansion between the copper conductor and stator winding insulation as a result of rapid thermal overheating.

3.2. Failure Due to Vibration on Stator Windings

This is a case where a ground fault occurred on the stator winding of an air cooled gas turbine generator manufactured with the global vacuum pressure impregnation (VPI) method. The global VPI type generator was manufactured with the side ripple spring, which holds the stator winding in position inside the slot, removed. The stator winding was inserted into the slot and immersed in resin without the ripple spring to reduce the size of the core of the generator for saving manufacturing time and cost. This method is widely used in the generator industry, to keep the power density high in the globally competitive market.

However, when the generator is started and stopped, the winding insulation is subject to thermal expansion and contraction because of Joule heating, and the resin that holds the stator winding in its position within the slots is separated. This creates a gap between the winding and the slot, and the stator winding starts to vibrate at 120 Hz because of the electromagnetic force of the rotor, further reducing the thickness of the insulation. This increases the length subject to vibration, accelerating the wear process and eventually causing dielectric breakdown. In Figure 6, it can be observed how the semiconducting layer is almost eliminated because of the slot vibration-induced wear on the surface of the stator winding. The semi-conductive layer is used to reduce the partial discharge between the stator winding and the slot [12].

Figure 6. *Wear in stator winding surface.*

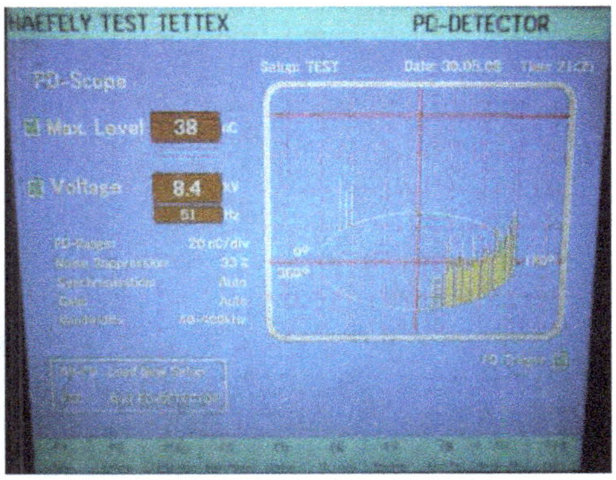

Figure 7. *PD pattern for slot discharge.*

Insulation diagnosis tests were performed on the remaining stator windings except for those that experienced dielectric breakdown. The PD measurement was carried out at rated phase voltage, and the measured PD magnitude and phase pattern is shown in Figure 7. The results show that the PD value was relatively high (38,000 pC), and the PD pattern showed that the positive discharge pulses, which occurred at a test voltage phase angle of 225 °, were produced more than negative discharge pulses, which occurred at 45 °, forming a slot discharge pattern [8, 9]. Most PDs are considered to occur at a gap between the stator winding and the slot, and it was also confirmed that the surface of the stator winding drawn from the slot was also significantly worn, as shown in Figure 6.

3.3. Internal Discharge Pattern of Ground-Wall Insulation

Figure 8 shows the PD measurements on a 22 kV steam turbine generator, which has been operating for approximately 20 years. According to the test results, the location of the main discharge is considered to be in the void inside the ground-wall insulation because the positive and negative PDs occur almost similarly. This is because the voids within the ground-wall insulation were not completely removed during the vacuum pressure impregnation process of the winding manufacturing or because voids were created inside the ground-wall insulation as a result of thermal deterioration. This discharge pattern due to the internal voids of the insulation material occurs during normal deterioration and is a frequent form of partial discharge pattern.

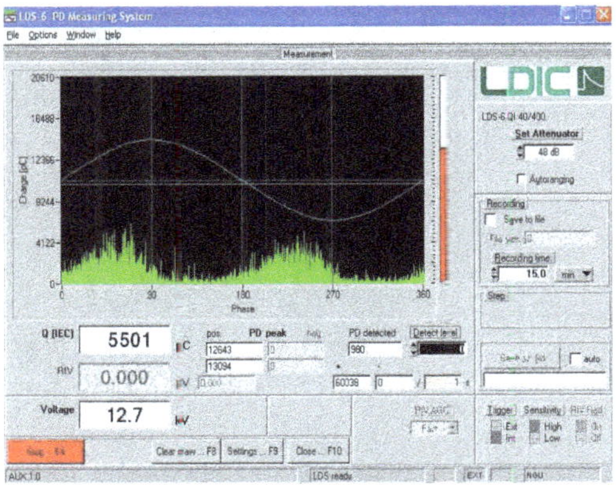

Figure 8. *PD pattern for internal discharge.*

3.4. Noise Occurrence Due to Gap Discharge

The screen of the PD measurement during a PD test on a generator that failed because of overheating in the copper conductor is shown in Figure 9. The location of the preliminary pulses are near 0° and 180°, and the size of the discharged pulses are almost identical. This pulse had a completely different shape with respect to the discharge pattern on insulation systems. This discharge pulse sizes being similar implies that the pulses occurred between particular electrodes. Because external noise pulses were detected in addition to the PD occurring at the electrical insulation, a visual inspection was performed to check the location that produced the noise. As a result, it was found that the contact between two isolated phase bus (IPB) bars, through which the generator output flowed for each phase, and the clip, which connected the two IPB bars to the common plate, was defective resulting in a discharge. This problem was rectified, and a test was conducted.

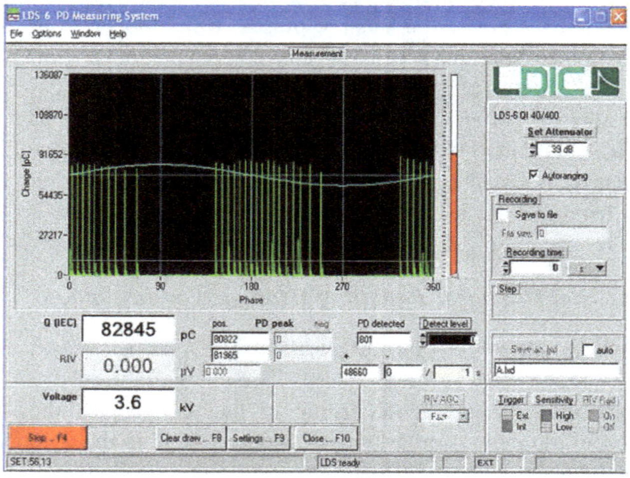

Figure 9. *PD pattern for gap discharge noise.*

Figure 10. *Gap discharge location.*

4. Conclusion

This study conducted PD measurements on four cases, two generators that had dielectric breakdown, one generator undergoing normal deterioration process, and the other generator with external noise discharge, which have shown different discharge patterns. The PD patterns were analyzed and the correlation between the causes of the defects and the discharge patterns were presented.

First, the discharge pattern at the time of the dielectric breakdown due to overheating of the winding, which resulted from the loss of coolant for the stator winding, was examined. In this case, the negative pulses were larger and more frequent than the positive pulses. This was a result of a separation phenomenon occurring between the insulating material and copper conductor because of rapid differential thermal expansion between the copper conductor and insulation. Furthermore, when compared to the measurement performed prior to failure, the AC current increasing ratio, dissipation factor, and PD levels all increased, which subsequently led to a rapid deterioration.

Second, a generator manufactured according to the global VPI. method without a side ripple spring that immobilizes the stator winding in a slot was studied. Repeated expansion and contraction occurred when the generator stopped operating, creating a crack between the winding and slot. The generated vibration in the generator impaired the semiconducting layer, which primarily resulted in negative pulses.

Third, the PD pattern of the generator with normal temperature aging appeared to be very similar for positive and negative pulses.

Fourth, the level and status of the discharge pulses occurring in a particular external gap instead of insulation system of the generator appeared to be almost identical.

It was found that the discharge patterns differed depending on the location of the defects in the stator winding. Hence, an analysis of the PD patterns of a generator will be useful in maintenance control and operation according to type of defect.

References

[1] Hee-Dong Kim, "Analysis of Insulation Aging Mechanism in Generator Stator Windings", Journal of the KIEEME, Vol. 15, No2, pp. 119-126, 2002.

[2] R. Morin, R. Bartnikas and P. Menard, "A Three-Phase Multi-Stress Accelerated Electrical Aging Test Facility for Stator Bars", IEEE Trans. on Electrical Conversion, Vol. 15, No. 2, pp. 149 ~ 156, 2000.

[3] H. Zhu, C. Morton and S. Cherukupalli, "Quality Evaluation of Stator Coils and Bars under Thermal Cycling Stress", Conference Record of the 2006 IEEE International Symposium on Electrical Insulation, pp. 384 ~ 387, 2006.

[4] H. G. Sedding, R. Schwabe, D. Levin, J. Stein and B. K. Gupta, "The Role AC & DC Hipot Testing in Stator Winding Ageing", IEEE Electrical Insulation and Electrical Manufacturing & Coil Winding Conference, pp. 455 ~ 457, 2003.

[5] "Recommended Practice for Insulation Testing of Large AC Rotating Machinery with High direct Voltage", New York : Institute of Electrical and Electronics Engineers, IEEE+ Std. 95-1977, pp. 13, 1977.

[5] IEEE Standard "IEEE Guide for Insulation Maintenance of Large Alternating-Current Rotating Machinery (10,000kVA and Larger)" IEEE Std 56-1997, pp. 12, 1997.

[7] Hee-Dong Kim, Tae-Sik Kong, Young-Ho Ju, Byong-Han Kim "Analysis of Insulation Quality in Large Generator Stator Windings", Journal of Electrical Engineering & Technology Vol. 6, No. 2, pp. 384-390, 2011.

[3] Claude Hudon and Mario Belec, "PD Signal Interpretation for Generator Diagnostics", IEEE Trans. on Dielectrics and Electrical Insulation, Vol. 12, No. 2, pp. 297 ~ 319, 2005.

[9] Y. Ikeda and H. Fukagawa, ""A Method for Diagnosing the Insulation Deterioration in Mica-Resin Insulated Stator Windings of Generator"", Yokosuka Research Laboratory Rep. No. W88046, 1988

[10] IEEE Standard "Trial-Use Guide to the Measurement of Partial Discharge in Rotaing Machinery", IEEE Std 1434-2000, pp. 40, 2000

[11] H. Yoshida and U. Umemoto, "Insulation Diagnosis for Rotating Machine Insulation", IEEE Trans. on Electric Insulation, Vol. EI-21, No. 6, pp. 1021-1025, 1986

[12] J.H. Dymond, N. Stranges, K. Younsi and J. E. Hayward, "Stator Winding Failures : Contamination, Surface Discharge, Tracking", IEEE Trans. on Industry Applications, Vol. 38, No. 2, pp. 577-583, 2002.

15

CSP-Biogas Combined Microgrid System for Rural and Remote Areas of Bangladesh

Atiqur Rahman, Miftah Al Karim

Department of Electrical and Electronic Engineering, American International University-Bangladesh (AIUB), Dhaka, Bangladesh

Email address:
atiqur160@gmail.com (A. Rahman), miftah.aiub@gmail.com (M. A. Karim)

Abstract: Power plays a vital role for a developing country like Bangladesh. Like the rest of the countries of the world, the demand of power is rising day by day in our country. But Bangladesh has been facing electricity shortage for many years. In our country, a major portion of total population still does not have the access to electricity. For becoming a developed country, Bangladesh has to overcome the problem of power crisis. Renewable energy can be a great source to solve this problem. Already some government and non-government organization are working on renewable sources like solar energy, bioenergy, wind energy, etc. In this paper we mainly focused to develop and implement microgrid system with CSP-biogas combined power plant for providing electricity in rural and remote areas of Bangladesh.

Keywords: Concentrating Solar Power, Heat Transfer Fluid, Concentrating Solar Power with Biogas Plant, Collectors

1. Introduction

Energy is a key to socio economic development of any country. Bangladesh is a small country of 1,47,570 square kilometers having more than 160 million people would require gigantic amount for its development. Nowadays almost 59.6 percent of Bangladeshi people are under the connection of the electricity power grid that means 41.4 percent people are does not access to power grid connection. The power supply is not sufficient to meet the peak demand in Bangladesh. In rural areas, only about 42 percent populations have grid electricity connection where about 58 percent of that out of grid connection [1]. Rural Electrification Board in Bangladesh (REB) provides electricity connection in many rural areas. Every year about 40000 new consumers are access to electricity; it would take about 40 years to provide electricity connection for all people in Bangladesh [2]. Many rural people who live in isolated areas, they are totally out of grid connection. This paper set to develop and implement the initiative step. Bangladesh is one of the sunniest parts of the world [3]. CSP-biogas combined power plant will be a best solution for isolated area of Bangladesh.

2. Concentrating Solar Power (CSP)

Concentrated Solar Power (CSP) is a technology which produces electricity by using mirrors or lenses to concentrate a large area of sunlight into a small area. In this system the concentrated light is converted to heat, which produces steam. Then stream drive a stream turbine that connected to an electric power generator. CSP are most promising system for microgrid application. CSP utilizes three technological approaches: Parabolic solar collectors systems, Power tower system and Dish engine systems.

2.1. Parabolic Trough System

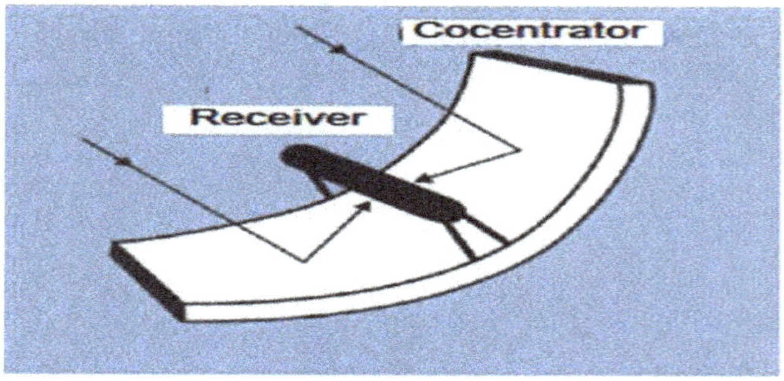

Figure 1. *Parabolic through system.*

These types of solar concentrators use large U-shaped reflectors. It contains oil filled pipes that running along their central point. The focus sunlight heats the oil inside the pipes. The hot oil, which make steam that run conventional steam turbines and generator. Fig. 1 shows the parabolic trough system.

2.2. Power Tower System

Power tower system is called a central receiver system. In this systems concentrate sunlight focus on a central receiver on top of a tower. The receiver contains a fluid deposit. The working fluid in the receiver is heated with high temperature. The hot fluid can be used to make steam for generation of electricity and stored for later use. So power tower systems can be produce electricity during peak periods on cloudy days or few hours after sunset. Fig. 2 shows the power tower system.

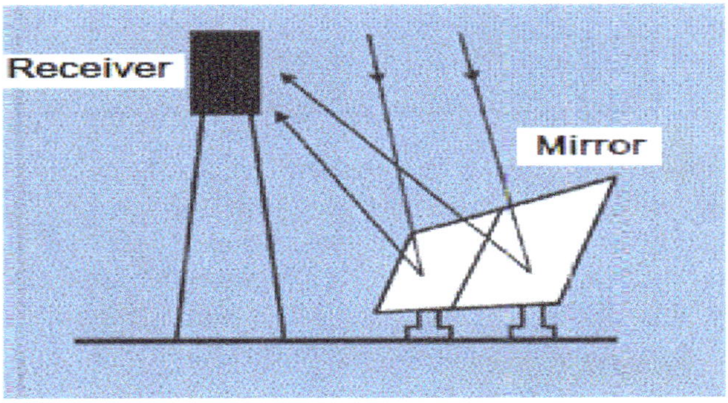

Figure 2. *Power tower system.*

2.3. Dish Sterling System

Dish sterling system use mirrored dishes to focus and concentrate light onto a receiver. The receiver embeds at the central point of the dish. The receiver is compact into an external combustion engine. The engine has thin tube that contains hydrogen or helium gas. Fig. 3 shows the dish sterling system.

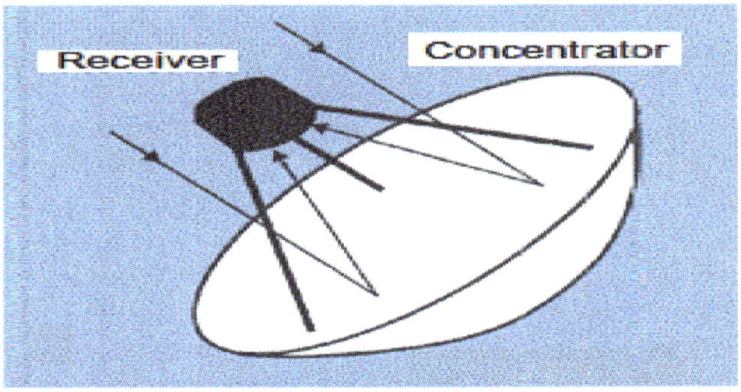

Figure 3. *Dish sterling system*

When concentrated light falls on the receiver, it heats the gas inside the tubes with very high temperature, which causes hot gas to expand inside the cylinder. This expanding gas runs the pistons. The piston rotates a crankshaft. Then crankshaft drives an electric generator.

3. Biogas

Biogas is an anaerobic digester that produces from organic matter by a consortium of bacteria. Biogas is mixture of methane (CH4), carbon dioxide (CO2), hydrogen sulphide (H2S) and varying quantities of water. Methane is the most valuable component. Natural gas consists of 80%-90% methane. Biogas can produce 55%-75% methane which can be increased to 80%-90% with free liquid [4]. Biogas can be used for heating, electricity production and many other operations.

4. Profile of Thanchi

For research purpose, a survey was done in a remote upazila Thanchi in the district of Bandarban. It is located at 21.78621°N 92.4278° E. Thanchi is located about 85 km away from Bandarban city. Total area of Thanchi is about 1020.82 sq.km. There are 4 unions and 178 villages in Thanchi. Total population of thanchi is about 27586. Population density of Thanchi is 27 per sq.km [5]. Thanchi is one of the most remote areas of Bangladesh. There are no electricity connections in Thanchi. The nearest grid electricity line is about 34 kilometers far from Thanchi. People need to alternative solution for energy, such as firewood and oil lamp. Now a few numbers of people are using solar home system and some diesel generators are using to serve electricity for only bazaar areas. All the diesel generators are runs from 6pm to 10pm. There is no biogas plant in Thanchi. There is a small river named sangu. It contains small amount of water over the year. So micro hydro is not possible at this area. The average air temperature is 24.10c. Wind speed is around 2.9m/s which is not possible to run the micro wind generator, finally it can't be installed at that area. Daily radiation of sun in Thanchi is about 4.69KWh/m2/c which is good for running CSP plant.

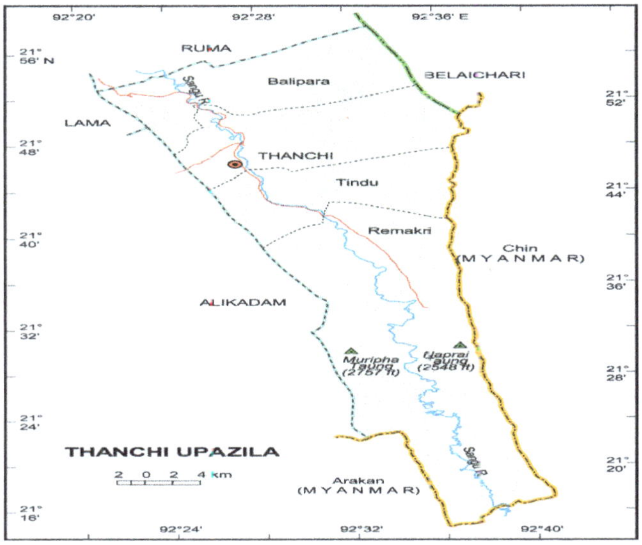

Figure 4. *The area map of Thanchi.*

Table 1. *Weather information of Thanchi [6].*

Month	Air temperature °c	Relative Humidity %	Daily adiation horizontal KWh/m²/c	Atmospheric pressure in Hg(0°c)	Wind speed /s	Earth temperature °c	Heating degree-days 18 °c	Cooling degree-days 10 °c
January	19.1	52.7	4.80	29.1	2.5	18.9	0	282
February	21.6	53.o	5.32	29.1	2.6	22.2	0	325
March	24.5	58.5	5.84	29.0	2.7	26.4	0	451
April	26.0	68.2	5.92	29.0	2.7	28.0	0	479
May	26.4	77.2	5.31	28.9	2.8	28.1	0	510
June	26.4	85.6	3.86	28.8	4.0	26.9	0	491
July	26.0	87.1	3.81	28.8	4.0	26.3	0	496
August	26.0	86.3	3.95	28.9	3.6	26.4	0	495
September	25.7	84.4	4.25	28.9	2.8	26.0	0	471
October	25.0	78.5	4.40	29.0	2.2	25.0	0	466
November	22.7	69.8	4.34	29.1	2.3	22.3	0	382
December	20.0	59.3	4.48	29.1	2.3	19.4	0	309
Annual	24.1	71.8	4.69	29.0	2.9	24.7	0	5157

Livestock Population of Thanchi

Total number of cows and buffaloes 5649, total number of goats 2628, total number of sheep 167, total number of hens 20081, total number of ducks 625 in Thanchi [7].

Table 2. *Total livestock in Thanchi.*

Species	Number
Cow&buffalo	5649
Goat	2628
Sheep	167
Hen	20081
Duck	625
Total	29150

Table 3. *Load calculation of Thanchi.*

Load type	Kilowatt(Kw)	Megawatt hour per day (Mwh/d)
Total load for house	1003.632	11.868
Total load for school	17.70	0.1416
Total load for shop	5.34	0.0772
Total load for hospital	0.89	0.00187
Total load for BGB and police station	2.225	0.02835
Total load of Thanchi	1029.787	12.117

5. Load Assumption of Thanchi

For designing a CSP-biogas combined system first task is to determine the approximate load that will be connected to the system. Each and every part of the CSP-biogas plant system will be designed according to the load requirement. There is 4872 house in Thanchi. Average members of each house are 4.73. In Thanchi there are 14 government primary school, 2 registered primary school, 22 non registered primary school, 77 NGO school, 2 government secondary school, 1 non-government secondary school, 4 BGB camp, 1 community health centre and 2 bazaar with 50 shops. In an average 4 pieces bulb and 2 fans are enough for each an every house at Thanchi. All bulbs are 14 watt energy saving bulb and fans are 75 watt. Each bulb will runs 6 hours per day and fan will runs 14 hours per day. All school will run 2 fans 8 hours per day. Each shop will runs 2 bulbs 5 hours and 1 fan 15 hours per day. In hospital 10 fans will run 24 hours and 10 bulbs will run 5 hour. Each BGB camp 5 fans will runs 14 hour and 5 bulbs will run 6 hours per day.

6. Proposed CSP-Biogas Power Plant

For the supplying of electricity in Thanchi we proposed some small scale CSP plant and some biogas power plant. Here we have considered some 60KW CSP plant and some 20KW biogas power plant. Then run CSP and biogas together for making combined power plant. Thanchi is a hilly area so firstly we have to select some plane area for installation CSP plant. For installation of biogas power plant we have to select some density area where number of cows is enough to run this type of power plant.

6.1. 60KW CSP Plant

The system used here includes parabolic trough collectors and an organic rankine cycle (ORC). This system has a storage system. For developing 60KW small scale plant, at first a steam engine has been considered. This steam engine needs to supply superheated steam that temperature is more than 300^0c. Instead an ORC in the right power range which need only an inlet temperature of about 170^0c. In this system 3 parallel collector rows and 12 rows in series collector have been selected [8]. Each collector with an aperture width of 2.37m and length of 5.95, that total of 428.4m for the 12 collectors with a distance about 7.2m. The absorber pipe diameter is about 38.4mm [9]. For installation of a 60 KW CSP plant a ground of 27000m^2 is needed.

6.2. 20KW Biogas Power Plant

Biogas can be converted directly into electricity using a fuel cell or by using combustion engine. Generally combustion engine are more useful for biogas power generation. The combustion engine used biogas as fuel.

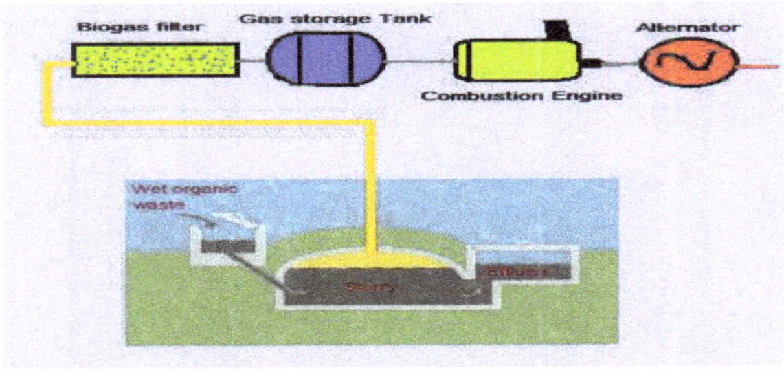

Figure 5. Biogas power plant.

In this system at first organic matter mixed with an equal quantity of water in mixing tank. That called slurry. Then slurry is feed into the digester through the inlet chamber. Then digester produces biogas. After filtering of this biogas it store in a gas storage tank. This gas storage tank supply biogas as fuel to combustion engine. The combustion engine run a generator and produces electricity. For running of a 20KW biogas power plant 18700L methane is needed [10]. One cow that can produce up to 250L methane [11]. So 75 cow enough for a 20 KW biogas power plant. This type of power plant can produce 320 KWh of electricity.

7. Microgrid Design for Thanchi

Thanchi is a hilly area so it is very difficult to buildup and maintains the grid connection. For hilly area the cost of transmission line is also very high. Table 4 shows the cost of transmission line.

Table 4. Cost of transmission line [12].

Voltage level	Cost per km (USD)	
	Flat surfaces	*Hilly areas*
11 KV	3200	5600
33 KV	11200	16000
66 KV	25600	38400
132 KV	32000	44000

Main grid connection is far away from Thanchi. In Bangladesh generally 132KV transmission line is used for long distance transmission of electricity. Table 5 shows the distance from nearest grid and transmission line cost for four union of Thanchi.

Table 5. Distance from nearest grid of Thanchi and its cost [13].

Name of union	Distance from nearest grid (km)	Cost for 132KV line (USD)
Balipara	34	1496000
Thanchi Sadar	52	2288000
Tindu	64	2816000
Remakri	77	3388000

Microgrid can be a great solution for supplying electricity in Thanchi. Single microgrid is also difficult and costly for this area. So five individual microgrid systems are proposed for removing of this problem. Those five microgrid can be covered about 90% of total population. Fig.6 shows the energy consumption of five microgrid systems.

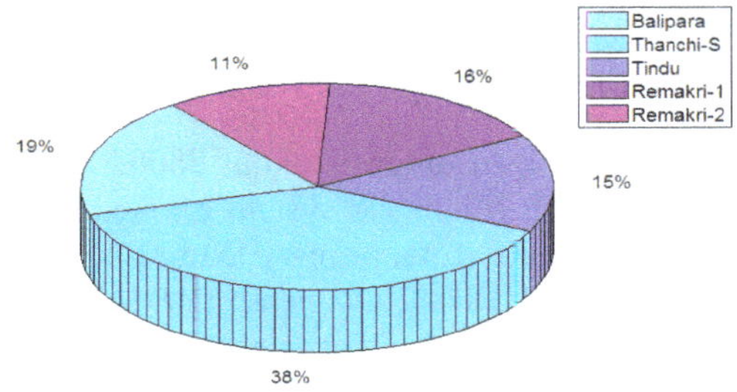

Figure 6. Energy consumption of Thanchi.

7.1. Microgrid Connection Design for Balipara Region

Total population of Balipara is about 5447.There are 56 school and one bazaar in Balipara. This microgrid design can cover about 4500 people. The primary load for Balipara is 2.76 MWh/d and 207.47 KW peak. Each CSP plant is 60 KW and biogas power plant is 20 KW.

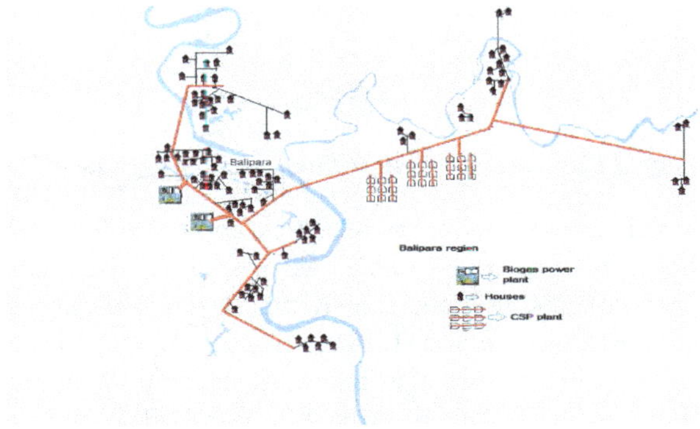

Figure 7. Microgrid design of Balipara region.

For 207 KW power supplies three CSP and two biogas power plant have to be considered. Here each house in the design is considered equal to 15 actual houses.

7.2. Microgrid Connection Design for Thanchi Region

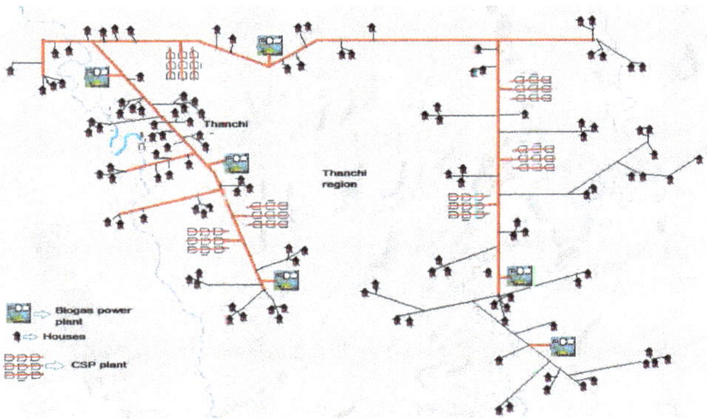

Figure 8. Microgrid design of Thanchi region.

Total population of Thanchi sadar union is about 8040.It contains six schools, one hospital, one police station and one BGB camp. Microgrid system of Thanchi region can cover total 7500 people of Thanchi and also can cover 3000 people of Tindu and 300 people of Balipara. Load of this region about 5.581MWh/d and 473.039KW peak. For 473.039KW power supplies six CSP plant and six biogas power plant have to be considered.

7.3. Microgrid Connection Design for Tindu Region

Total population of Tindu union is about 7800.Tindu has only one primary school. Microgrid system of Tindu region can cover total 4300 people. Total primary load of this region is about 2.22MWh/d and 187.42KW peak. For 187.42 KW power supplies three CSP and one biogas power plant have to be considered.

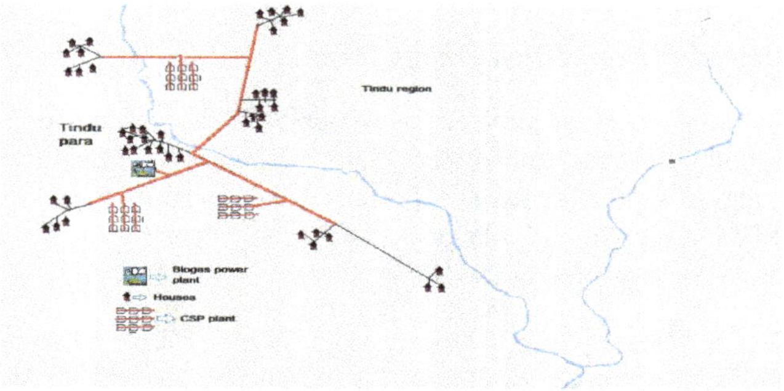

Figure 9. *Microgrid design of Tindu region.*

7.4. Microgrid Connection Design for Remakri Region-1

Total population of Remakri union is about 7800.It has three school and one BGB camp. Microgrid for Remakri region-1 can cover 4000 people of Remakri and 500 people from Tindu. Remakri region 1 can cover about total 4500 people. Total primary load for Remakri region-1 is about 2.327MWh/d and 196.878KW peak. For 196.878KW power supplies three CSP plant and one biogas power plant have to be considered.

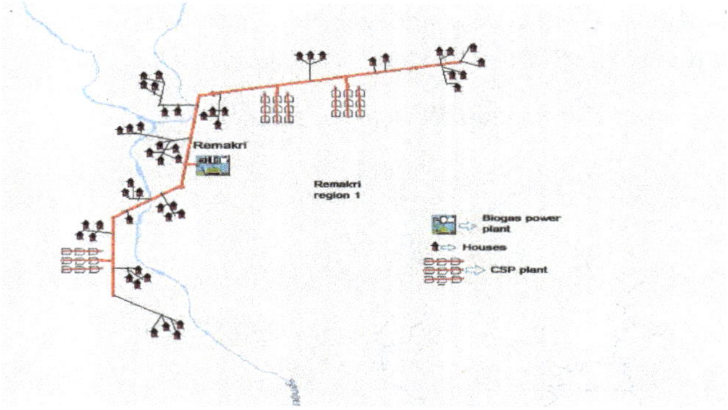

Figure 10. *Microgrid design of Remakri region-1.*

7.5. Microgrid Connection Design for Remakri Region-2

Total population of Remakri region-2 is about 3200.Total load for Remakri region 2 is 1.64MWh/d and 139.36KW peak. For total 139.36 KW power supplies two CSP plant and one biogas have to be considered.

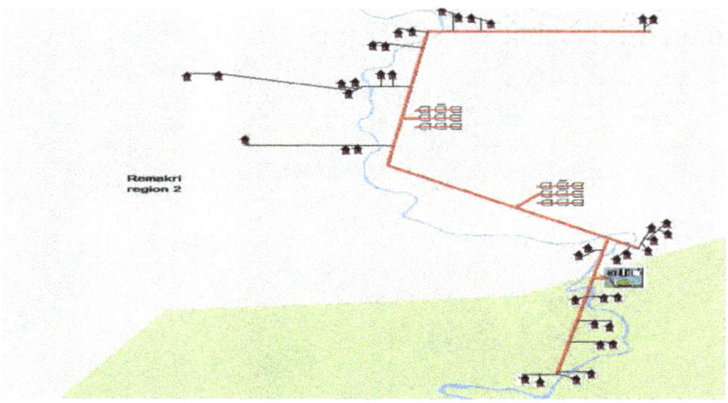

Figure 11. *Microgrid design of Remakri region-2.*

8. Cost Calculation of CSP-Biogas Power Plant

It is very important to understand the information of the relative costs and benefits of renewable energy technologies to arrive at an accurate assessment of most appropriate renewable technologies. Cost can be measure including financing cost, implement cost (solar reflectors, biogas plant), total installation cost, fixed and variable operation and maintenance costs.

8.1. Cost Calculation of 60KW CSP Plant

Initial cost of CSP plant is higher than fossile fuel plant. The initial cost of CSP plants approximately four-fifths of the total cost [14]. CSP plant with thermal energy storage is more expensive, but allowed higher capacity factors. The operation costs of CSP plants are lower than fossil fuel power plants. The operation and maintenance cost depend on replace mirrors and receiver, cost of mirror washing, etc. This microgrid system used 17 CSP plant. Each CSP plant has considered 60KW plant.

Table 6. *60KW CSP plant cost calculation by using RETScreen 4.*

RETScreen Energy Model-Powerproject		
Proposed case power system		**Incremental initial costs**
Technology Solar thermal power	Solar thermal power	
Power capacity	60 KW	
Manufacturer		
Model		$300000
Capacity factor	50%	
Electrcity exported to grid MWh	263	
Electricity exported rate $ per MWh	100	

According to the RETScreen 4 total installation cost for 60KW power plant is about $300000. Operation annual cost for CSP plant can be between 0.02%-0.035% of total installation cost [15]. Here from RETScreen 4 the operation and maintenance cost is about $3000. Electricity generation cost is about $0.10/KWh. One 60KW CSP plant can generate upto 263 MWh electricity per year that means

720.5 KWh of electricity per day.

8.2. Cost Calculation of 20KW Biogas Power Plant

Cost of biogas power plant depends on cost of installation of plant, cost of maintenance, cost of combustion engine, etc. Total installation cost of a 20 KW biogas power plant is about $63011[16]. Generation cost of biogas power is about $0.11/KWh [17].

9. Result Analysis

The current generation cost of electricity in Bangladesh is about $0.079/KWh [18]. For gas based power plant cost varies from $0.062 to $0.070/KWh [19]. In PV power plant generation cost is about $0.51/KWh [20]. Government purchases electricity at the rate of $0.29/KWh from Quick Rental Power Plant (QRPP)[21]. For CSP-biogas combined system, cost is found $0.10/KWh for CSP and $0.11/KWh for biogas power plant, which is better than PV plant and quick rental power plant. Fig. 12 shows the cost comparison.

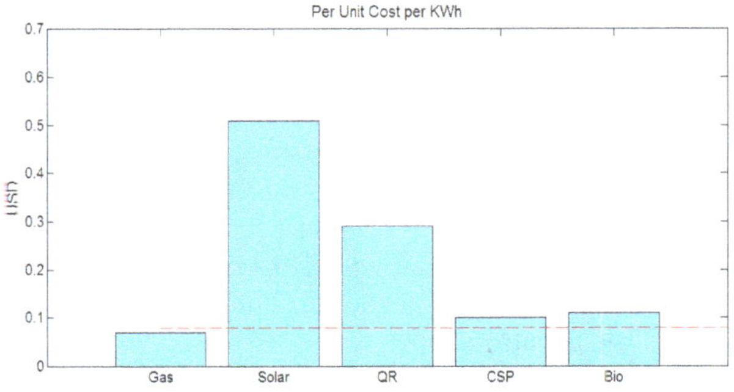

Figure 12. *Cost comparison of CSP and biogas plant with other power plant.*

10. Conclusion

The performance of CSP-biogas combined microgrid system is more economical. This microgrid system can be a great solution for removing electricity crises in rural and remote area of Bangladesh. Thanchi is a remote area and CSP-biogas combined microgrid can be very useful. Population density of Thanchi is not same in all areas. Some areas are highly density and some areas have no population. So cost of single microgrid is more for coverage total area. We have taken some population areas and proposed CSP-biogas combined microgrid system. But for implementation of this project some suitable places have to be selected for installation CSP and biogas power plant.

Acknowledgment

Authors would like to thank American International University Bangladesh (AIUB) for their financial and logistical support and for providing necessary guidance.

References

[1] Bangladesh Energy Situation - energypedia.info.

[2] Expanding Renewable Energy in Bangladesh – CLIMATE HIMALAYA.

[3] K. Anam, H. A. Bustam, "Power Crisis & Its Solution Through Renewable Energy in Bangladesh," in Multidisciplinary Journal in Science and Technology, September edition, 2011.

[4] https://en.wikipedia.org/wiki/Biogas.

[5] Bangladesh National information window. Bangladesh.gov.bd.

[6] RETScreen plus software "Climate database".

[7] Agriculture Census 2008.

[8] D. Kruger,A. Kenissi, S. Dieckmann, C. Bouden, A. Baba, A. oliveira, H. Soares, E. Rojas Bravo, R. Ben cheikh, F. Orioli, D. Gasperini, K. Hennecke, H. Schenk,"Pre-Design of a Mini CSP plant"International Conference on Concentrating Solar Power and Chemical Energy System, Solar PACES 2014.

[9] Leonel Reyes Ochoa "Engineering Aspects of a Parabolic Though Collector Field with Direct Steam Generation and Organic Ranking Cycle" Cologne, Germany, October 2014.

[10] A model biogas plant that produce 320 units of power a day.

[11] How much methane does a cow produce in one day.

[12] Standards/Manual/guidelines for small hydro development, ministry of new and renewable energy govt. of India, June 2011.

[13] Off grid mapping of Bangladesh using gis tools.

[14] IRENA "RENEWABLE ENERGY TECHNOLOGIES: COST ANALYSIS SERIES".

[15] RENEWABLE ENERGY TECHNOLOGY: COST ANALYSIS SERIES, Volume 1: Power sector Issue 2/5, June 2012.

[16] THE HINDU, http://www.thehindu.com/todays-paper/tp-national/tp-karnataka/a-model-biogas-plant-that-produces-320-units-of-power-a-day/article3879626.ece.

[17] RETScreen International 50 kW - Biogas / Canada.

[18] The Daily Star "Power utility press for big hike in tariff".

[19] Mustafa, K. Mujeri, Tahreen Tahrima Chowdhury, "Quick Rental Power Plants in Bangladesh: An Economic Appraisal," Bangladesh Institute of Development Studies, June 2013.

[20] RETScreen International Power-Photovoltaic-1000kw/Germany.

[21] Quick Rental power plant project failing in Bangladesh, https://bangladeshstudies.wordpress.com/2012/04/23/quick-rental-power-plant-project-failing-in-bangladesh/.

Trapezoidal control of a coiled synchronous motor optimizing electric vehicle consumption

Aicha Khlissa, Houcine Marouani, Souhir Tounsi

School of Electronics and Telecommunications of Sfax, Sfax university (B.P. 1163, 3018 Sfax-Tunisie, Sfax, Tunisia

Email address:
aichakhlissa@gmail.com (A. Khlissa), Houcine.marouani@isecs.rnu.tn (H. Marouani), souhir.tounsi@isecs.rnu.tn (S. Tounsi)

Abstract: In this paper, we present a systemic trapezoidal control methodology of a coiled rotor axial flux synchronous motor dedicated to electric traction, taking into account of several constraints such as the speed limit, the energy saving, the cost of the power chain and the reliability of the whole system. Indeed the control law developed allows to impose the electromotive forces in phase with the phase currents, which reduces vehicle consumption. Also based on the technique of overfluxing during periods of high acceleration to reduce the phase current of the motor and then the vehicle consumption. This optimization technique is based on the increase of the excitation current for a given torque, thereby increasing the electric motor constant. Therefore, the phase current is lowered leading to a reduction in consumption. During the phases of constant speed operation and high decelerations, the value of the excitation current is calculated iteratively to minimize the consummation. Finally, the results obtained are with good level which encourages the electronic integration phase of this control law.

Keywords: Trapezoidal Control, Coiled Rotor Motor, Controlling Parameters, Systemic Control, Electric Vehicles

1. Introduction

The production of electric vehicles in large series suffers from their low autonomy, their high costs relative to combustion vehicles, and the infrastructure of batteries charging problem [1-3]. In this context, this paper addresses the problems of reducing the cost and increasing autonomy for a given stored energy. Indeed, our choice was directed towards the axial wound rotor motors with smooth pole, which is in a low-cost structure of production. Regarding the static converter, our choice fell on a structure to electromagnetic switches to low cost of production compared to its equivalent structure with IGBTs [1-3]. A trapezoidal control approach optimizing the consumption seen in increasing autonomy is developed. This approach can impose control current in phase with the back electromotive forces and use the over-fluxing technique during phases of strong accelerations to reduce consumption. The excitation current is optimized for the operating phases at constant speed and during deceleration phases as to optimize autonomy.

The paper briefly describes the choice and the design principle of the motor-converter at first time and describes the trapezoidal control strategy optimizing autonomy.

2. Power Chain Structure

Several configurations of power chain are shown in the literature. We cite as examples:

- The four-engine wheels configuration to direct mechanical linkage or gears.
- The configuration with two motors front or rear to direct connection or with gear.
- The single-engine configuration with mechanical differential transmission more gears or gearless. This configuration is chosen for our application because it offers the advantage of low cost, because the manufacture of a single motor is less expensive than many engines. This configuration also avoids the problem of slippage since it is impossible to control several motors at the same speed.

The power chain structure is illustrated by the figure 1 [4].

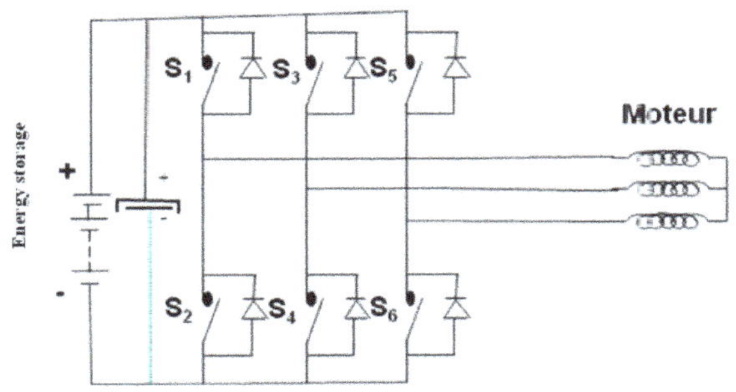

Figure 1. Power chain structure.

Usually the static converter is with IGBTs. In our case we have replaced the transistor IGBT by electromagnetic switch to push the multiple disadvantages of IGBTs.

During the phases of acceleration and constant speed operation, the motor is driven by the power converter with electromagnetic switch according to the scalar control strategy imposing the motor phase current in phase with the electromotive force, leading to a minimization of energy consumption [5], [6].

3. Systemic Interactions

The sizes of the motor and the converter as well as the parameters of order of the power chain are calculated according to one methodology of conception holding in account of the following interactions systems [1-12]:

- The sizing torque is calculated to the starting of the vehicle on a slope, iteratively by the method of the genetic algorithms while respecting the thermal constraints of the whole motor-converter.
- The magnetic inductions in iron are chosen close to the bends saturation to reduce the mass and thereafter the cost of the motor.
- The speed amplification report of the reducer is calculated in seen to reach the maximal speed of the vehicle with a good interpolation of the reference voltages and a reduction of the ripple torque.
- The continuous bus voltage is calculated to maximal speed to have the possibility to cover a large beach of speed.

4. Converter Structure

The static converter is a two-level inverter voltage. This structure is the least expensive compared to others and offers good quality of voltages and currents wave-forms, which leads to a good dynamic characteristic of EVs. Two inverter types are studied, the IGBT converter structure and the electromagnetic switches converter structure. The latter structure has the disadvantage of low switching frequency (Below 150 Hz), but it is less expensive and does not pose the problem of floating potential, since each inverter arm is controlled by a single electro-magnet. Against, the IGBT structure offers the possibility to achieve a switching frequency of 8000 Hz which leads to a good quality of the dynamic characteristic of EVs, it present a lot of disadvantages which can be cited as examples :

- Energy losses leading to a reduced range for a stored energy also establishes the temperature rise in the transistors and diodes leading to the incorporation of a cooling system in most cases.
- The problem of potential-floating leading to a complication of the electronic control circuit.
- The intervention of capacity Trigger-emitter, Trigger-collector and Collector-emitter. These capacities occur especially at high frequencies leading to a deterioration of the quality of control signals and subsequently to performance degradation of the overall drive system
- The problem of static and dynamic Luch-up generally leading to the deterioration of the converter. In this paper we only present the design process of the converter with electromagnetic switches. Design methods of the IGBT inverter are highly processed and presented in the literature [3].
- The design parameters of the generator coil are shown in figure 2.

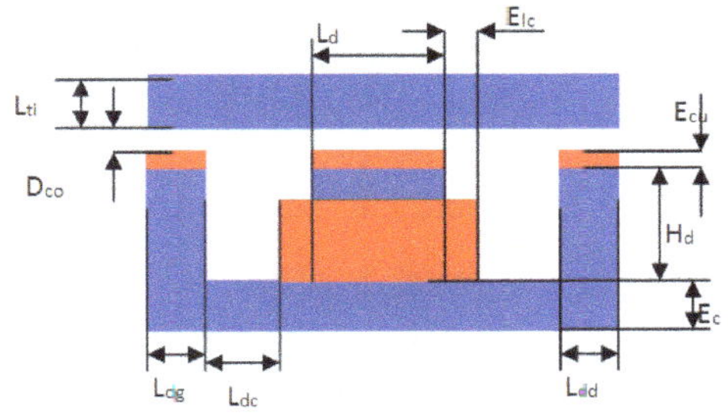

Figure 2. Design parameters of the generator coil.

For the electromagnetic switches converter structure, the electromagnet is a modular structure. Indeed, several modules can be stacked either in series or in parallel to increase the attraction force, and thereafter the opening and closing frequency of the contacts of the static converter.

The stack in parallel has the advantage that the frequency of closing and opening of the switches is clearly higher in regard to the stack in series structure, because these actions are performed by action of the attraction force of two generator coils.

The parallel stack of two modules is illustrated in figure 3 [3].

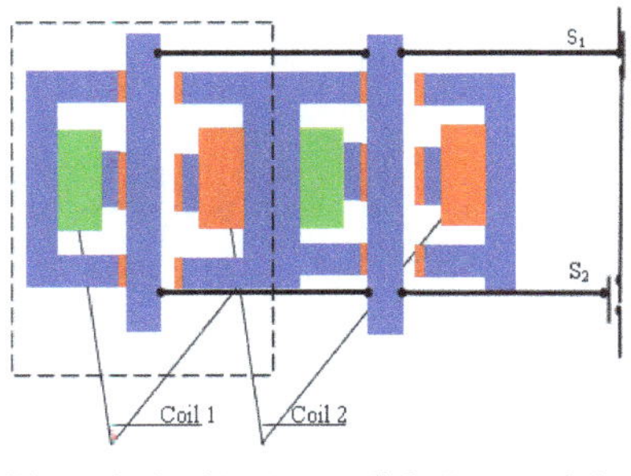

Figure 3. Stacking in parallel of two modules.

The equation of movement of the stem is deducted from the fundamental relation of the dynamics [3]:

$$\text{nmp} \times M_t \times \frac{dv}{dt} = \text{nmp} \times \frac{\mu_0 \times \mu_r}{4} \times \frac{I^2 \times N_{sb}^2}{(E_{cu}+D_{co}-X_t)^2} \quad (1)$$

$$v = \frac{dx_t}{dt} \quad (2)$$

Where v and M_t are respectively the velocity and the mass of the moving rod, nmp is the number of modules, N_{sb} is the spires number of the generating coil, E_{cs} is the thickness of the copper layer, D_{co} is the opening of the movable stem, x_t displacement of the mobile stem and μ_r is the relative permeability of copper.

This structure presents the time of opening and the one of closing weakest, since closing and opening of the power contacts take place joint-stock of two generating coils. The increase of the number of module increases the frequency of closing and opening of contacts, what brought us to choose only one module to push the problem of increase of the cost of the structure, slightly.

For only one module, the time of closing and opening is estimated to:

$$T_{on}=T_{off} = 3.74\text{ e-3 s}$$

From where the frequency of closing and to the opening is estimated to:

$$F_{ri} = \frac{1}{T_{on}+T_{off}} = \frac{1}{2\times 3.74e-3} = 133.7\text{ hz} \quad (3)$$

In conclusion, the structure to parallel stacking with only one module is chosen for the continuation of survey.

5. Motor-Converter Model

The motor is supplied by a voltage inverter at two levels. IGBT transistors are replaced by ideal electromagnetic switches, to reduce losses and sink multiple disadvantages of IGBTs, such as:

- Tail current.
- The problem of floating mass leading to a complication of the control circuit.
- Static and dynamic Luch-up phenomenon leading to a deterioration of the converter in the majority of cases.
- Switching and conduction losses significant.
- Heat IGBT transistors, leading to the need for the integration of a cooling system in the most cases.
- Each phase of the motor is equivalent to a resistor in series with an inductance and a back electromotive force. The model of the three phases is described by the following equations [4], [5]:

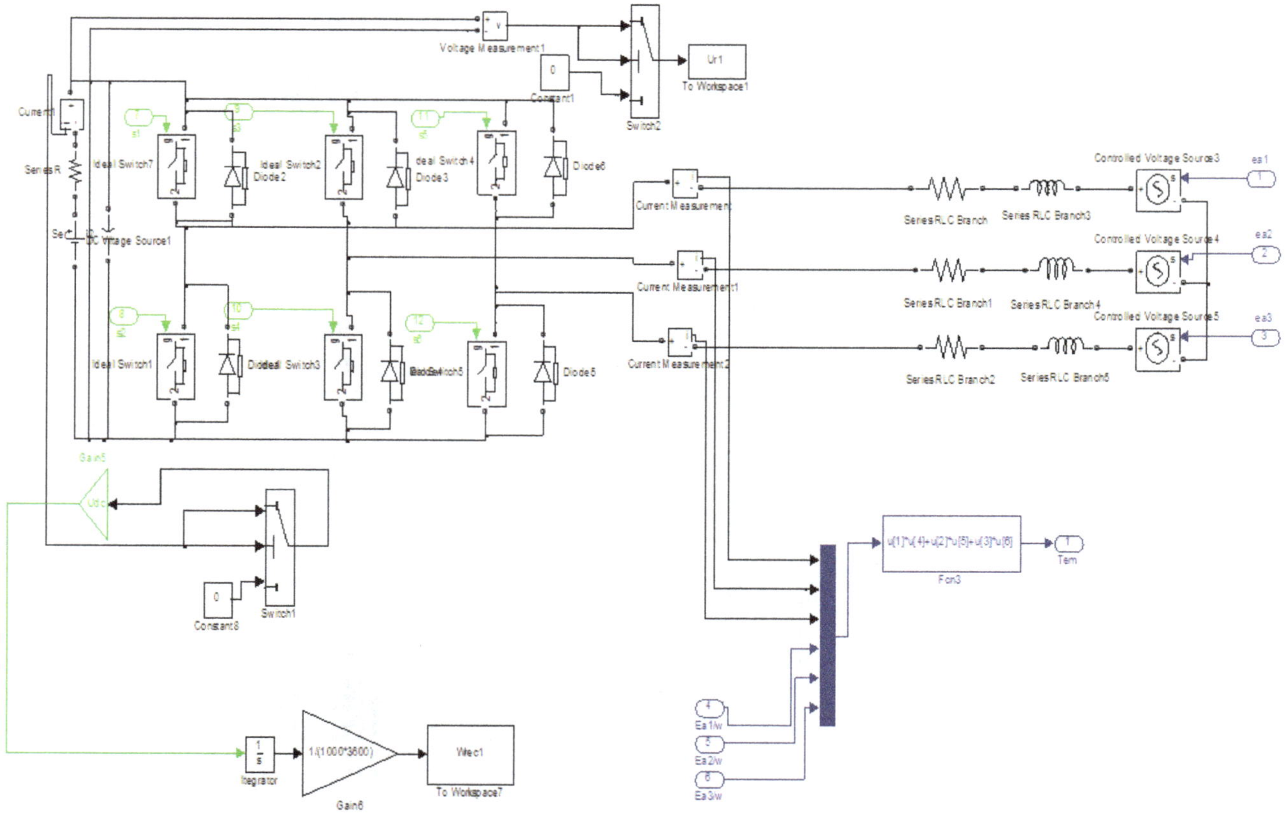

Figure 4. Simulink model of the motor-converter.

The equations of phase voltages are expressed as follows:

$$u_1 = R_{t.} \times i_1 + L_t \times \frac{di_1}{dt} + e_1 \quad (4)$$

$$u_2 = R_{t.} \times i_2 + L_t \times \frac{di}{dt} + e_2 \quad (5)$$

$$u_3 = R_{t.} \times i_3 + L_t \times \frac{di_3}{dt} + e_3 \quad (6)$$

$$L_t = L - M \quad (7)$$

Where e_1, e_2 and e_3 are respectively the electromotive forces of the phases 1, 2 and 3, L is the phase inductance and M is the mutual inductance.

The electromagnetic torque is given by the following equation:

$$T_{em} = \frac{1}{\Omega}(e_1 \times i_1 + e_2 \times i_2 + e_3 \times i_3) \quad (8)$$

Motor model is implanted under Matlab / Simulink as shown in the figure 4:

6. Speed Regulator

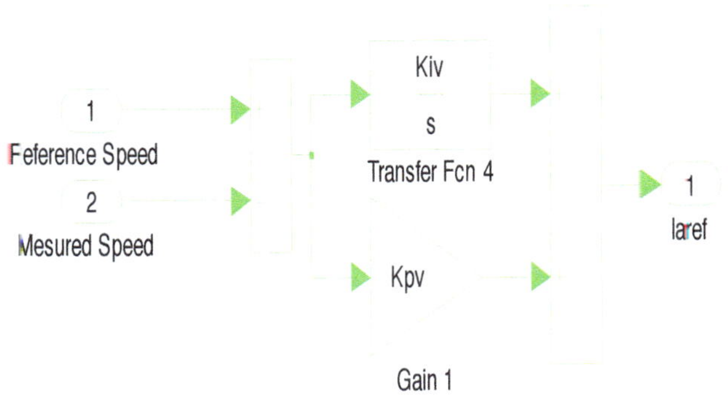

Figure 5. *Simulink model of the speed regulator.*

The comparison of the reference speed and the response speed provides the amplitude of the reference currents minimizing the error between the reference speed and the speed of response. Indeed, the reference speed is compared to the response speed. The comparator attacks a proportional-integral-type (PI regulator) to provide the amplitude of reference currents minimizing the speed error. The Simulink model of speed regulator is illustrated by the figure 5.

7. Currents Regulators

The reference current generator allows generating three currents with trapezoidal shapes and phase shifted relative to each other by an angle equal to 120 ° electrical. This three phase currents are out in phase with electromotive forces to minimize consumption and its amplitudes are controlled by the speed controller. Three control loops are used to convert currents to the three reference voltages of the motor.

The model of the reference current generator is implanted under the environment of Matlab / Simulink as shown according to the figure 6:

8. Model of the Back Electromotive Forces

The three back electromotive forces are estimated from the following three equations:

$$a = \cos\left(p \times \Omega \times t + \frac{\pi}{2}\right) \quad (9)$$

$$b = \cos\left(p \times \Omega \times t - \frac{2 \times \pi}{3} + \frac{\pi}{2}\right) \quad (10)$$

$$c = \cos\left(p \times \Omega \times t - \frac{4 \times \pi}{3} + \frac{\pi}{2}\right) \quad (11)$$

The models of the back electromotive forces (e_1, e_2, e_3) are estimated from the following algorithm:

```
{Begin
if
a>1/2;
a1=1/2.Ke.Ω:
else
a1=0;
if
 a<-1/2;
a2=-1/2.Ke.Ω;
else
a2=0;
e1=a1+a2;

if
b>1/2;
b1=1/2.Ke.Ω:
else
b1=0;
if
b<-1/2;
b2=-1/2Ke.Ω;
else
b2=0;
e2=b1+b2;

if
c>1/2;
c1=1/2.Ke.Ω:
else
c1=0;
if
c<-1/2;
c2=-1/2.Ke.Ω;
```

else
c2=0;
e_3=c1+c2;
end}.

With K_e is the back electromotive constant and Ω is the motor angular speed.

The Simulink model of the back electromotive forces is illustrated by the figure 7.

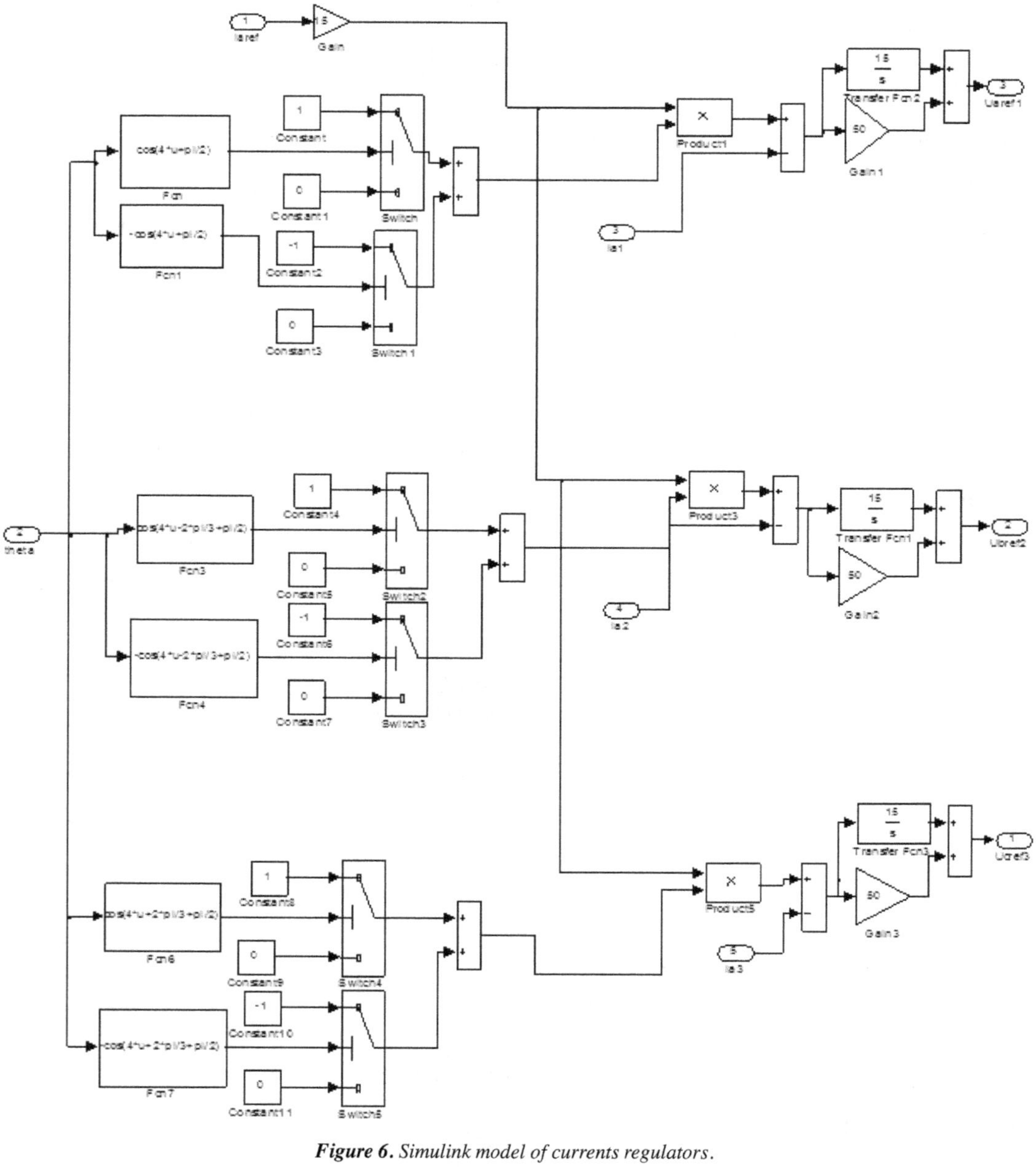

Figure 6. *Simulink model of currents regulators.*

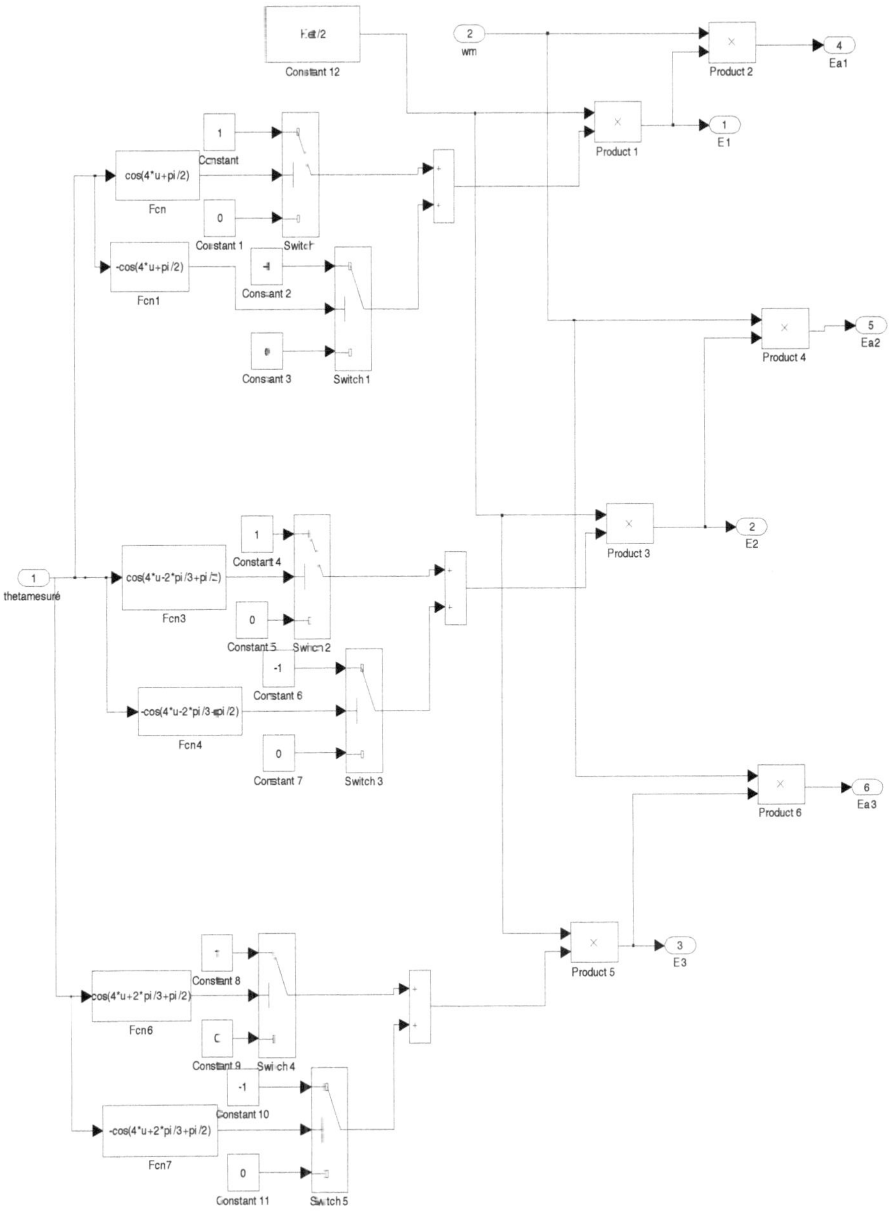

***Figure 7**. Simulink model of the back electromotive forces.*

9. Control Signals Generator

The control signal generator compares the three reference voltages to a triangular signal with frequency significantly higher than the frequency of the voltages provided by the currents regulators. The output of each comparator attack an hysteresis variant between 0 and 1 for outputting the signals for controlling the switches S1 , S3 and S5 . The speed controller and current controller adjusts the pulse width of the control signals so as to impose currents in phase with the electromotive forces against and minimize the error between the reference speed and the speed of response. Signals for controlling the switches S2, S4 and S6 are respectively complementary to the signals S1, S3 and S5. To prevent short circuits, control pulses S2, S4 and S6 are shortened to avoid duplication between two signals control arm. The delay to the closing of the power switch is held in account by the proposed model. The Simulink model of the generator control signals is shown in figure 8:

10. Dynamic Equation

The dynamic equation of the vehicle is derived from the fundamental relationship of dynamics:

$$(M_v \times R_r)\times\frac{dv}{dt} = r_d \times T_m - (F_r + F_a + F_c)\times R_r \qquad (12)$$

The equation of motion of the vehicle is implemented under the environment of Matlab/ Simulink as shown in to the figure 9:

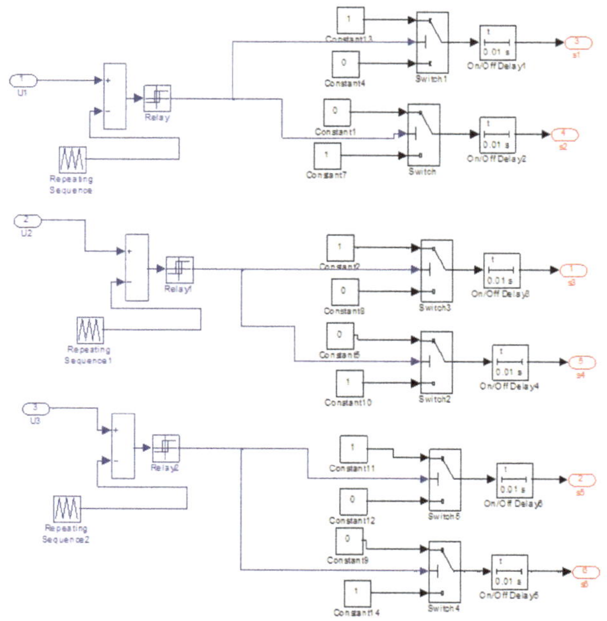

Figure 8. Simulink model of control signals generator.

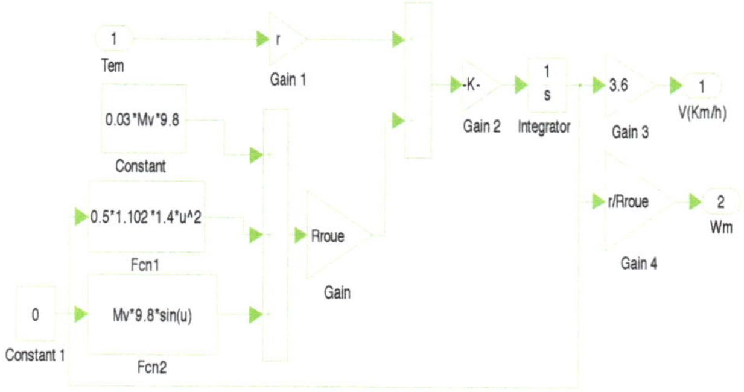

Figure 9. Simulink model of the dynamic equation.

11. Excitation System Optimizing Consumption

The excitation system use the over-fluxing technique during phases of strong accelerations to reduce consumption (Excitation current augmentation). The excitation current is optimized for the operating phases at constant speed and during deceleration phases as to optimize autonomy (figure 10).

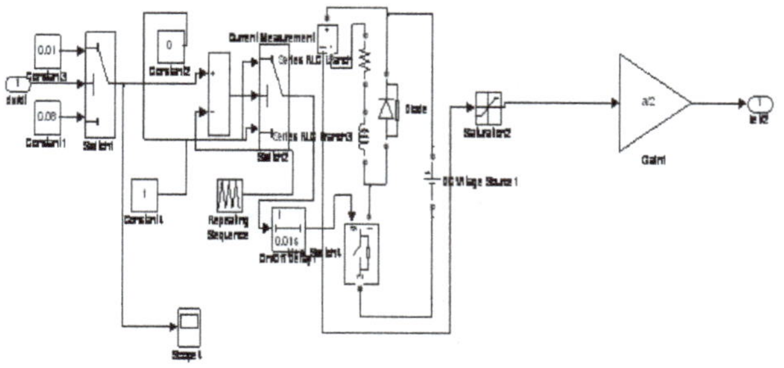

Figure 10. Excitation system.

12. Global Model of the Power Chain

The coupling of different models of the electric vehicle power chain leads to overall model implemented under the environment of Matlab/ Simulink as shown in to the figure 11 [9-12]:

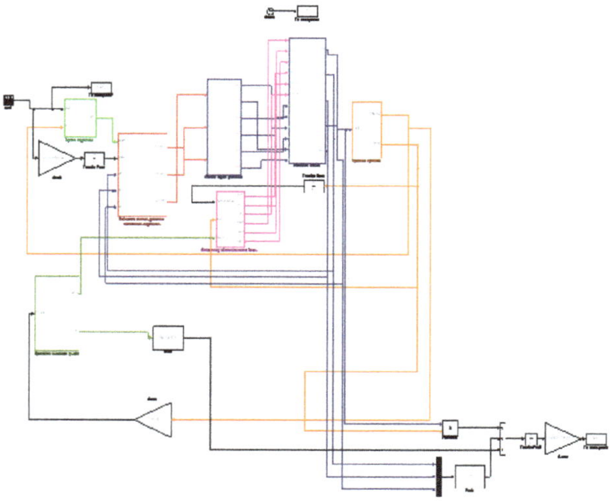

Figure 11. Global model of the power chain.

13. Descriptions of the Simulations Results

Scalar control has the disadvantage of convergence at startup, but for the electric car application, the acceleration during start-up is reduced to avoid the problem of abrupt starting and to increase the reliability of the vehicle. This property makes it very robust the scalar control for electric car application. This feature is illustrated in Figure 12. Also this figure shows that the speed of response precisely follows the reference speed, which shows the performance of the control technique chosen. This characteristic validates design process of the power chain.

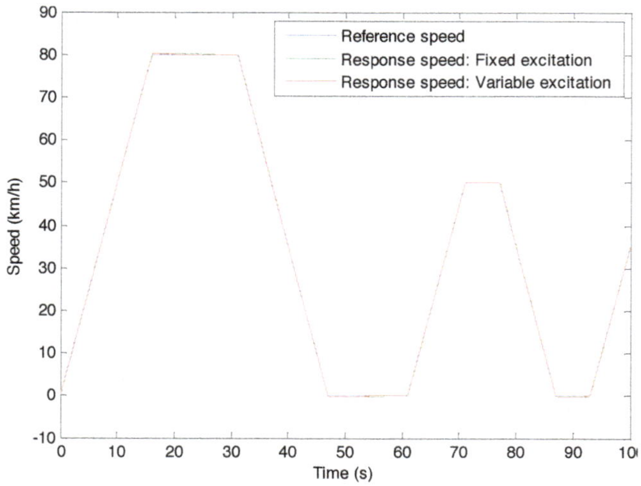

Figure 12. Speed response.

Figure 13 shows that the starting current is significantly reduced, which shows the good choice of the parameters of speed and currents regulators.

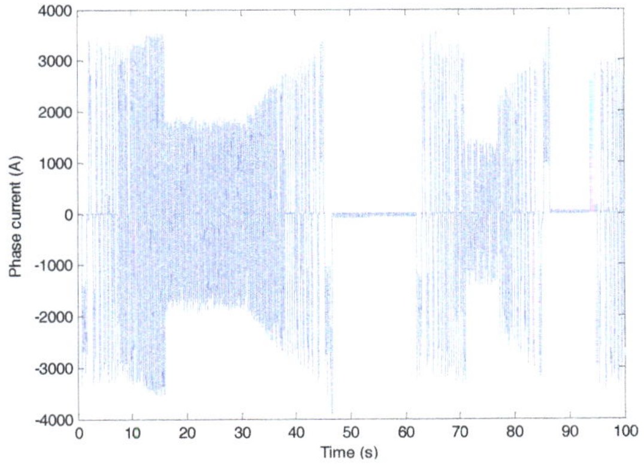

Figure 13. Phase current.

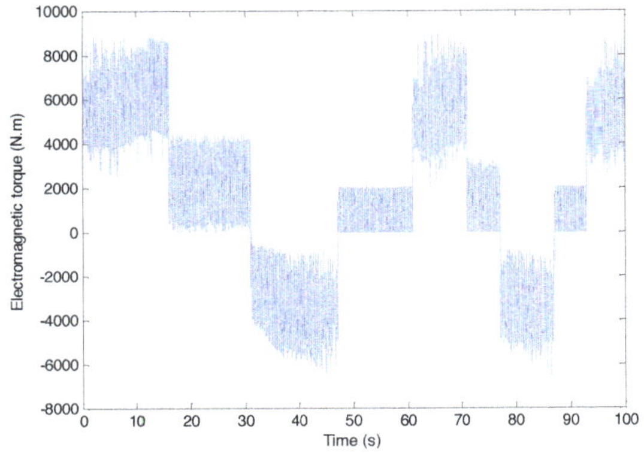

Figure 14. Torque on the motor shaft.

Figure 14 shows that the torque on the motor shaft is negative during deceleration, since the inertia of the drive power is negative.

The electric vehicle recovers energy hang deceleration phases. This property is illustrated by figure 15.

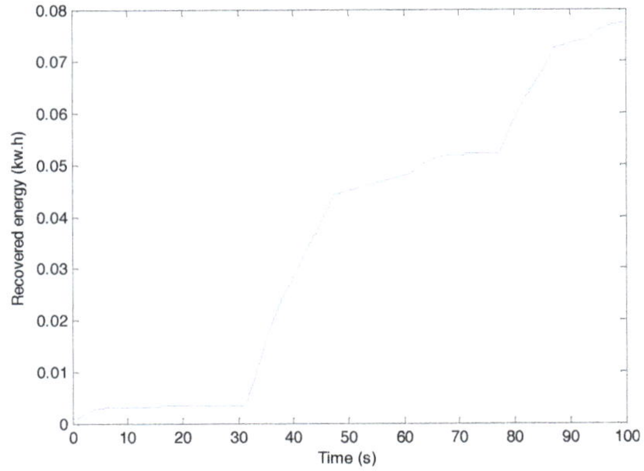

Figure 15. Energy recovered.

The control law imposes a phase shift ideally zero to minimize consumption, but in reality this phase shift is close to zero since the time constant of the motor is not zero. This property is illustrated by figure 16:

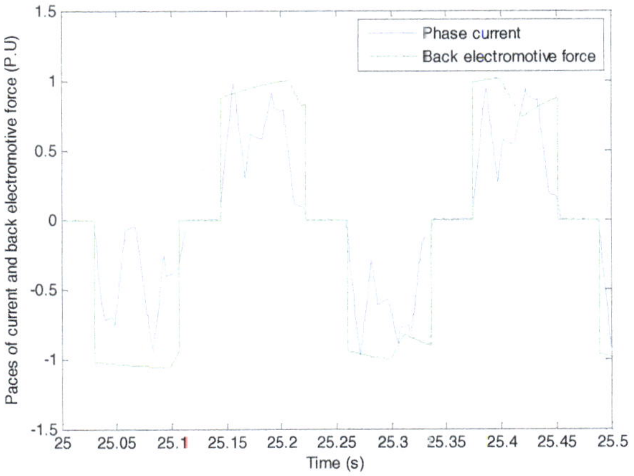

Figure 16. Paces of current and back electromotive force.

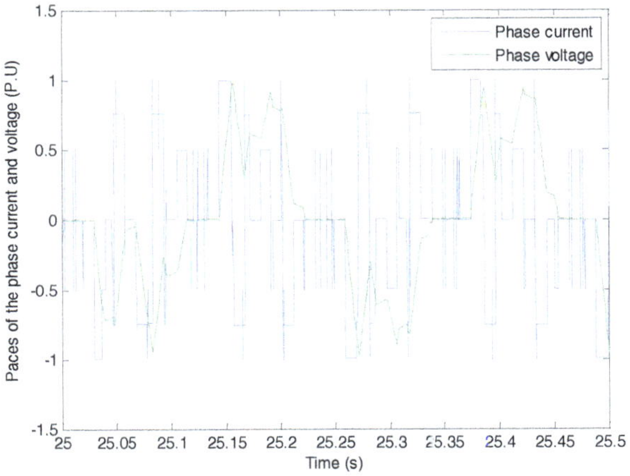

Figure 17. Paces the phase voltage and back electromotive force.

The figure 17 shows that the shape of the current is near to trapezoidal shape, which shows the effectiveness of the selected control law. The distortion in the shape of the current is essentially owed to the delay to the closing of the power contacts.

The current of excitation presents its elevated values during the phases of strong acceleration to minimize electric vehicle consumption (figure 18)

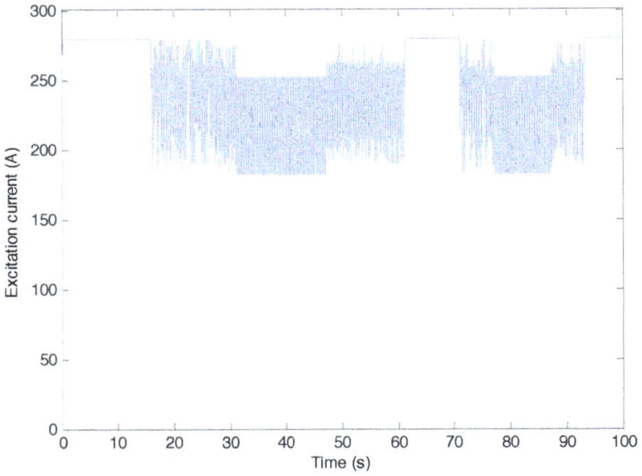

Figure 18. Excitation current.

The electric vehicle recovers energy hang deceleration phases. This property is illustrated by figure 19 and 20.

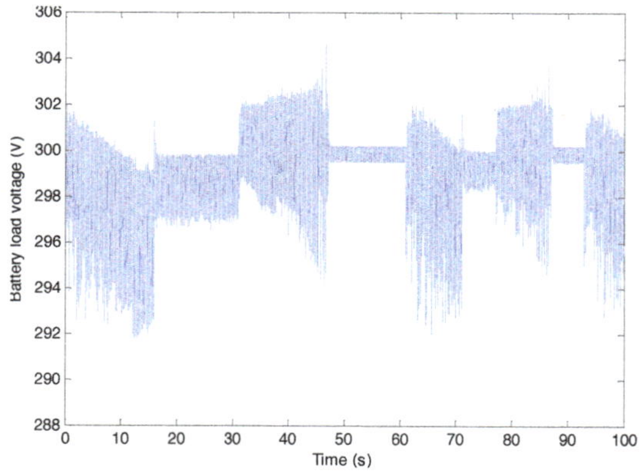

Figure 19. *Battery load voltage.*

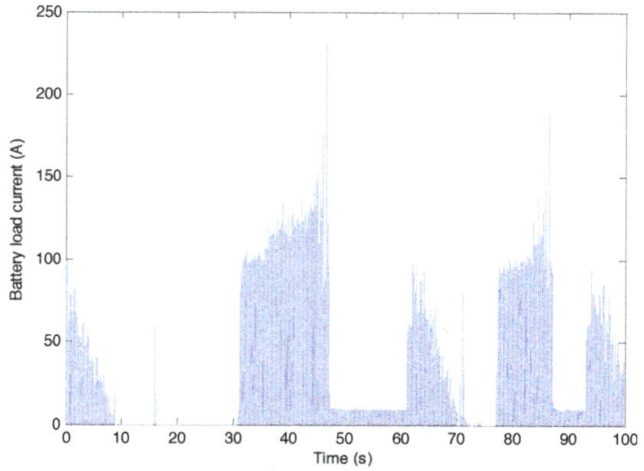

Figure 20. *Battery load current.*

The figure 21 shows that the system with optimized excitation permits a non negligible reduction of the consummation.

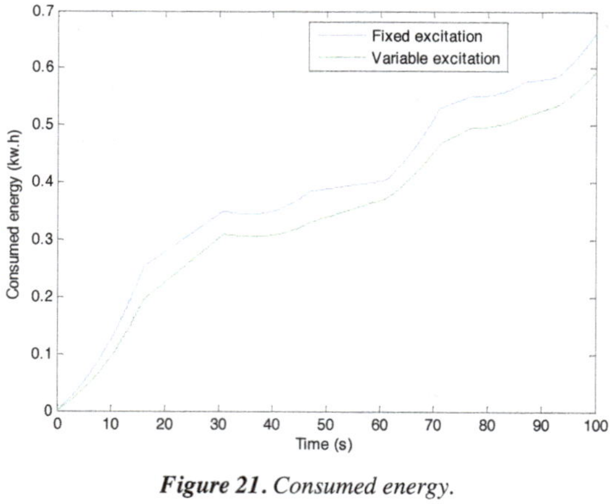

Figure 21. *Consumed energy.*

14. Conclusion

In this paper we present a systemic trapezoidal control a coiled rotor synchronous motor with reduced pruction cost. This methodology takes into account the interactions between the control and the design of the motor-converter. The system modeling along with a trapezoidal control law under the environment of Matlab / Simulink, validates this design approach and leads to a scientific results to good level. As future work, it will be interesting to start the problem of excitation system parameters optimization.

References

[1] S. TOUNSI, «Modélisation et optimisation de la motorisation et de l'autonomie d'un véhicule électrique», Thèse de Doctorat 2006, ENI Sfax.

[2] S. TOUNSI et R. NEJI: «Design of an Axial Flux Brushless DC Motor with Concentrated Winding for Electric Vehicles», Journal of Electrical Engineering (JEE), Volume 10, 2010 - Edition: 2, pp. 134-146.

[3] S. TOUNSI, M. HADJ KACEM et R. NEJI « Design of Static Converter for Electric Traction », International Review on Modelling and Similations (IREMOS) Volume 3, N. 6, December 2010, pp. 1189-1195.

[4] S. TOUNSI « Losses modelling of the electromagnetic and IGBTs converters », International Int. J. Electric and Hybrid Vehicles (IJEHV), Vol. 5, No. 1, 2013, pp:54-68.

[5] S. TOUNSI « Comparative study of trapezoïdal and sinusoïdal control of electric vehicle power train», International Journal of Scientific & Technology Research (IJSTR), Vol. 1, Issue 10, Nov 2012.

[6] [M. HADJ KACEM, S. TOUNSI, R. NEJI «Systemic Design and Control of Electric Vehicles Power Chain », International Journal of Scientific & Technology Research (IJSTR), Vol. 1, Issue 10, Nov 2012.

[7] S. TOUNSI « Control of the Electric Vehicles Power Chain with Electromagnetic Switches Reducing the Energy Consumption», Journal of Electromagnetic Analysis and Applications (JEMAA) Vol.3 No.12, December 2011.

[8] S. TOUNSI, M. HADJ KACEM et R. NEJI « Design of Static Converter for Electric Traction », International Review on Modelling and Similations (IREMOS) Volume 3, N. 6, December 2010, pp. 1189-1195.

[9] S. TOUNSI et R. NEJI: "Design of an Axial Flux Brushless DC Motor with Concentrated Winding for Electric Vehicles", Journal of Electrical Engineering (JEE), Volume 10, 2010 - Edition: 2, pp. 134-146.

[10] S. TOUNSI, R. NEJI, and F. SELLAMI: "Design Methodology of Permanent Magnet Motors Improving Performances of Electric Vehicles", International Journal of Modelling and Simulation (IJMS), Volume 29, N° 1, 2009.

[11] A. Moalla, S. TOUNSI et R. Neji: "Determination of axial flux motor electric parameters by the analytic-finite elements method", 1nd International Conference on Electrical Systems Design & Technologies (ICEEDT'07), 4-6 Novembre, Hammamet, TUNISIA.

[12] N. Mellouli, S. TOUNSI et R. Neji: "Modelling by the finite elements method of a coiled rotor synchronous motor equivalent to a permanent magnets axial flux motor", 1nd International Conference on Electrical Systems Design & Technologies (ICEEDT'07), 4-6 Novembre, Hammamet, TUNISIA.

Techno-Economic Comparative Assessment of Asvt Versus Conventional Sub-stations for Rural Electrification

Kitheka Joel Mwithui[1], David Murage[1], Michael Juma Saulo[2]

[1]Department of Electrical and Electronic Engineering, Jomo Kenyatta University of Agriculture and Technology, Nairobi, Kenya

[2]Department of Electrical and Electronic Engineering, Technical University of Mombasa, Mombasa, Kenya

Email address:

kithekajoelmwithui@tum.ac.ke (K. J. Mwithui), dkmurage25@yahoo.com (D. Murage), michaelsaulo@tum.ac.ke (M. J. Saulo)

Abstract: The overall electricity access rate is still very low in most sub-Saharan African (SSA) countries. The rate is even lower in rural areas where most of the population in these countries lives. In Kenya about 8% of rural communities lives at close proximity to High voltage transmission lines yet they have no electricity. One of the main obstacles to rural electrification (RE) is the high cost of laying the distribution infrastructure owing to the dispersed nature of loads and low demand. Thus, electrifying the rural areas needs to be considered holistically and not just on the financial viability. To reduce cost, it is important that auxiliary service voltage transformer (ASVT) sub-station, which are cheaper than the conventional sub-station be explored. This research aimed at carrying out the techno-economic assessment of Auxiliary service voltage transformer sub-station and the conventional sub-station that can be used to step down 132kv supply from transmission line to 240v to supply single phase loads in rural areas where there is no any nearby conventional sub-station but there are trunks of high voltage transmission lines at close proximity. The research further explored the maximum number of ASVT sub-stations that can be terminated on 132kv within a specified distance beyond which it would be economically viable to use a conventional sub-station. In this research local prices and the life cycle costing of sub-stations were used.

Keywords: Auxiliary Service Voltage Transformer (ASVT), Techno-Economic Assessment (TEA), Transmission Line (TL), Life Cycle Costing (LCC), Conventional Sub-station (CS)

1. Introduction

Many people in rural areas in developing countries do not have access to electricity and even electrification of the metropolitan areas and suburbs is incomplete or unreliable. It has been reported that more than 1.6 billion people, mostly in developing countries, do not have access to electricity and that most of them live in rural areas. This trend is even highly pronounced in rural areas of Sub-sahara Africa. [1, 2]

The high rate of low power connectivity has been amplified by the low concentration of electricity users and the cost of setting up conventional sub-stations being very high, thus the power utility company cannot generate an adequate return on investment. [3, 5], on the other hand there are large number of rural communities living around or in close proximity to high transmission line but are not supplied with electricity. The main obstacle being, these transmission lines have very high voltages that cannot be directly and cheaply be used for electrification. [4, 6]

To address the prohibitive costs incurred with the use of conventional substations, non conventional sub-station namely; Auxiliary Service Voltage transformer (ASVT) substation is explored in this journal.

The auxiliary service voltage transformer also known as station service voltage transformer (SSVT) combines the characteristics of instrument transformer with power distribution capability. In this transformer, the high voltage side is connected directly to the overhead transmission line of either 220kV or 132kV, while the secondary side may be of typical voltage ratings of 240V, 480V, 600V or any other voltage level supplies designed on order. One step down principle is applied to achieve the low voltages just like in instrument transformers [7, 8].

The Auxiliary service voltage transformer can either be used with its low voltage output to directly supply needed power near transmission lines or simply step up the ASVT low voltage output through distribution transformer for a local distribution network.

In developing countries where transmission line infrastructure is already in place but a wide spread distribution infrastructure is lacking, the non conventional distribution substation technologies can be used to greatly reduce the electrification costs for small villages [4].

2. The Auxiliary Service Voltage Transformer

The ASVT, sometimes known as a station service voltage transformer (SSVT) is insulated in sulfur hexafluoride (SF6) gas and combines the characteristics of instrument transformer with power distribution capability [9]. All the dielectric characteristic of the conventional instrument transformer are applicable to ASVT even though these are hybrid apparatus which are between an instrument transformer and a distribution transformer. These transformers fulfill the standards for both types, i.e. IEEEC 57.13.1993 and IEEE C57.12.00 [10]. This inductive transformer has a very high thermal power in comparison with conventional instrument transformer, in general from 20 up to 60 times more than the design of new generation, without reaching the capacity of a power transformer. [8]

The ASVT is capable of tapping either 220kv or 132kv and step it down to 240v in one step. This eliminates the number of transformers required to step down high voltage using the conventional method i.e. 132kv/66kv/33kv/11kv, hence cutting the cost required to set up ASVT sub-station. [10, 12].

ASVTs were originally designed to suit supply for auxiliary services within the sub-station such as lighting loads, motor loads and instrument purposes [11]. In developing countries where transmission lines infrastructure is already in place but a wide spread distribution infrastructure is lacking, the ASVT sub-station technologies can be used as a compact transformer to greatly reduce the electrification cost for rural electrification. The ASVT can be used to supply loads directly with its low voltage or simply step up the ASVT low voltage output through distribution transformer for local distribution network.

Tapping the high voltage transmission line and connecting an ASVT with a small foot print sub-station will provide affordable, readily available electricity to many rural dwellers in close proximity to high voltage lines and presently without power [11].

2.1. The Penetration Point of Asvt Sub-station on 132kv Transmission Line

A research was carried out to investigate whether termination of the ASVT sub-station on 132kv transmission line could lead to violation of the voltage profile of the line. The point of ASVT termination was also varied to investigate whether there was a specific penetration point of the ASVT sub-station that was to be adhered to. The researcher further investigated whether variation of the of the 132kv transmission line from 440KM to 600KM could lead to violation of the transmission line voltage profile. The study results were as shown in Fig. 1 graph. [13]

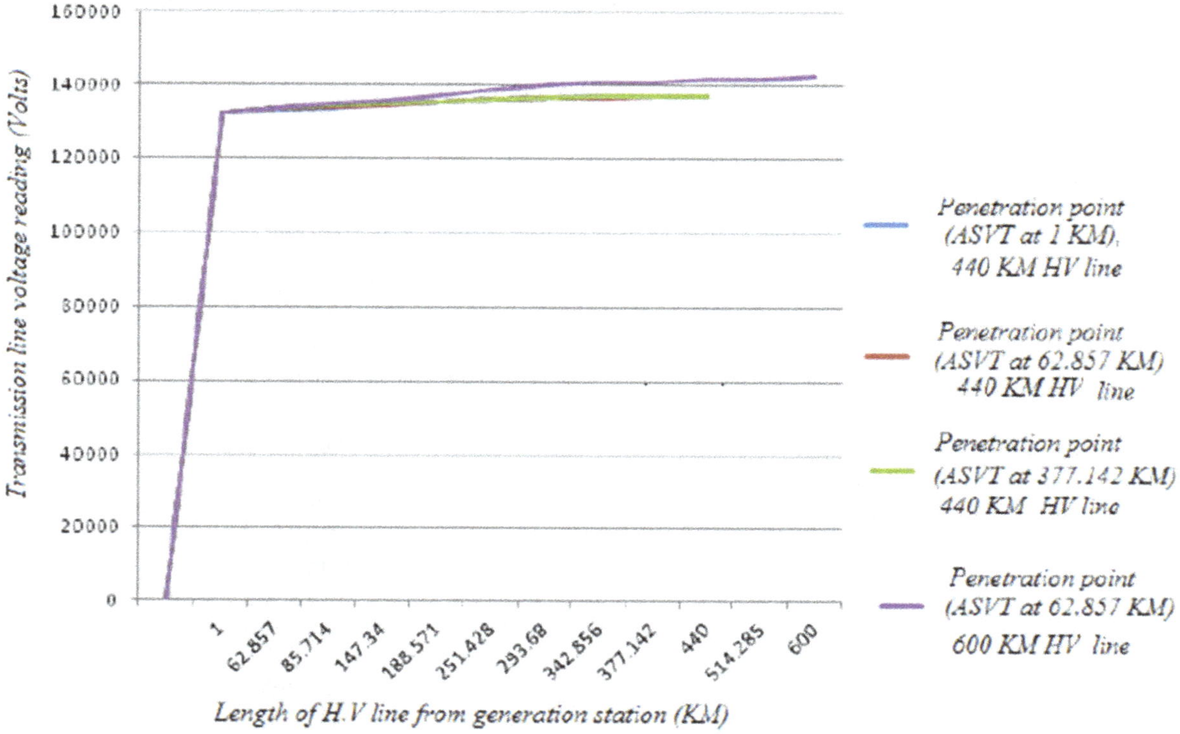

Fig. 1. *Penetration point of ASVT sub-station on 132kv transmission line.*

The above study revealed that the ASVT sub-station can be terminated from any point of the 440KM, 132kv transmission line without the violation of the line voltage profile. This was a positive realization in the fact that the termination of the ASVT sub-station would depend on the location of the village at close proximity to the H.V line.

Further study was carried out to investigate whether a longer transmission line would lead to voltage profile violation. This line was increased to 600KM. This lead to violation of the transmission line as displayed by blue line in Fig. 1. The voltage profile was to be maintained at ±6% of the transmission line voltage.

2.2. The Optimum Penetration Level of Asvt Sub-station on 132kv Transmission Line

A further study was carried out to investigate the maximum number of ASVT sub-stations that can be terminated on a 440KM, 132kv transmission line without voltage profile violation.

The results were as captured in Fig. 2 [8]

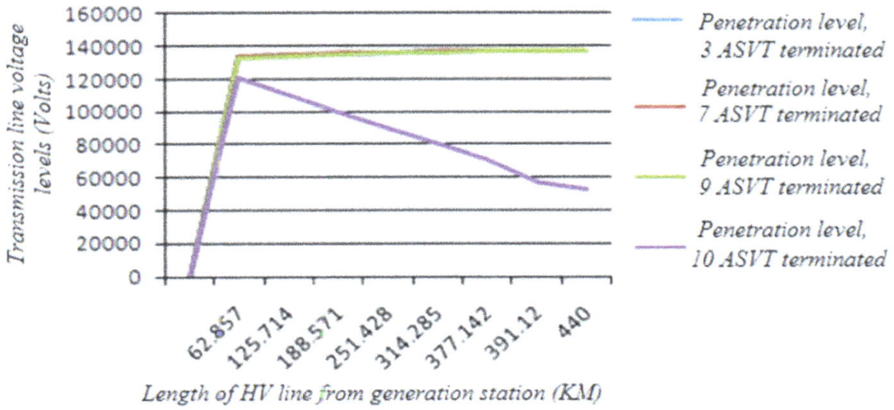

Fig. 2. The penetration level of AVST sub-station on 132kv transmission line.

The ASVT sub-stations were terminated on the 440KM, 132kv transmission line then the voltage profile of the line was monitored.

When 3, 7, 9 ASVT sub-stations were terminated on the 132kv line, the transmission line voltage as maintained within 132±6% kv, Which is the Kenya Power and Lighting Company recommended transmission line voltage profile.

When the tenth ASVT sub-station was terminated, the transmission line voltage levels reduced drastically as captured by the blue line of Fig. 1.

This showed that the surge impedance loading of the transmission line was affected which led to violation of the line voltage profile.

3. Methodology

3.1. Asvt Sub-station to Supply Maungu Village

A case study of Maungu village located along Nairobi - Mombasa highway was used to analyze the cost of setting up an ASVT sub-station to supply the village. The features of the village are as follows: [13]

(i) Has 18 households and a shopping centre.
(ii) Located 300 metres from the 132kv transmission line
(iii) 40KM away from the conventional sub-station.
(iv) Sparsely populated.
(v) The village is not supplied with electricity.

The ASVT sub-station tap power directly from the over head transmission lines through high voltage connectors without interrupting the power flow.

The ASVT sub-station use one transformer and a bus bar to interconnect transformers. A disconnector switch and a circuit breaker are required, but to further reduce the cost of components the circuit breaker can be replaced by a disconnector switch.

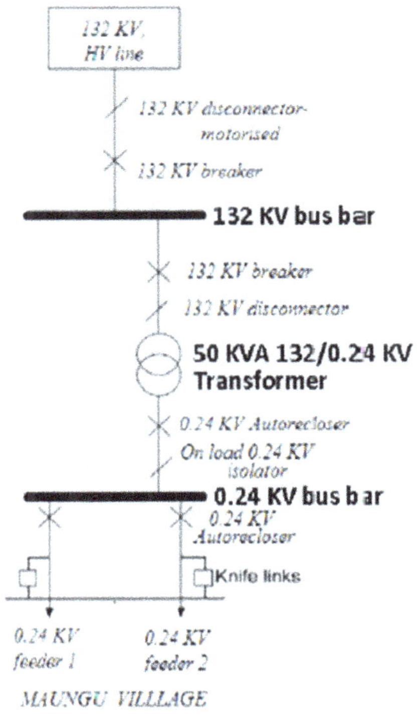

Fig. 3. 132/0.24kV ASVT sub-station single line diagram.

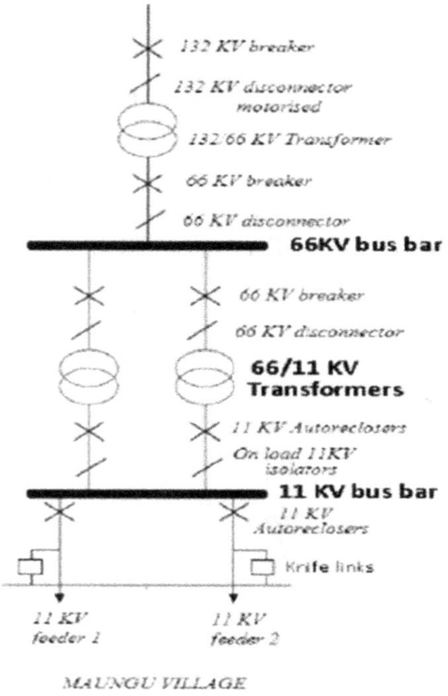

Fig. 4. *132/66/11kv conventional sub-station single line diagram.*

The ASVT sub-station removes all the back up that is found in the conventional sub-station and as such when one component of the sub-station fails customers will definitely experience power outage. This is not a serious problem to the customers in the rural villages in comparison to having no electricity supply as long as the cost of electricity connection is affordable. [14]

The ASVT sub-station single line diagram is as shown in Fig. 3 and ASVT sub-station layout in Fig. 5.

3.2. Conventional Sub-station to Supply Maungu Village

Another study was carried out to assess the cost to be incurred to set up and maintain a conventional sub-station to supply Maungu village.

A conventional sub-station is a fully flashed sub-station that terminates the transmission line and as a result maintains a high level of service.

A conventional sub-station is designed with a large amount of redundancy in terms of transformers, disconnect switches, circuit breakers, bus bars in order to provide continued operation under failure or high load.

This distribution sub-station has many components involved, thus being very large and spreads over large ground area. A single line diagram of the conventional sub-station is as shown in Fig. 4.

3.3. Life Cycle Costing of Sub-stations.

A life-cycle cost analysis (LCC) is geared at determining the total cost of a sub-station and expenses incurred in maintaining it.

The main reasons for carrying out LCC analysis are: [14]

(i) Comparison of the costs incurred in setting up a sub-station.

(ii) To determine the most cost-effective sub-station.

The life cycle costing of ASVT sub-station and conventional sub-station were carried out. The study considered the prices of the equipment required to set up the sub-station, its protection and one year equipment.

The local prices were considered and in cases where the equipment were not locally available, the quotation prices used by Kenya power and lighting company (KPLC) were used.

The techno-economic comparative assessment of ASVT versus conventional sub-station is tabulated in table 1. [13]

The life-cycle cost of a project can be calculated using the formula:

$$LCC = C + Mpw + Epw + Rpw - Spw$$

Where;

C= capital cost of a project i.e. initial capital expense for the equipment, the system design, engineering, and installation.

M = maintenance i.e. is the sum of all yearly scheduled operation and maintenance (O and M) costs.

Pw = the present worth of each factor.

E = Energy cost of the system i.e. the sum of the yearly fuel cost.

R = replacement cost i.e. the sum of all repair and equipment replacement cost anticipated over the life of the system.

S = salvage value of a system i.e. its net worth in the final year of the life-cycle period.

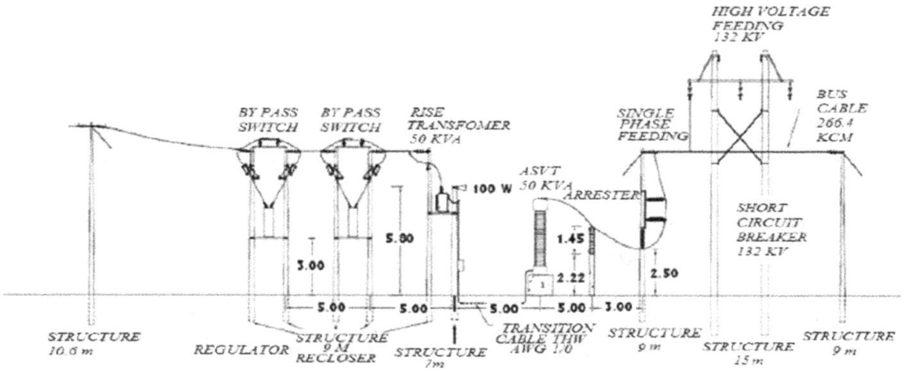

Fig. 5. *ASVT sub-stations layout.*

Table 1. *Life cycle costs foe a fully flashed conventional versus ASVT sub-station.*

PARTICULARS		CONVENTIONAL SUB-STATION		ASVT SUB-STATION	
CAPITAL (C)	**PRICE PER ITEM (Kshs)**	**Quantity**	**Total Price (Kshs.)**	**Quantity**	**Total Price (Kshs.)**
132kv disconnector switch motorized	300,000	2	600,000	1	300,000
132kv breaker	1,250,000	2	2,500,000	1	1,250,000
66 kv disconnector switch	300,000	3	900,000		
66kv breaker	1,250,000	3	3,750,000		
66kv busbar	1,600,000	1	1,600,000		
11kv auto recloser	1,640,000	3	4,920,000		
ON load 11kv isolator	225,000	3	675,000		
240V busbar	100,000			1	100,000
240V recloser	1,000,000			3	3,000,000
ON load 240V isolator	225,000			1	225,000
Knife link	58,000	2	116,000	2	116,000
11kv busbar	100,000	1	100,000		
11kv tie bar	100,000	1	100,000		
240V tie bar	100,000			1	100,000
Civil works			4,000,000		2,000,000
Earthing			1,750,000		750,000
132/66kv, 10MVA Transfomer	50,000,000	1	50,000,000		
66/11kv. 5MVA Transformer	49,000,000	2	98,000,000		
SUB-TOTAL			115,911,000		56,841,000
Operation and maintenance (M)					
Transformer and switch gear service (per year)	15,000,000	2	30,000,000	1	15,000,000
SUB-TOTAL			115,911,000		56,841,000
REPLACEMENT (RPw) (per year)					
ON load kv isolator	225,000	3	675,000	1	225,000
66kv breaker	1,250,000	3	3,750,000		
66kv breaker	1,250,000	3	3,750,000		
SUB-TOTAL			4,425,000		1,475,000
SALVAGE (SPw) (per year.)					
20% of original			6,954,660		2,500,000
SUB-TOTAL			6,954,000		2,500,000
TOTAL LCC = C+ MPw +RPw - SPw			196,481,340		70,816,000

3.4. Results and Discussion

Fig. 3 and 4 shows single line diagram of ASVT and conventional sub-station technologies respectively. Table 1 displays the life cycle cost analysis of both conventional and ASVT sub-stations. The cost of setting up a conventional sub-station to supply Maungu village with electricity is kshs. 196,481,340 (1,926,284 Us Dollars) while the cost of setting up an ASVT sub-station to supply the same village with electricity is kshs.70,816,000 (694,274 Us dollars).

From table 1, it was observed that ASVT sub-station is three times cheaper than the conventional sub-station.

The research also intended to identify the maximum number of ASVT sub-stations that can be used to supply villages in close proximity to high voltage transmission line with electricity beyond which a conventional sub-station will be more economically viable.

3.5. Research Recommendation

The techno-economic comparative assessment of the non-conventional sub-stations should be carried out to draw conclusion on the most economically viable non-conventional sub-station that can be used to supply Maungu village with electricity and the power utility company realize return on investment.

4. Conclusion

The data analysis of table 1 led to a conclusion that it is more economical to set up ASVT sub-station to supply Maungu village with electricity than to set up a conventional sub-station to supply the same village with electricity.

The life cycle cost comparative assessment of table 1 also led to a conclusion that a maximum of three ASVT sub-stations can be used to supply electricity to villages living at close proximity to 132kv transmission line for a stretch of 40KM, beyond which a conventional sub-station will be more economically viable.

References

[1] World energy outlook, SBN: 978-92-6412413-4.

[2] G. Dagbjartsson, C. Gaunt., 'Rural electrification,' A scoping report, 2013.

[3] M. Saulo, M. Mbogho, "Implication of capacitor coupling substation on rural electrification planning in kenya," in Procceedings of 3rd international Kenya Society of Electrical and Electronics Engineers Conference, KSEEE, 2014.

[4] M. J. Saulo, Penetration level of unconventional rural electrification technologies on power networks. PhD thesis, University of Capetown, May 2014.

[5] M. Saulo, C. Gaunt, M. Mbogho, 'Comparative assessment of capacitor coupled substation and auxiliary service voltage transformer for rural electrification' in 2nd Annual International Conference in Kabaraka University, 2012.

[6] R. Gomez, A. Solano, C. Gaunt, 'Rural electrification project development, using auxiliary service voltage transformer," pp. 1-6, 2010.

[7] Arteche Instrument Transformer Manual (2010): ASVT245 and ASVT-145 manual and technical brochures.

[8] Kitheka J. Saulo J. Murage D. 'Determination of the penetration level of ASVT sub-station on 132kv line without voltage profile violation.' International Journal of energy and power engineering vol 5-1, pp 22-28, Feb 2016.

[9] Omboua A. Application report "the high voltage line becomes a power distributor: A successful test in Congo – Brazzaville" Congo. 2006.

[10] Omar C., Gomez R. Solano A. Acosta E. (2010) Eradicating energy poverty "Rural Electrification in Chuahua, Mexico at one third of the cost versus a conventional substation" Mexico.

[11] Saulo M. J, Gaunt C. T "implication of using Auxiliary Service Voltage Transformer substation for Rural Electrification." *International journal of energy and power engineering.* Vol 4-1) pp 1-11. 2014.

[12] Kitheka J. Saulo J. Murage D. 'The penetration level of Auxiliary service voltage transformer sub-station on a power network for rural electrification,' in Kabarak University 5th Annual conference, July 2015.

[13] Kitheka J. 'The optimum penetration level of Auxiliary service voltage transformer sub-station on 132kv transmission line without voltage profile violation.,' Msc thesis, JKUAT, 2016.

[14] G. Anderson, K. Yanev, 'Non conventional substation and distribution system for rural electrification," in 3rd IASTED Africa PES 2010, 2010.

Analysis and Investigation of Direct AC-AC Quasi – Resonant Converter

Mihail Hristov Antchev

Department of Power electronics, Technical university-Sofia, Sofia, Bulgaria

Email address:
antchev@tu-sofia.bg

Abstract: The present article reports an analysis and investigation of direct AC-AC quasi-resonant converter. A bidirectional power device, whose switching frequency is lower than the frequency of the current passing through the load, is used for its realisation. A mathematical analysis of the processes has been made and comparative results from computer simulation and experimental study have been brought. The converter can find application in wide areas of power electronics: induction heating, wireless power transfer, AC-DC converters.

Keywords: Quasi – Resonant Converter, Constant Frequency

1. Introduction

The standard method of converting AC to AC power is by rectifier, supplying power to inverter, to whose output is connected the load. The so called "direct converters" are used to increase the energy efficiency. There are matrix converters for direct AC to AC power conversion [1,2]. Resonant converters are also used for the same purpose [3,4,5]. In [6] a direct AC-AC resonant converter has been considered, using two bidirectional power devices, including an active-inductive load while switching to the power supply. The converter described in [7,8] uses 4 power devices, and during the active intervals a serial oscillator circuit is connected to the input voltage source. The converter presented in [9] uses two bidirectional power devices, an additional capacitor for soft switching and a serial oscillator circuit connected to the input voltage at specified intervals. The induction heating converter described in [10] is most similar to the converter presented in this paper. An analysis has been made on the assumption that the converter is supplied with DC voltage, whose value is equal to the effective value of the AC input voltage, which does not correspond to the actual physical action. The "multi-cycle modulation" used in the study provides only discrete power regulation of the load and the use of variable switching frequency deteriorates the electromagnetic compatibility.

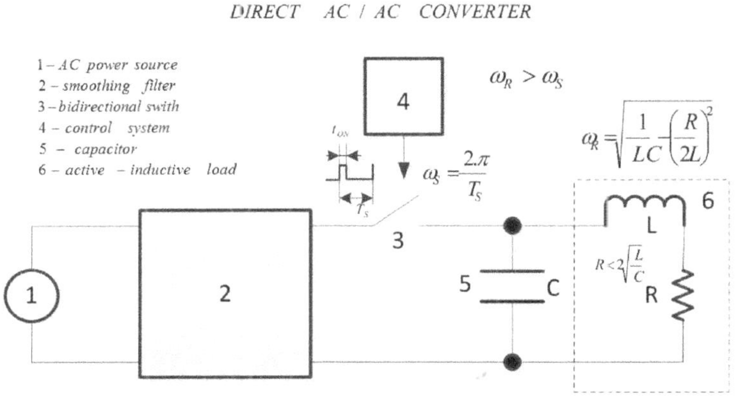

Figure 1. Block diagram of the new AC-AC quasi-resonant converter.

The purpose of the present paper is to give more accurate mathematical analysis, suitable for testing the converter's operation in random mode, and not only in the so called "multi-cycle modulation", which is a particular case of the present investigation. Constant frequency and pulse width modulation operation is proposed for the power regulation.

Fig.1. shows a block diagram to illustrate the implementation of the considered converter.

Its effect is similar to that of the so called "class E-inverters" [11,12]. When switching on the bidirectional switch 3 for time t_{ON} the capacitor 5 is quickly charged to the momentary value of the AC input voltage. This voltage is applied also to the load 6 and the current passing through it increases in absolute value. When turning off the switch for time $T_S - t_{ON}$, damped oscillations develop in the oscillator circuit, composed of load and capacitor, with frequency ω_0, determined by the elements values. The process of switching is repeated at a period $T_S > T_0$, and each switching on in the circuit adds energy. This energy can be regulated via change in time t_{ON} at constant switching frequency ω_S . This frequency is much bigger than the frequency of the source 1 AC input voltage. With the help of the smoothing filter 2 the higher harmonics of the switch current are removed. Thus the current from the AC voltage source has an almost sinusoidal shape.

2. Mathematical Description

Fig.2 shows the implemented power circuit, which will be used also for mathematical analysis. The timing diagrams of the basic values are shown on fig.3 for one switching period T_S. The analysis in one switching period is divided in two intervals: the first with duration t_{ON} , and the second - $T_S - t_{ON}$. The aim is to obtain expressions for the basic values during both intervals: the current through the inductance, capacitor's voltage and the current through the bidirectional switch.

For the first interval with duration t_{ON} the current through the load inductance changes from I_{t_0} to I_{LtON} . The value of the capacitor's voltage is constant,

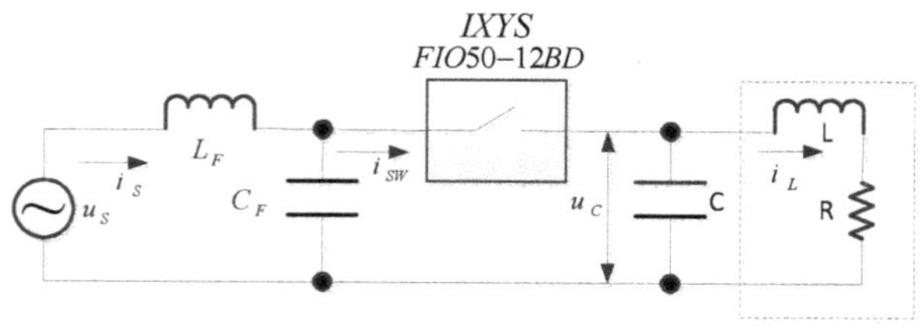

Figure 2. Practical design of the converter.

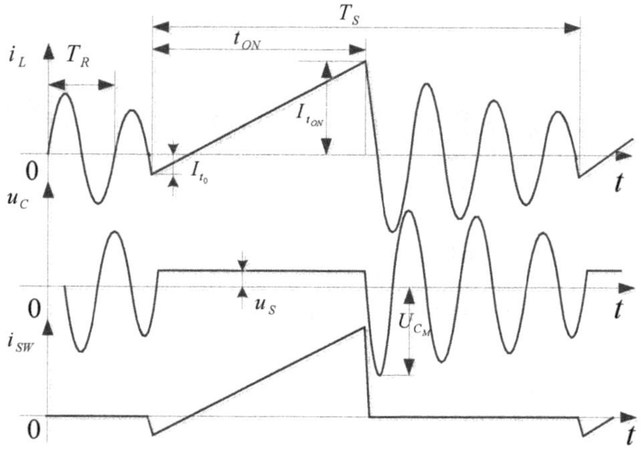

Figure 3. Waveforms for analysis: top- inductor current i_L , middle-capacitor voltage u_C , bottom – bidirectional switch current i_{SW}.

equal to the momentary value of the input voltage (due to the high switching frequency of the switch).

$$u_C = u_S \tag{1}$$

The differential equation has the form

$$u_S = L.\frac{di_L}{dt} + R.i_L \tag{2}$$

From the initial condition

$$i_L(t=0) = I_{t_0} \tag{3}$$

is determined the integration time constant and this gives

$$i_L(t) = \frac{u_S}{R} + \left(I_{t_0} - \frac{u_S}{R}\right).e^{-\frac{t}{\tau}}, \tag{4}$$

where $\tau = \frac{L}{R}$.

The current value at the end of the interval is:

$$i_L(t_{ON}) = I_{LtON} = \frac{u_S}{R} + \left(I_{t_0} - \frac{u_S}{R}\right).e^{-\frac{t_{ON}}{\tau}} \tag{5}$$

The average current value through the switch is determined by:

$$I_{SW_{AVR}} = \frac{1}{T_S}\int_0^{t_{ON}} i_L(t).dt \ , \tag{6}$$

where substitution of (4) and transformation leads to the following result:

$$I_{SW_{AVR}} = \frac{u_S}{R}.\frac{t_{ON}}{T_S} + \frac{\tau}{T_S}.\left(I_{t_0} - \frac{u_S}{R}\right).\left(1 - e^{-\frac{t_{ON}}{\tau}}\right) \tag{7}$$

For the second interval with duration $T_S - t_{ON}$ the current changes from I_{LtON} to I_{t_0} . The last value is used as initial in the next switching period. The differential equation has the form:

$$L.\frac{di_L}{dt} + R.i_L + \frac{1}{C}\int i_L.dt = 0 \tag{8}$$

The solution is:

$$i_L(t) = e^{-\delta t}.\left(C1.\cos\sqrt{\omega_R^2 - \delta^2}.t + C2.\sin\sqrt{\omega_R^2 - \delta^2}.t\right), \quad (9)$$

where $\delta = \frac{R}{2L}; \omega_R = \frac{1}{LC}$

$$\text{Accepting that } \omega_0 = \sqrt{\omega_R^2 - \delta^2} \quad (10)$$

$$i_L(t) = e^{-\delta t}.(C1.\cos\omega_0 t + C2.\sin\omega_0 t) \quad (11)$$

The integration constants are determined by the following conditions:

$$i_L(t = 0) = I_{LtON} \rightarrow C1 = I_{LtON} \quad (12)$$

$$i_L\left(t \approx \frac{T_R}{4}\right) = 0 \rightarrow C2 = -I_{LtON}.\cot\omega_0.\frac{T_R}{4} \quad (13)$$

The final value of the current in this interval I_{t_o} is obtained using (11) for time $T_S - t_{ON}$.

Through integration is obtained the law on capacitor voltage variation:

$$u_C(t) = \frac{1}{C}\int i_L(t).dt \quad (14),$$

i.e.

$$u_C(t) = \frac{e^{-\delta t}}{C.\omega_R^2}[(C1.\omega_0 - C2.\delta).\sin\omega_C t - (C2.\omega_0 + C1.\delta).\cos\omega_0 t] \quad (15)$$

Its maximum value for this switching cycle is:

$$u_C\left(t \approx \frac{T_R}{4}\right) = u_{CM} \rightarrow u_{CM} = \frac{e^{-\delta\frac{T_R}{4}}}{C.\omega_R^2}.\left[(C1.\omega_0 - C2.\delta).\sin\omega_0.\frac{T_R}{4} - (C2.\omega_0 + C1.\delta).\cos\omega_0.\frac{T_R}{4}\right] \quad (16)$$

The maximum value of the capacitor's voltage corresponds to the switching cycle around the maximum value of the power supply voltage. The last values of the variables in each interval are used as initial values in the next one. For the first switching period after the input voltage passes through zero $I_{t_0} = 0$. Thus all values can be determined by sequential calculation from the moment the input AC voltage passes through zero to the end of its half-period.

3. Computer Simulation

An investigation has been made of the proposed converter's operation using the PSIM program. Fig.4 shows the circuit for computer simulation. The value of the resonant inductance is $100\mu H$, of the resistor in the oscillator circuit - 0.01Ω. The value of the resonant capacitor is $10nF$. The values of the smoothing input filter elements are: $L_F = 680\mu H, C_F = 20\mu F$. The maximum value of the input voltage is $70V$, and its frequency -$50Hz$. The switching frequency of the switch is set to$10kHz$, changing the time t_{ON} . The results from the simulation at $t_{ON} = 30\mu S$ are shown on fig.5, fig.6 and fig.7 in different time scale, in order to track the operation in one switching cycle, as well as for a longer period.

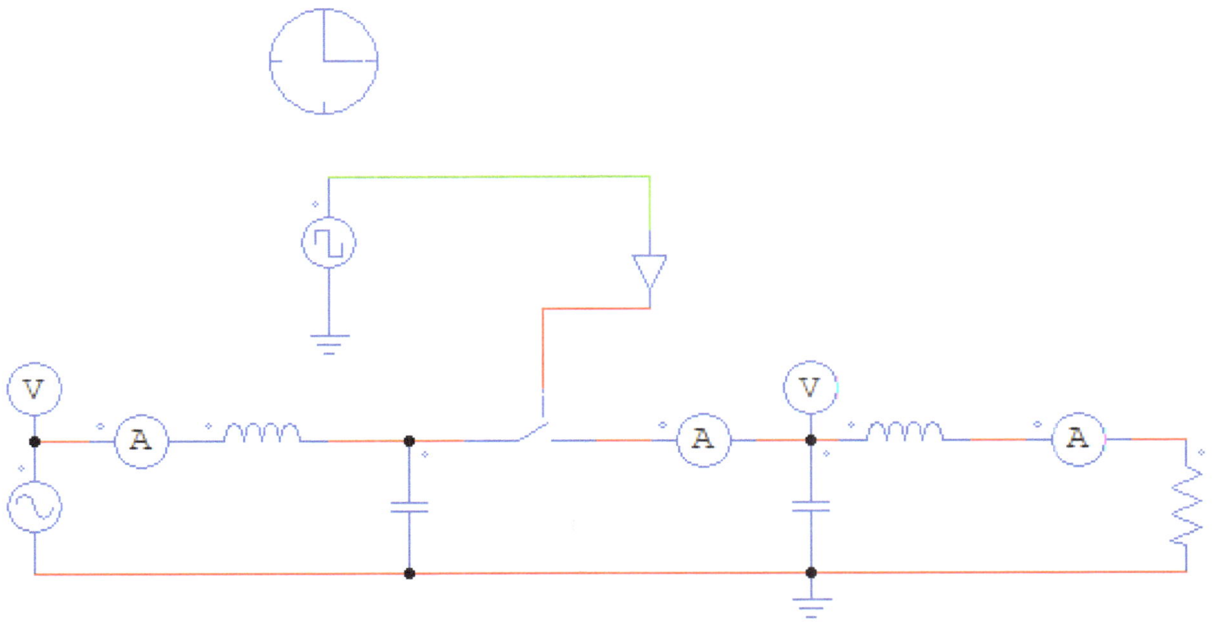

Figure 4. Circuit for computer simulation.

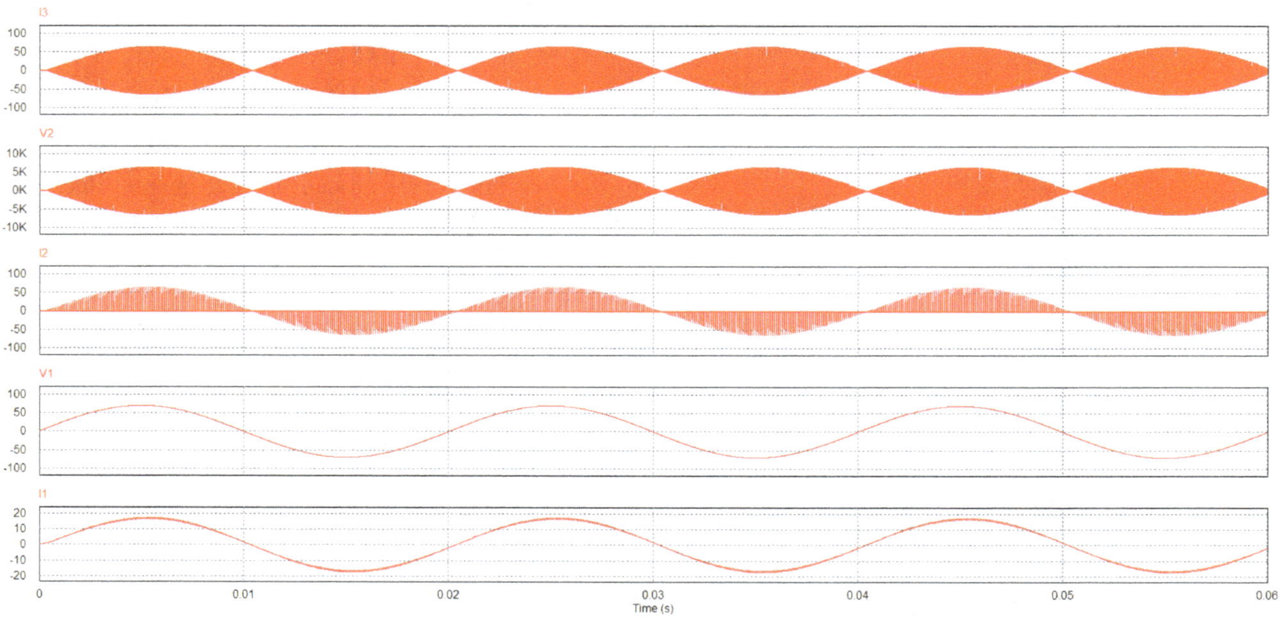

Figure 5. *Results from simulation: from top to bottom inductor current* i_L*, capacitor voltage* u_C*, bidirectional switch current* i_{SW}*, input voltage* u_S*, input current* i_S *;X-axis – from 0 to 60mS.*

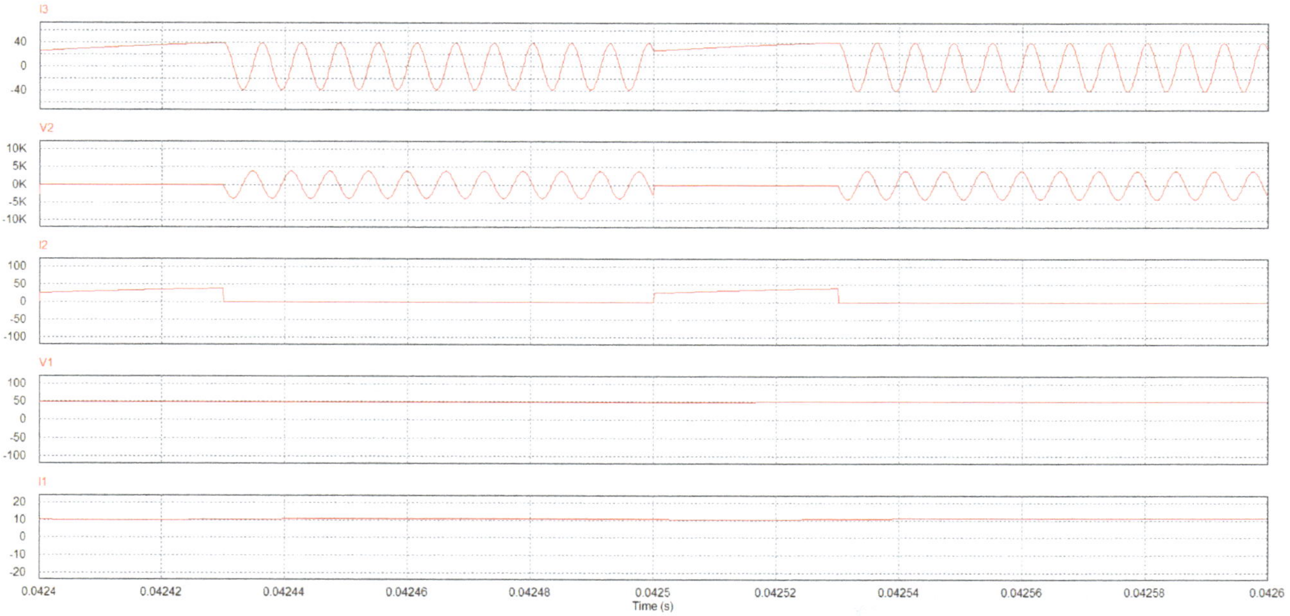

Figure 6. *Results from simulation: from top to bottom inductor current* i_L*, capacitor voltage* u_C *, bidirectional switch current* i_{SW} *, input voltage* u_S*, input current* i_S *; X axis – from 42.4 mS to 42.6mS.*

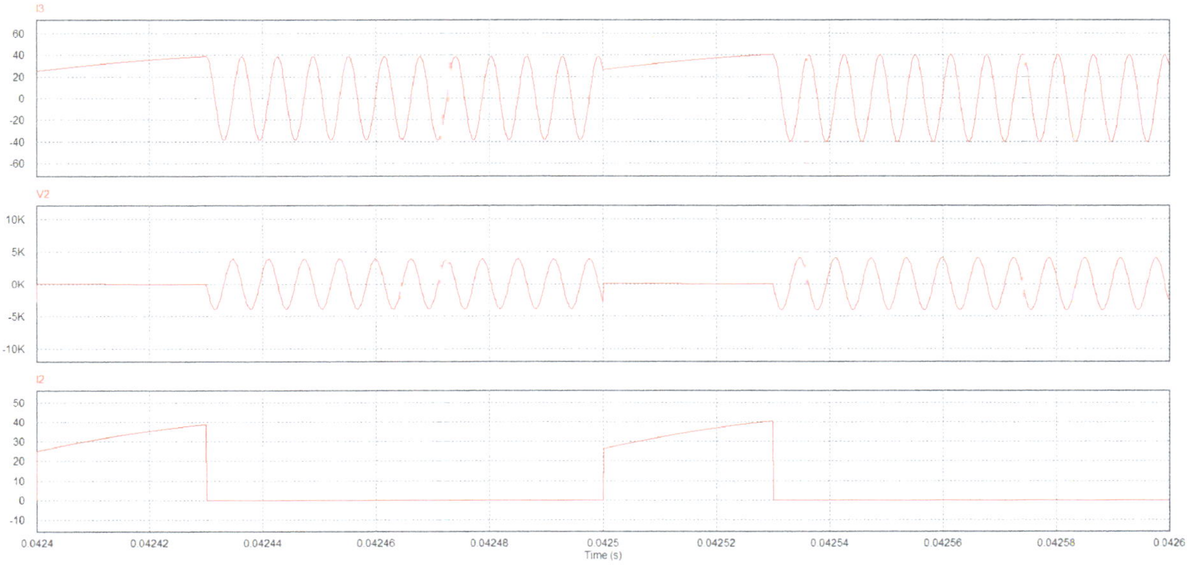

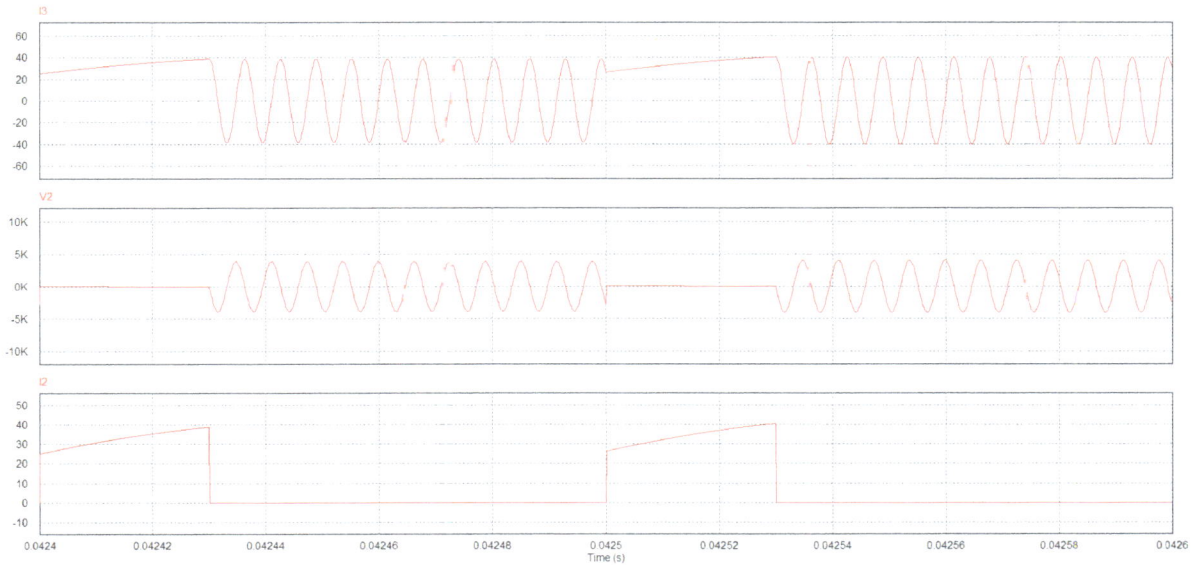

Figure7. *Results from simulation: from top to bottom inductor current* i_L *, capacitor voltage* u_C *, bidirectional switch current* i_{SW} *; X axis – from 42.4 mS to 42.6mS.*

For the cycle shown on fig. 7 a comparison has been made between the results from the computer simulation and those from the calculations according to the mathematical analysis formulas. From fig.7 are taken the readings from the simulation for the current $I_{t_0} = 26.46A$ and the voltage in the beginning of the interval $u_S = 63.1V$. Following formula (5) is calculated $I_{LtON} = 45.31A$, and fig.7 reports a value of $40.28A$. The value obtained from the simulation is smaller, as at the end of the first interval the voltage decreases and the result from fig. 7 reports a value for $u_S = 28.37V$. This decrease accounts for the difference in the average values of the current through the switch – calculated according to (7) $8.03A$, and according to fig.7 with the corresponding PSIM function $10.23A$. The constants $C1 = 40.28$ and $C2 = 0.032$ are calculated using the formulas (12) and (13), and formula (16) is used to obtain the maximum value of the capacitor's voltage $u_{CM} = 4027.56V$. Fig.7 reports a value of $4007V$.

4. Experimental Investigation

An experimental investigation of the converter has been made, carried out according to the scheme shown on fig. 2. The element's values are the same as those from the computer simulation except for the resistor in the oscillator circuit. The measurement shows, that the inductance used in the circuit has a value equal to $100\mu H$ within the range from $5kHz$ to $120kHz$. The value of its resistance from the serial equivalent circuit at $100kHz$ is 1Ω. Therefore the damping in the oscillator circuit will be greater in comparison with that in the computer simulation. The results from the experimental investigation are shown on fig.8, fig.9, fig.10, fig.11, fig.12 and fig.13 in different time scales.

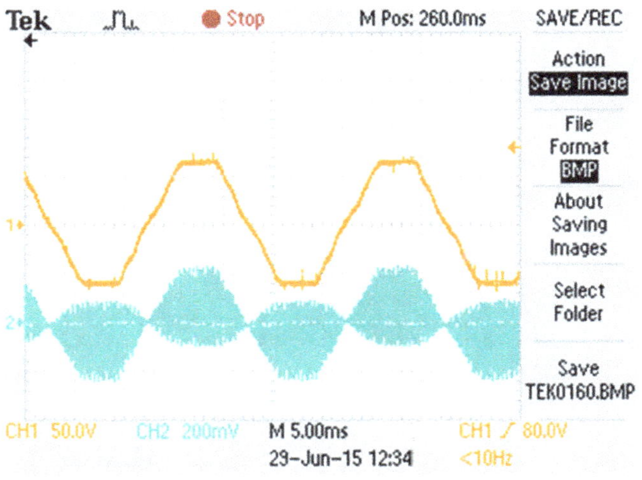

Figure 8. *CH1- input voltage, CH2 – inductor current – 2A/div.*

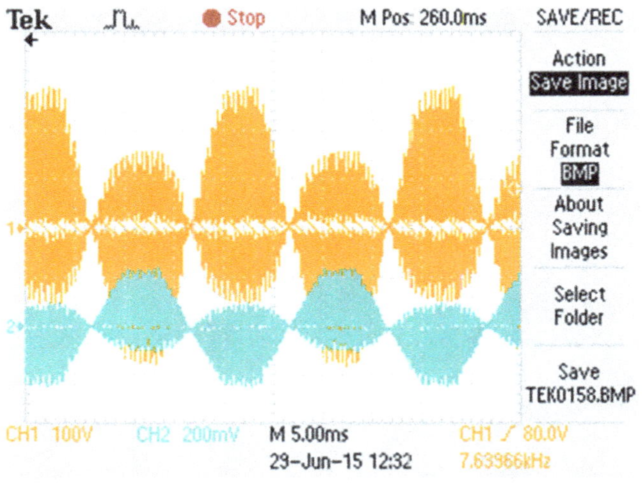

Figure 9. *CH1- capacitor voltage, CH2 – inductor current – 2A/div.*

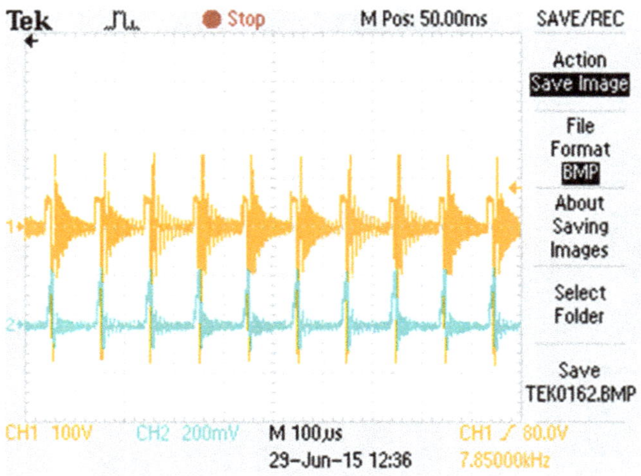

Figure 10. *CH1- capacitor voltage, CH2 – inductor current – 2A/div.*

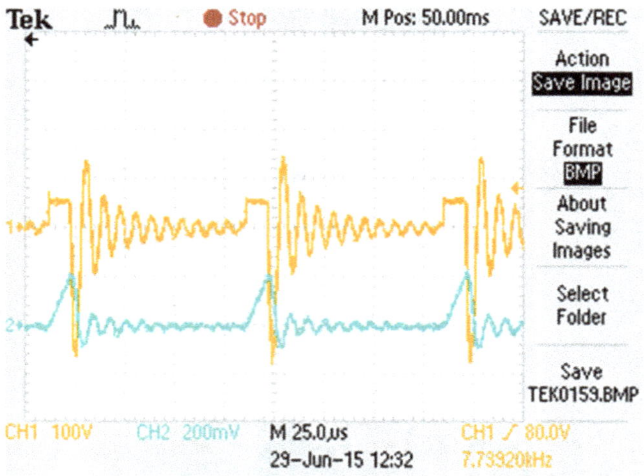

Figure 11. *CH1- capacitor voltage, CH2 – inductor current – 2A/div.*

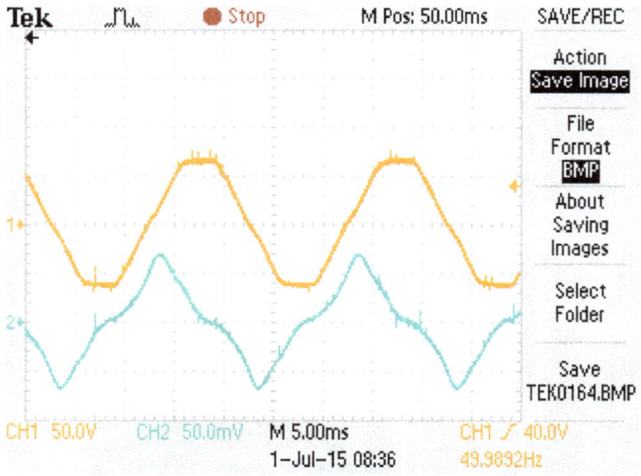

Figure 12. *CH1- input voltage, CH2 – input current – 0.5A/div.,* $t_{ON} = 10\mu S$.

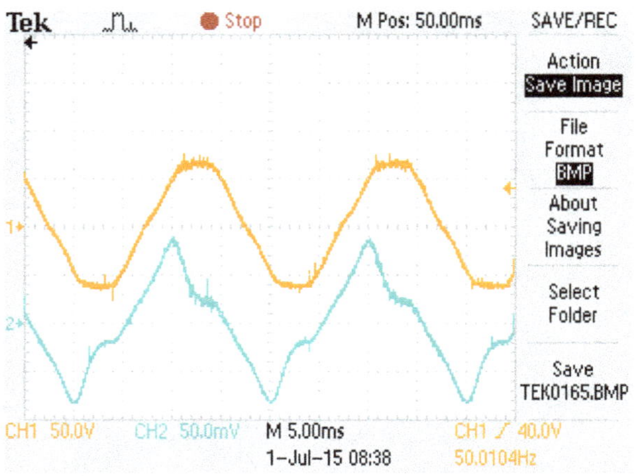

Figure 13. *CH1- input voltage, CH2 – input current – 0.5A/div.,* $t_{ON} = 15\mu S$

A comparison between the results from the experimental investigations and those from the calculations according to the mathematical analysis formulas has been made for the cycle shown on fig. 11. On fig.11 the reported current is $I_{t_0} \approx 0$ and $u_S = 45V$. Using formula (5) is calculated $I_{LtON} = 3.04A$, and from fig.11 the reported value is $2.4A$. Formulas (10) and (11) are used for the calculation of the constants $C1 = 2.4$ and $C2 = 0.0017$, and according to formula (14) is calculated the maximum value of the capacitor's voltage $u_{CM} = 238V$. Fig.11 gives a value equal to $280V$.

Fig.11 shows also that the damping in the experimental circuit is considerably greater than that in the computer simulation. Fig.12 and fig.13 show, that the shape of the current from the source is near to the sinusoidal. The first harmonic current is ahead of the voltage, which can be seen on fig.5. This is due to the greater value of the input filter capacitor. Increasing the switching time of the bidirectional power device reduces the dephasing, and the converter is approaching the active load with respect to the source. A power factor close to 1 could be achieved by appropriate design of the input filter and power device control system synchronisation.

5. Conclusions

The present paper provides a mathematical analysis of a direct AC-AC quasi-resonant converter. The analysis is based on a sequential calculation of the main values in each switching cycle from the beginning to the end of the input AC voltage half cycle. The comparison of the analysis results with those from the computer simulation and the experimental investigations shows good coincidence in the values of the key variables.

References

[1] Chlebis P., P.Simonik, M.Kabasta, The Comparision of Direct and Indirect Matrix Converters, Progress In Electromagnetics Research Symposium Proceedings Proceedings, Cambridge, USA, July, 2010, pp. 310-313 .

[2] Trentin A., P. Zanchetta, J. Clare, P. Wheeler, Automated Optimal Design of Input Filters for Direct AC/AC Matrix Converters, IEEE Transactions on Industrial Electronics, Vol.59, No.7, July 2012, pp.2811-2822.

[3] Bland M., L.Emprinham , J.Clare, P.Wheeler, A New Resonant Soft Switching Topology for Direct Ac-AC Converters, Power Electronics Specialist Conference proceedings, 2002, pp.72-77.

[4] Sornago H., O. Lucia, A. Mediano, J. Burdio, Direct AC-AC Resonant Boost Converter for Efficient Domestic Induction heating Application, IEEE Transactions on Power Electronics, Vol.29, No.3, March 2014, pp.1128-1140

[5] Sornago H., O. Lucia, A. Mediano, J. Burdio, Efficient and Cost – Effective ZCS Direct AC-AC Resonant Converter for Induction Heating , IEEE Transactions on Industrial Electronics, Vol.61, No.5, May 2014, pp.2546-2556.

[6] Moghe R., R.P.Kandula, A.Iyer, D.Divan, Losse in Medium – Voltage Megawatt-Rated Direct AC/AC Power Electronics Converters, IEEE Transactions on Powe Electronics, Vol.30, No. 7, July 2015, pp.3553-3562.

[7] Li H.L., A.Hu, G. Covic, A Direct AC-AC Converter for Inductive Power- Transfer Systems ,IEEE Transactions on Power Electronics, Vol.27, No. 2, February 2012, pp.661-669.

[8] Li H.L., A.Hu, G. Covic Current Fluctuation Analysis of a Quantum ac-ac Resonant Converter for Contactless Power Transfer, Energy Conversion Congress and Symposium Proceedings, 2010, pp.1838-1843.

[9] Sigimura H., S. Mun, S. Kwon, T. Mishima, M. Nakaoka, High Frequency Resonant Matrix Converter using One-Chip Reverse Blocking IGBT-Based Bidirectional Switches for Induction Heating Power Electronics Specialist Conference Proceedings, 2008, pp.3960-3966.

[10] Sornago H., O. Lucia, A. Mediano, J. Burdio, A Class-E Direct AC-AC Converter With Multicicle Modulation for Induction Heating Systems, IEEE Transactions on Industrial Electronics, Vol.61, No.5, May 2014, pp.2521-2531.

[11] Aldhaher S., P. Luk, A. Bati, Wireless Power Transfer Using Class E Inverter with Saturable DC-Feed Inductor, IEEE Transactions on Industry Applications, Vol.50, No.4, 2014, pp.2710-2718.

[12] Kaczmarczyk Z., A high-efficiency Class E inverter – computer model, laboratory measurement and SPICE simulation, Bulletin of the Polish Academy of Sciences, vol.55, No.4, 2007, pp.411-417.

Determination of the parameters of the synchronous motor with dual excitation

Moez Hadj Kacem[1], Souhir Tounsi[2], Rafik Neji[1]

[1]Electrical Engineering Department, National School of Engineers of Sfax , Sfax University, Sfax, Tunisia
[2]Industrial Informatic Department, National School of Electronics and Telecommunications of Sfax, Sfax University, Sfax, Tunisia

Email address:
Moez_haj_kacem@voila.fr (Moez H. K.), souhir.tounsi@isecs.rnu.tn (Souhir T.), rafik.neji@enis.rnu.tn (Rafik N.)

Abstract: This paper describes the electric parameters determination for Synchronous Motor with Dual Excitation, using the joined method analytic/finite elements. Indeed several models of mutual and principal inductances and electric motor constant are developed analytically and validated by the finite elements method. These models are fortunately parameterized allowing to the formulation of several optimization problems such as the motor ripple torque.

Keywords: Synchronous Motor, Dual Excitation, Electric Vehicles, Torque

1. Introduction

For electric traction applications, synchronous or asynchronous motors [1] with radial or axial fluxes [2, 3], can be used. In order to increase the torque generation capability, these motors can be modulated. Moreover, the consequent progress of the permanent-magnet technology makes permanent magnets synchronous motors more and more utilized for variable speed and high performance systems.

At the same time, in a context where public opinion is in caressingly sensitive to the problems of pollution and the depletion of fossil fuels (oil, gas, etc ....), a growing number of industrial organizations and public or private researches trying to find innovative technological solutions for better energy management. Machines double excitation (MSDE) fully meet requirements more stringent. Indeed, it appears that the authors' concept of double excitation offers the possibility of obtaining innovative and efficient electric machines.

The principle of double excitation [2] is to associate in electrical machines synchronous permanent magnet excitation and an excitation coil allowing operational flexibility and a good design of the converter assembly machine.

Possible applications of this type of machine are many and varied. The MSDE can be used in the production of electrical energy which they function as generators. But their application is preferred especially in the field of electric or hybrid drive.

2. Generalities about Synchronous Motor with Dual Excitation

The excitation flux is created by two different sources, one is permanent magnet, another wound (in most cases) [4] or permanent magnets (the mechanical weakening) [5], the aim being to use the second excitation source to control the flow in the gap

Depending on how the two circuits are arranged excitement there are several types of machines with double excitation. We have classified into two categories:

- double excitation synchronous parallel machines
- double excitation synchronous series machines

2.1. Parallel Excitation Synchronous Motor

The double excitation parallel offers many opportunities arrangement of the two excitation circuits. A varied panorama of solutions exists in the literature [4,-9], the following is an example for illustration.

We can see the diagram of a double parallel excitation in

Figure 1 [2-9]. For this type of excitation, the permanent magnets and the coil are at the rotor.

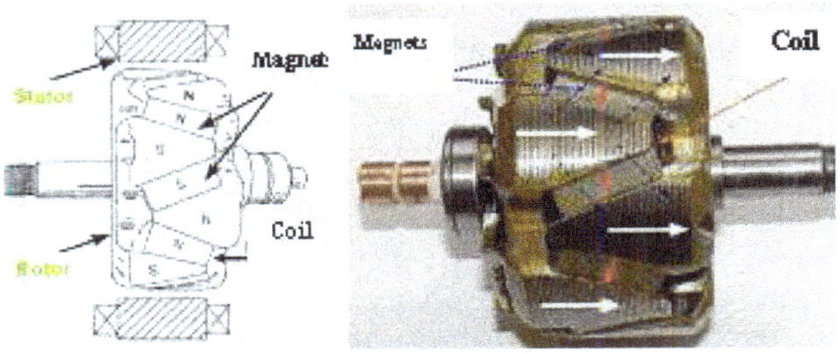

Figure 1. Synchrones machine with parallel double excitation.

This machine is more like a wound rotor machine, on which would add magnets to improve performance, to a machine magnet which we want to control the flow. In terms of performance, it is not certain that this solution can compete with magnet machines.

2.2. Serial Double Excitation Synchronous Machine

The schematic diagram of this type of double excitation is given in figure 2 [5]. For this type of double excitation, the flow through the excitation coil magnets. The flow of excitation coil and the magnets on the same path. Reducing the flow in the gap is carried out by injecting into the coils a current which creates a DMF opposite to that of the magnets. The power of the coil current is bidirectional, which increases the cost of food and therefore the overall cost. An iron loss for this type of machine down, this is one of the additional benefits specific to this structure [7].

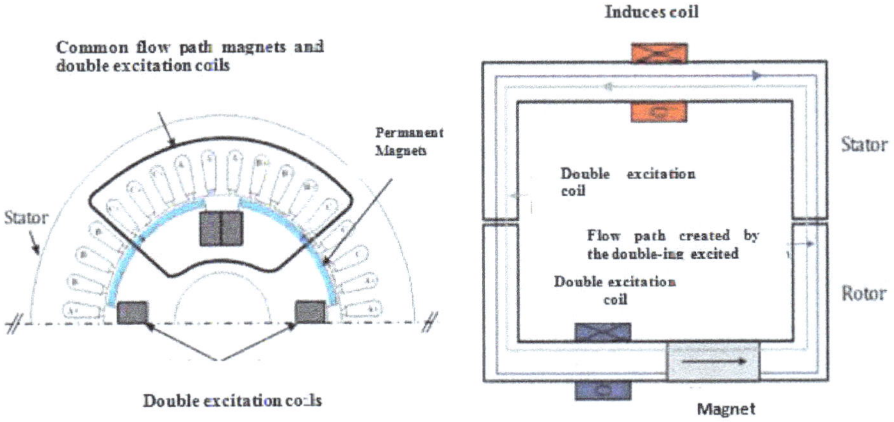

Figure 2. Serial double excitation synchronous machine.

The disadvantage of this type is that the double excitation coils are magnetically in series with the magnets. These have permeability close to that of air; the coils are therefore a high magnetic reluctance, which greatly reduces the efficiency of weakening by the coils.

An appropriate choice of the permanent magnet used is paramount to avoid the problem of irreversible demagnetization. The Iron Neodymium magnets Bor are generally recommended for this type of applications.

3. Stator Sizing of the all Configurations

All structures have the same motor stator. They therefore have the same dimensions.

The slot width of these structures is given by the following relationship [9]:

$$L_{enc} = \left(\frac{D_e + D_i}{2}\right) \times \sin\left(\frac{1}{2} \times \left(\frac{2\times\pi}{N_d} - \alpha\times\beta\times\frac{\pi}{p}\times(1 - r_{did})\right)\right) \quad (1)$$

Where N_d is the number of main teeth and r_{did} is the report between angular width of an inserted tooth and a the angular width of main tooth.

For the configurations to trapezoidal shapes of waves, the height of a tooth is given by the following relation [9]:

$$H_d = \frac{3\times2\times N_s}{2\times N_d} \times \frac{I_{dim}}{\delta} \times \frac{1}{K_f} \times \frac{1}{L_{enc}} \quad (2)$$

For the configurations to sinusoïdal shapes of waves, the height of the teeth is expressed by the following relation [9]:

$$H_d = \frac{3\times2\times N_s}{2\times N_d} \times \frac{I_{dim}}{\sqrt{2}\times\delta} \times \frac{1}{K_f} \times \frac{1}{L_{enc}} \quad (3)$$

Where K_f is the fill factor of slots, δ is the allowable current density in the slots and I_{dim} is the dimensioning current of copper conductors.

This current is given by the following relationship:

$$I_{dim} = \frac{C_{dim}}{K_e} \quad (4)$$

Where K_e is the back electromotive force constant.

4. Determining of Electric Parameters of the Motor

The study of the two-dimensional double-excitation synchronous machine is to etude conventional synchronous machine with permanent magnets. We study the influence of geometrical parameters on the machine performance (flux and electromagnetic torque load).

We will study here the influence of permanent magnets dimensions (height and width) and teeth static (width and height) on the maximum performance of the machine studied.

These effects will be evaluated analytically and by a 2D finite element method. To carry out this study, it is necessary to have an initial geometry. We will present the parametric study of the structure itself: it is to change the original structure so that it meets the specifications. This process has allowed us to scale a machine meets these specifications.

4.1. Structures with Permanent Magnets

The magnetic induction in the air gap is calculated for a position of maximum overlap (the edges of a magnet for MSAP configurations or the opening of the rotor coil for MSRB configurations, series excitation and those parallel excitation merge with the edges of the main tooth relative to the stator pole), since this position is relative to the maximum induction in the different active parts of the stator magnetic pole of the motor. The distribution of the field lines to this position is illustrated by figure 3.

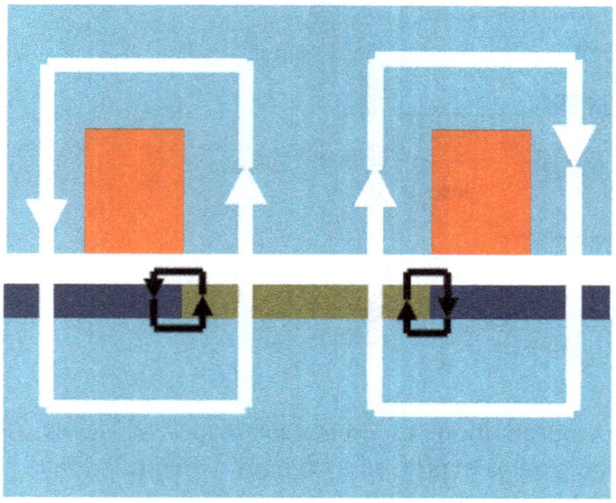

Figure 3. *Distribution of the field lines to a position of maximum recovery.*

This figure shows that the flux is divided into useful flux for the traction of the rotor and leakage flux between the magnets.

Applying Ampere's theorem at stator pole to a supply of the stator coil by the maximum current of the motor, regardless of the magnetic flux due to the magnets, used to calculate the magnetic induction caused by the current.

$$\int_{\text{Filed lines}} \overline{H} \times \overline{dl} = \frac{N_s}{\frac{N_d}{3}} \times I_{max} = 2 \times (H_{ri} \times H_a + H_{ri} \times e) \tag{5}$$

Where I_{max} is the maximal current of the motor, H is the magnetic field, H_{ri} is the magnetic field in the air-gap, H_a is the height of a magnet, e is the thickness of the air-gap, μ_0 is the permeability of air, N_s is the number of spires by phase and N_d is the number of main teeth.

$$B_{ri} = \mu_0 \times H_{ri} \tag{6}$$

Where B_{ri} is the magnetic induction due to the energize of the motor by the maximal current (I_{max}):

$$B_{ri} = \frac{\mu_0}{2 \times \frac{N_d}{3}} \times \frac{N_s \times I_{max}}{H_a + e} \tag{7}$$

This induction is negligible compared to the induction created by the magnets, since the flux generated by the coil through two times the thickness of the magnets and the air gap having both a very low magnetic permeability. Accordingly, only the magnetic flux generated by the magnets is used for sizing the motor.

This induction is derived from the application of Ampere law on a closed contour of the field lines:

$$\int_{\text{Filed lines}} \overline{H} \times \overline{dl} = 0 = 2 \times (H_m \times H_a + H_e \times e) \tag{8}$$

The magnetic induction in the air gap is linear in function of the magnetic field in the gap:

$$B_e = \mu_0 \times H_e \tag{9}$$

While applying the theorem of conservation of flux to the level of the air-gap, we deducts the expression of the induction in the entrefer according to the induction in the magnets and the coefficient of leakage flux:

$$B_a \times S_a \times K_{fu} = B_e \times S_d \tag{10}$$

The magnetic induction of the magnets takes the following relation:

$$B_a = \frac{S_d}{S_a} \times \frac{B_e}{K_{fu}} \tag{11}$$

The induction in the magnets is approached by the following linear equation:

$$B_a = \mu_0 \times \mu_r \times H_m + B_r \tag{12}$$

Where μ_r is the relative permeability of the magnets and B_r is the residual magnetic induction of magnets.

From the equations (8), (9), (10), (11) and (12), we deducts the height of the magnets imposing an induction in the air-gap B_e. This induction is chosen of a manner to have magnetic inductions in the different active parts of the motor near of the bends saturation of the characteristic B-H, leading to a minimal mass of the motor and a working in the linear regime [9]:

$$H_a = \mu_r \times \frac{B_e}{B_r - \frac{S_d \times B_e}{S_a \times K_{fu}}} \times e \tag{13}$$

Where K_{fu} is the leakages flux coefficient.

To avoid the demagnetization of the magnets, the current of phase must be lower than demagnetization current I_d [9]:

$$I_d = \left(\frac{B_r - B_c}{\mu_r} \times H_a - B_c \times K_{fu} \times e \right) \times \frac{p}{2 \times \mu_0 \times N_s} \qquad (14)$$

Where B_c is the demagnetization induction, B_r is the residual induction of magnets and μ_0 is the permeability of air.

The height of the rotor yoke and the stator yoke is derived by application of the theorem of conservation of flux between the magnet and a rotor yoke, and between the stator tooth and the stator yoke [9]:

$$H_{cr} = \frac{B_e}{B_{cr}} \times \frac{Min(S_d, S_a)}{2 \times \left(\frac{D_e - D_i}{2} \right)} \times \frac{1}{K_{fu}} \qquad (15)$$

$$H_{cs} = \frac{B_e}{B_{cs}} \times \frac{Min(S_d, S_a)}{2 \times \left(\frac{D_e - D_i}{2} \right)} \qquad (16)$$

Where B_{cr} and B_{cs} are respectively the induction in the rotor yoke and in the stator yoke, S_d and S_a are respectively the section of a tooth and the one of a magnet and K_{fu} is the coefficient leakages flux.

4.2. Structures with Coiled Rotor

The distribution of the field lines at a stator pole is shown in figure 4:

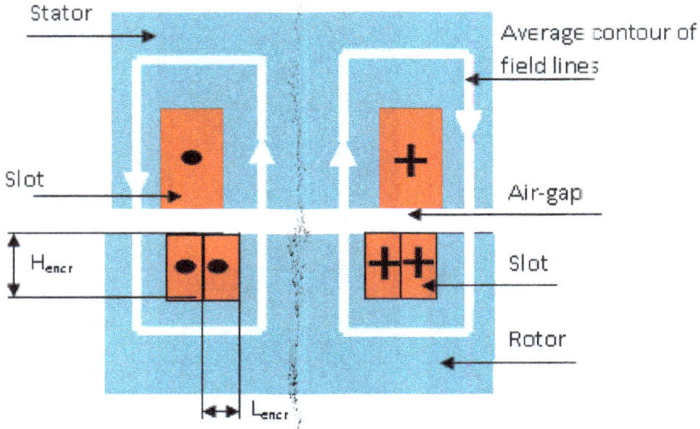

Figure 4. *Distribution of the field lines to a position of maximum recovery.*

The magnetic induction in the air gap is derived from the application of Ampere law on a closed contour of field lines:

$$\int_{\text{Filed lines}} \vec{H} \times \vec{dl} = 2 \times N_{sr} \times I_e = 2 \times e \times H_e \qquad (17)$$

The magnetic induction in the air gap is linear as a function of magnetic field:

$$B_e = \mu_0 \times H_e \qquad (18)$$

We deduce from relationships (17) and (18), expression of the induction in the air gap:

$$B_e = \mu_0 \times \frac{N_{sr}}{e} \times I_e \qquad (19)$$

The width of a rotor slot depend on the factor of shortening of the main tooth γ in relation to the polar step. In general, this coefficient doesn't pass 20% of the opening of the polar step:

$$L_{encr} = \frac{D_e - D_i}{4} \times \beta \times \frac{\pi}{p} \times \gamma \qquad (20)$$

The height of the rotor slots depends on the density of the current allowable in the copper δ and the fill factor of slot K_{rr}.

$$H_{envr} = \frac{e \times B_e}{\mu_0 \times K_{rr} \times \delta \times L_{encr}} \qquad (21)$$

4.3. Structures with Serial Excitation

The distribution of the field lines to the level of a stator pole is illustrated by figure 5.

The induction in the air gap generated by the magnet is derived by application of Ampere law on a closed contour of the field lines and the theorem of conservation of the flux between a magnet and a main tooth:

$$B_{ea} = \frac{H_a \times B_r}{\left(\frac{H_a \times S_d}{S_a \times K_{fu}} + \mu_r \times e \right)} \qquad (22)$$

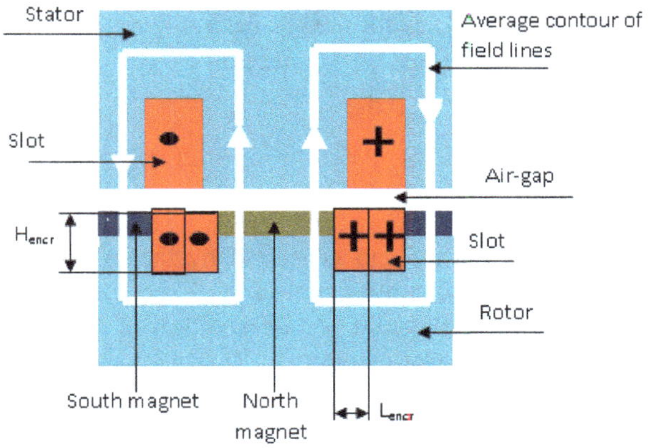

Figure 5. *Distribution of the field lines to a position of maximum recovery.*

The induction in the air-gap generated by the excitation of the motor by a current I_e is deducted by application of the ampere theorem on a middle contour of the field lines:

$$B_{ee} = \mu_0 \times \frac{N_{sr} \times I_e}{H_a + e} \qquad (23)$$

The application of the superposition field theorem in the air gap leads to the following expression for the induction in the air gap:

$$B_e = \frac{H_a \times B_r}{\left(\frac{H_a \times S_d}{S_a \times K_{fu}} + \mu_r \times e \right)} + \mu_0 \times \frac{N_{sr} \times I_e}{H_a + e} \qquad (24)$$

The height of the magnets is calculated iteratively from this equation, of a manner to have a magnetic induction B_e in the air-gap minimizing the mass of the motor (Inductions in the different active parts of the motor close to the elbow of saturation of the B-H curves) and guaranteeing a working in a linear régime.

The width of a rotor slot depend on the factor of shortening of the main tooth γ in relation to the polar step. In general this coefficient doesn't pass 20% of the opening of the polar step:

$$L_{encr} = \frac{D_e - D_i}{4} \times \beta \times \frac{\pi}{p} \times \gamma \tag{25}$$

The height of the rotor slot depend on the density of the admissible current in the copper δ and the slot filling factor K_{rr}.

$$H_{encr} = \frac{e \times B_e}{\mu_0 \times K_{rr} \times \delta \times L_{encr}} \tag{26}$$

4.4. Structures with Parallel Excitation

The distribution of the field lines to the level of a stator pole is illustrated by the figure 6.

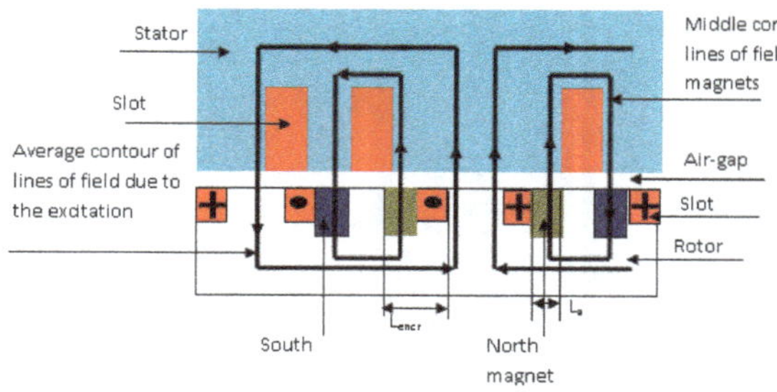

Figure 6. *Distribution of the field lines to a position of maximum recovery.*

The induction in the air gap generated by the magnet is derived by application of Ampere law on a closed contour of the field lines and the theorem of conservation of the flux between a magnet and a main tooth:

$$B_{ea} = \frac{H_a \times B_r}{\left(\frac{H_a \times 2 \times S_d}{S_a} + \mu_r \times e\right)} \tag{27}$$

The induction in the air gap generated by the motor by an excitation current I_e is derived by application of Ampere's law on an average contour of the field lines:

$$B_{ee} = \mu_0 \times \frac{N_{sr} \times I_e}{e} \tag{28}$$

This induction corresponds to the maximal induction in the air-gap.

The height of the magnets is calculated iteratively from the equation (31), of a manner to have a magnetic induction in the air-gap B_e minimizing the mass of the motor (Inductions in the different active parts of the motor near of the elbow of saturation of the B-H curves) and guaranteeing a working in a linear régime.

The width of a rotor slot depend on the collar shortening factor of the main tooth γ_c, doesn't pass 50% of the opening of the polar step in general:

$$L_{encr} = \frac{D_e - D_i}{4} \times \beta \times \frac{\pi}{p} \times \gamma_c \tag{29}$$

The height of the rotor slots depends on the density of the current allowable in the copper δ and the fill factor of slot K_{rr}:

$$H_{encr} = \frac{e \times B_e}{\mu_0 \times K_{rr} \times \delta \times (L_{encr} - L_a)} \tag{30}$$

4.5. Choice of Structure that Better Meets to Our Specifications

To refine the choice of best structure suited to the specifications we have taken as a criterion for performance comparison. And after a review of the literature [4, 8, 9] depth on the subject, it seemed that the permanent magnet synchronous machine is the best choice.

Finally, after a comparative study on these structures we opted for the double excitation synchronous machine series "MSDE" because of its good performance compared to the machines of the same type and for the power density, congestion the cost of reduced production and have a simple excitation coil (figure 7)

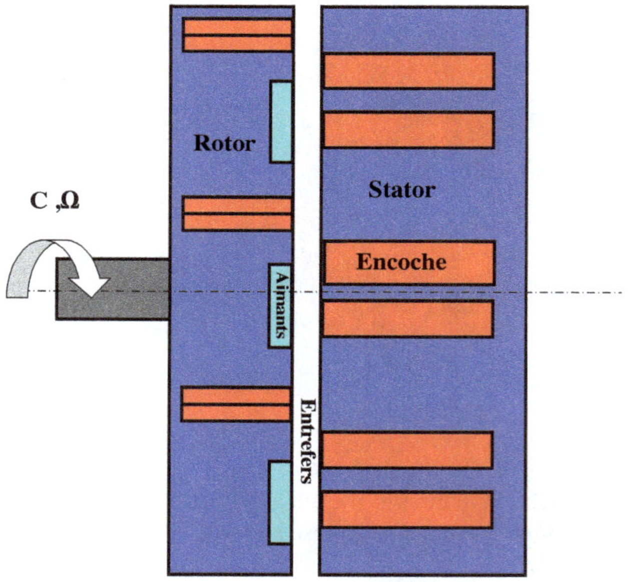

Figure 7. *Synchronies' machine excitation double.*

5. Validation of the Model

5.1. Distribution of Field Line

The distribution of field lines to load and load is shown in Figures 8 and 9:

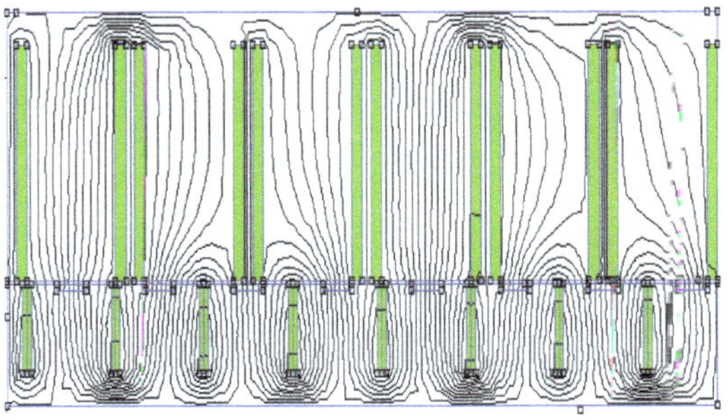

Figure 8. *Field lines of the motor at no load.*

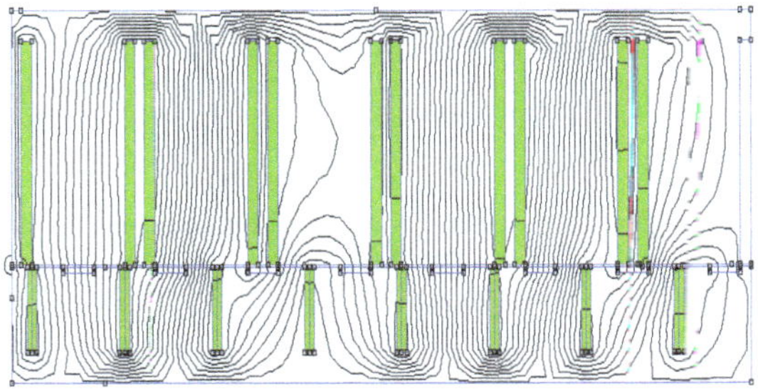

Figure 9. *Field lines of the motor at load.*

5.2. Flux

The figure 10 and 11 illustrate the trend of the flow of three phases to empty and loaded according to θ.

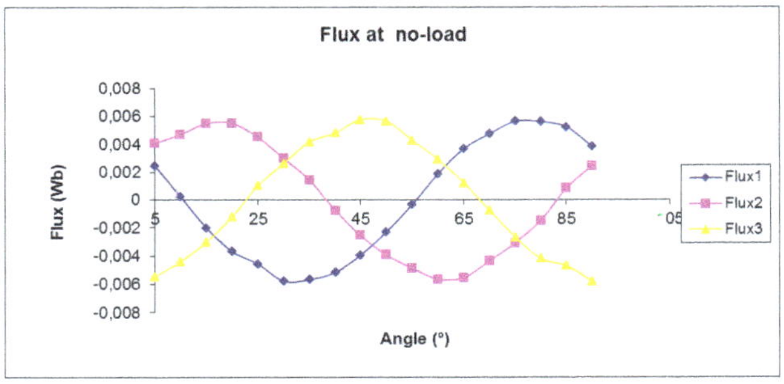

Figure 10. *Flux at no-load.*

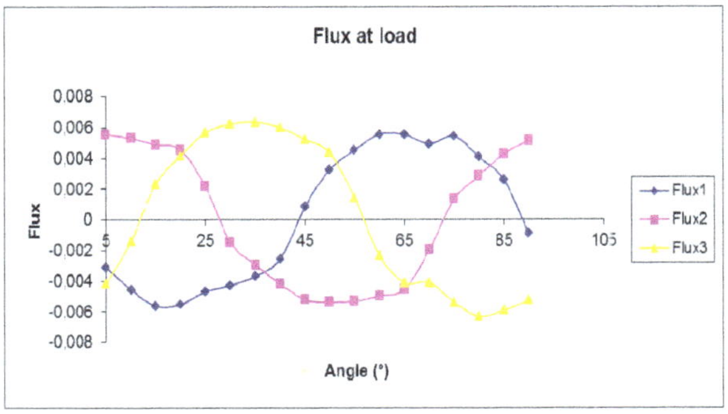

Figure 11. *Flux at load.*

We note from figures 13 and 14 that the flow undergoes a slight deformation load. This is due to the magnetic armature reaction. The flow reaches the value calculated analytically, which validates the analytical design approach.

5.3. Electromotive Forces Against

The figure 12 and 13 show the results obtained from the back M.F.E:

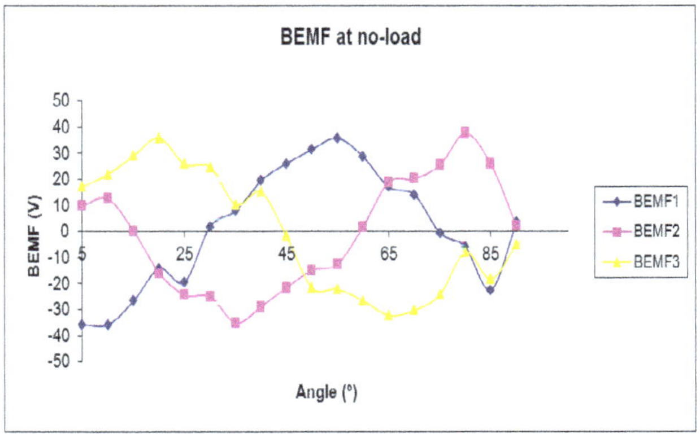

Figure 12. *Back M.F.E.at no-load.*

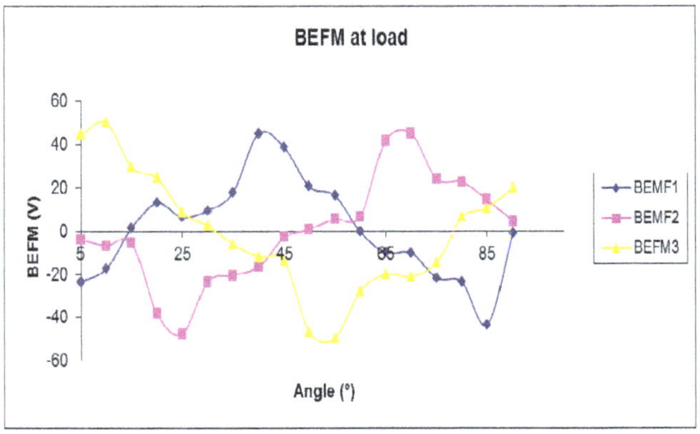

Figure 13. *Back M.F.E.at load.*

5.4. Electromagnetic Torque

The figure 14 illustrates the torque variation of motor according to the swing angle at load and at no load.

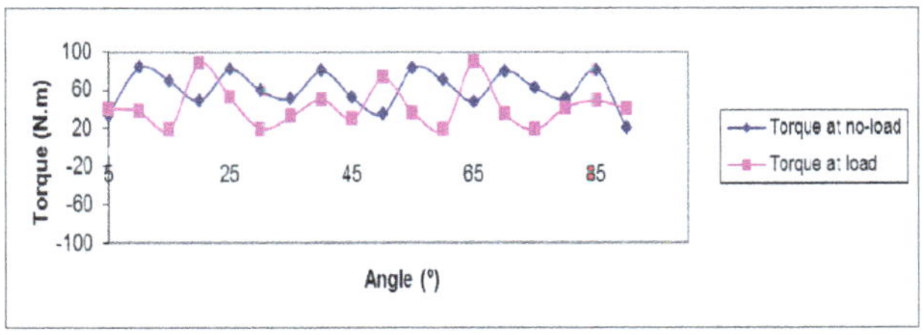

Figure 14. *Load torque and load.*

The torque ripple is mainly due to the presence of cogging torque

6. Conclusion

In this article, we presented a method for calculating the parameters of a synchronous dual actuator excitation by finite element method.

The results validate the analytical model. Indeed, in a first step, we determined the analytical values of flux, electromotive force against the torque calculated by the finite element simulations; in a second step we validate the model by finite element simulations.

This work will allow us to develop to allow control and optimization studies for these types of engines.

References

[1] M. HADJ KACEM, " Conception des Composants Electriques de la Chaîne de Puissance d'un Véhicule Electrique"; Thèse de Doctorat 2013. ENIS Tunisie.

[2] S. TOUNSI, M. HADJ KACEM et R. NEJI « Design of Static Converter for Electric Trac-tion », International Review on Modelling and Similations (IREMOS) Volume 3, N. 6, De-cember 2010, pp. 1189-1195.

[3] M.HADJ KACEM, S.TOUNSI et R. NEJI: « Systemic Design and Control of Electric Vehicles Power Chain »; IJSTR, volume1, n° 2012,pp.73-81.

[4] S.TOUNSI, R.NÉJI, F.SELLAMI: Design of a Permanent Magnet Actuator for Electric Vehicles. Revue Internationale de Génie Électrique volume 9/6 2006 - pp.693-718.

[5] Y. Amara, J. Lucidarme, M. Gabsi : A new topology of hybrid synchronous machine. IEEE Trans.Ind.Appl., vol .37, Issue 5, pp. 1273-1281.

[6] S. TOUNSI, N. BEN HADJ, R. NEJI et F. SELLAMI, "Optimization of Electric Motor Design Parameters Maximizing the autonomy of electric vehicles", International Review of Electrical Engineering (IREE), ISSN 1827-6660, Volume 2 N° 1, January-February 2007, pp.118-126..

[7] Y. Amara : « Contribution à la conception et à la commande des machines synchrones à double excitation. Application au véhicule hybride », Thèse de Doctorat, Université de Paris, 2001.

[8] S.TOUNSI: «Conception et Optimisation Systématiques de la Chaîne de Puissance des Véhicules Electriques»; Habilitation universitaire, Ecole Nationale d'Ingénieurs de Sfax - Tunisie, Mai 2012.

[9] D. FODOREAN: Design and implementation of a synchronous machine double excitation: Application to the direct drive: Doctoral thesis, 2005.

Design and optimization of axial flux brushless DC generator dedicated to generation of renewable energy

Souhir Tounsi

National School of Electronics and Telecommunications of Sfax, Sfax University, SETIT Research Unit, Sfax, Tunisia

Email address:
souhir.tounsi@isecs.rnu.tn

Abstract: In this paper, we present a design model of permanent magnet generator dedicated to generate renewable energy, taking in account of several systemic and physical constraints. Being couple to a model of the losses of the power chain and to a model of the mass of the generator, this analytic model puts a problem of conjoined optimization of the recovered energy and the cost of the generator. This problem is solved by genetic algorithms method.

Keywords: Renewable Energy, Design, Generator, Converters, Optimization

1. Introduction

A modular axial generator structure with permanent magnet reducing the cost of manufacture is chosen to generate renewable energy [1], [2], [3], [4] and [5].

We choice the analytic method to conceive the permanent magnet generator seen its compatibility to optimization approaches. Indeed, it's fast and product results quickly and without iterations.

The coupling of power chain losses model and the model of the generator mass to the program dimensioning the generator, pose an optimization problem. This last is solved by the software of optimization based on the Genetic Algorithm method.

2. Renewable Energy System

The system generating renewable energy comprises:

- A propeller attached to the rotor transmitting the mechanical energy caused by the movement of the air to the stator.
- A synchronous generator with permanents magnets to convert mechanical energy from the rotor in an alternating electrical energy.
- AC-DC converter to convert the alternative energy into continuous energy.
- DC-DC converter to elevate voltage loading batteries to optimize the recovered energy.
- An energy accumulator for energy storage.

3. Structures of the Electric Generator

3.1. Manufacturing Cost Reduction

The electric generator structure is modular i.e. it can be with several stages. This technology allows the reduction of the production cost of these generators types. The slots are right and open what facilitates the coils insertion and reduces the generator manufacturing cost. The concentrated winding is used because of its advantages:

- Reduction of the manufacturing time of this generator (insertion of coils in one block).
- Reduction of the end-windings.
- Reduction of the generator bulk.

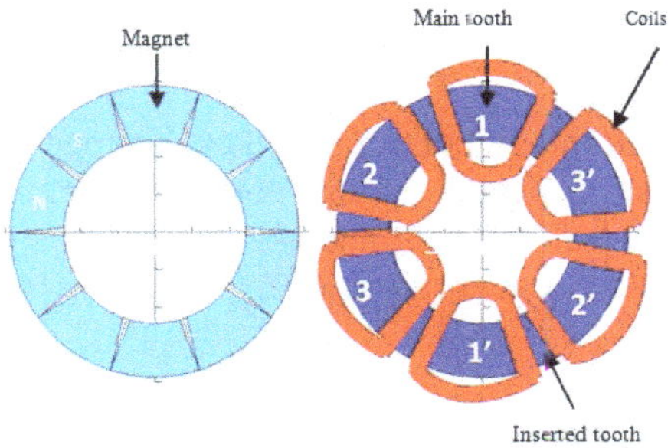

Figure 1. *5 pairs of poles, 6 main teeth, axial flux and trapezoidal configuration.*

Figure 1 illustrates the first trapezoidal configuration (n=1) with axial flux only one stage [6].

Five configurations with a trapezoidal wave-form are found while being based on optimization rules of the ripple torque and cost. Each configuration is characterized by a variation law of the pole pairs number (p) according to an integer number n varying from one to infinity, the ratio (r) of the number of main teeth (N_t) by the number of pole pairs, the ratio (ν) between the angular width between two main teeth and that of a principal tooth, the ratio (α) between the angular width of a principal tooth and that of a magnet and the ratio (β) between the angular width of a magnet and the polar step. Table .1 gives these ratios for these configurations [6], [7], [8], [9].

Table 1. *Found configurations.*

Trapezoidal configurations	p	r	ν	α	β
1	2.n	1.5	1/3	1	1
2	5.n	1.2	2/3	1	1
3	7.n	6/7	4/3	1	1
4	4.n	0.75	5/3	1	1
5	5.n	0.6	7/3	1	1

3.2. Design Methodology

We choose the analytic modelling of the generator, because it's compatible to the optimizations approaches [10], [11], [12], [13].

The worksheet computes the geometrical dimensions of rotor and stator as well as windings, temperature, inductance, leakages and efficiency for different operating points.

A sizing program is developed with equations detailed below. The program inputs are:

1. Generator specifications.
2. Materials properties.
3. Configuration, i.e. magnet number and teeth number.
4. Inner and outer diameter of the motor.
5. Notebook data.
6. Current density in coils δ.
7. Rotor yoke B_{ry}, stator yoke B_{sy}, flux density in the air-gap B_g and number of phase turn N_s.

When inputs 3. and 4. are set, magnet shapes, teeth and slots are fixed. Then, the area of one tooth A_t and the average length of a spire L_{sp} are calculated from geometric equations.

This model is validated by finite elements method. Indeed, the generator is drawn according to its geometrical magnitudes extracted from analytical model with the software Maxwell-2d, and is simulated in dynamic and static in order to compare the results obtained with those found by the analytical method.

The coupling of this model to a model evaluating the power train losses and generator mass, poses an optimization problem with several variables and constraints. This latter is solved by the genetic algorithms (GAs) method [10], [11], [12], [13].

4. Dimensioning Torque

The generator constant is defined by [10], [11], [12], [13]:

$$K_e = 2 \times N_s \times A \times B \times B_g \tag{1}$$

For the axial flux structures A and B are given by:

$$A = \frac{D_e - D_i}{2} \tag{2}$$

$$B = \frac{D_e + D_i}{2} \tag{3}$$

The dimensioning torque is given by the following relation:

$$T_{dim} = \frac{1.137 \times r \times v_{max}^3}{\Omega_{max}} \tag{4}$$

where V_{max} is the maximum air velocity, r is the radius of the rotor and Ω_{max} is the maximum angular speed of the generator.

The dimensioning current is expressed as follows:

$$I_{dim} = \frac{T_{dim}}{2 \times N_s \times A \times B \times B_g} \tag{5}$$

5. Generator Sizing

The air-gap flux density is calculated for a maximal recovery position, or the magnet is in front of a main tooth. At this position the air-gap flux density is maximal. The distribution of the field lines to the level of a pole is illustrated by the figure 2:

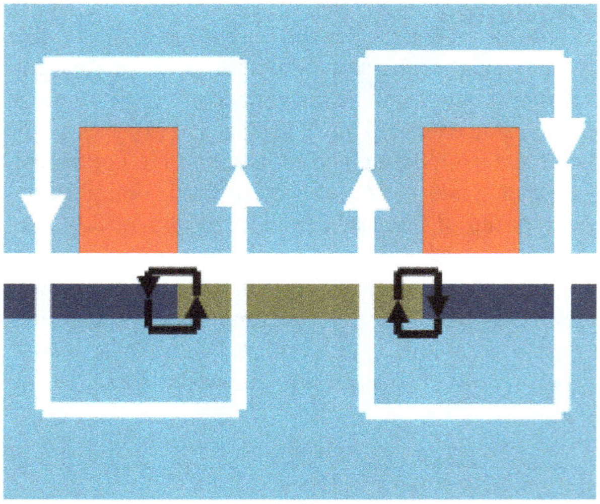

Figure 2. *Flux lines distribution at maximal recovery position.*

The flux decomposes itself in main flux and in leakages flux between magnets.

As applying the Ampere theorem to the level of a stator pole, we can deduct the flux density due to the stator current.

$$\int_{\text{flux lines}} \vec{H} \times \vec{dl} = \frac{N_s}{2} \times I_{max} = 2 \times (H_{ri} \times t_m + H_{ri} \times g) \tag{6}$$

where I_{max} is the stator maximal current, H is the magnetic field, H_{ri} is the air-gap magnetic field, t_m is the magnet thickness and μ_0 is the air permeability.

$$B_{ri} = \mu_0 \times H_{ri} \quad (7)$$

where B_{ri} is the flux density in the air-gap due to the stator current.

$$Bri = \frac{\mu_0}{4} \times \frac{N_s \times I_{max}}{t_m + g} \quad (8)$$

While applying the Ampere theorem, we can deduct the magnet thickness imposing a fixed flux density in the different zones of the motor while disregarding the flux density due to the stator current circulation, since the flux must cross two times the air-gap thickness and magnet with permeability very close to the air permeability.

$$\int_{\text{flux lines}} \vec{H} \times \vec{dl} = 0 = 2 \times (H_m \times t_m + H_g \times g) \quad (9)$$

The air-gap flux density is linear according to the magnetic field for this working regime:

$$B_g = \mu_0 \times H_g \quad (10)$$

While applying the flux conservation theorem to the level of the air-gap, we deduct the value of the air-gap flux density in function of the magnet flux density and the coefficient of the leakages flux.

$$B_m \times S_m \times K_{fu} = B_g \times S_m \quad (11)$$

The magnet flux density becomes:

$$B_m = \frac{B_g}{K_{fu}} \quad (12)$$

The magnet flux density is approached by the following linear equation:

$$B_m = \mu_0 \times \mu_m \times H_m + B_r \quad (13)$$

where μ_m is the magnet's relative permeability, B_r is the remanence.

From the equation (10), (11), (12) and (13), we deduct the magnet thickness fixing the air-gap flux density equal to B_g:

$$t_m = \mu_m \times \frac{B_g}{B_r - \frac{B_g}{K_{fu}}} \times g \quad (14)$$

where $K_{fu} < 1$ is the magnet's leakage coefficient and g is the air-gap thickness. To avoid demagnetization, the phase currents must be lower then the demagnetization current I_d [8]:

$$I_d = \left(\frac{B_r - B_{min}}{\mu_m} \times t_m - B_{min} \times K_{fu} \times g \right) \times \frac{p}{2 \times \mu_0 \times N_s} \quad (15)$$

where Bmin is the minimum flux density allowed in the magnets and µ0 is the air permeability. The rotor yoke thickness try and stator yoke thickness tsy derive from the flux conservation [9]:

$$t_{ry} = \frac{B_g}{B_{ry}} \times \frac{Min(A_t, A_m)}{2 \times A} \times \frac{1}{K_{fu}} \quad (16)$$

$$t_{sy} = \frac{B_g}{B_{sy}} \times \frac{Min(A_t, A_m)}{2 \times A} \quad (17)$$

where A_t is the tooth area, A_m is the area of one magnet, B_{ry} and B_{sy} are respectively the flux densities in rotor and stator yokes. For the axial flux and trapezoidal wave-form motor configurations the slot height is [9]:

$$h_s = \frac{3.2.N_s}{2N_t} \frac{I_{dim}}{\delta} \frac{1}{K_f} \frac{1}{A_s} \quad (18)$$

where N_t is the number of principal teeth, δ is the current density in slots, K_f is the slot filling factor, A_s is the slot width and I_{dim} is the dimensioning current:

$$I_{dim} = \frac{T_{dim}}{K_e} \quad (19)$$

The slot width is expressed as follows:

$$A_s = B \times SIN\left(\frac{1}{2} \times \left(\frac{2 \times \pi}{N_t} - \alpha \times \beta \times \frac{\pi}{p} \times (1 - r_{did}) \right) \right) \quad (20)$$

where r_{did} is the ratio between the angular width of the inserted tooth and that of the principal tooth. This ratio is optimized by finite elements simulations in order to reduce the flux leakages and to improve the electromotive force wave-form.

6. Optimization Problem

The optimization problem consists on the determination of the generator sizes minimizing its mass and the power train losses, while respecting the technological constraints of the application.

The generator weight is expressed as follows:

$$W_m = W_{sy} + W_t + W_c + W_{ry} + W_m \quad (21)$$

For the axial flux configurations the weight of stator yoke W_{sy}, tooth W_t, copper W_c, rotor yoke W_{ry}, and magnets W_m are expressed as follows:

$$W_{sy} = n \times d \frac{\pi}{4} \times (D_e^2 - D_i^2) \times t_{sy} \quad (22)$$

$$W_t = n \times d \times N_t \times A_t \times h_s \quad (23)$$

$$W_c = 3 \times n \times N_s \times L_{sp} \times \frac{I_{dim}}{\delta} \times d_c \quad (24)$$

$$W_{ry}=\pi\times\left(\left(\frac{D_e}{2}\right)^2-\left(\frac{D_i}{2}\right)^2\right)\times t_{ry}\times d \qquad (25)$$

$$W_m=2\times n\times p\times A_m\times t_m\times d_m \qquad (26)$$

where d is the density of the metal sheet, d_c is the density of copper, d_m is the magnet density, A_a is the magnet angular width, A_d is the angular width of principal teeth and A_e is the slot angular width.

For the trapezoidal wave-form configurations, the copper losses are expressed by the following relation:

$$P_c= 2\times R\times I^2 \qquad (27)$$

The phase resistance is given by the following expression:

$$R=r_{cu}(T_b)\times\frac{N_s\times L_{sp}}{S_c} \qquad (28)$$

where r_{cu} is the copper receptivity, L_{sp} is the average length of spire, T_b is the copper temperature and S_c is the active section of one conductor:

$$S_c=\frac{I_{dim}}{\delta} \qquad (29)$$

The iron losses are expressed by the following relation [14], [15], [16], [17], [18], [19]:

$$P_{fer}= C\times f^{1.5}\times\left(n\times W_t\times B_g^2 +n\times W_{sy}\times B_{sy}^2\right) \qquad (30)$$

where c is the core loss, f is the motor supplying frequency, W_t is the teeth weight, W_{sy} is the stator yoke weight, B_g is the ai-rgap flux density and B_{sy} is the flux density in stator yoke. The mechanical losses are expressed by the following relation [11]:

$$P_m=\left(T_b+T_{vb}+T_{fr}\right)\times\Omega \qquad (31)$$

where Ω is the angular speed of the electric generator.

The losses in the static converter are nearly hopeless, they are not held in account in the model of power train losses calculation.

$$P_{ptl}= P_c+P_{fer}+P_m \qquad (32)$$

7. Genetic Algorithms Optimization of the Generator Mass and the Power Train Losses

The function to optimize is expressed by the following expression:

$$F_o= W_m +a\,P_{ptl} \qquad (33)$$

Where "a" is a coefficient fixing the influence degree of P_{ptl} at the global objective function compared to W_m. Indeed, "a" brings closer the value of (a P_{ptl}) to the value of W_m.

The optimization problem consists in optimizing the F_o with respect to the problem constraints. In fact, Genetic Algorithms (GAs) are used to find optimal values of the internal diameter D_i, the external diameter D_e, the flux density in the air-gap B_g, the current density in the coils δ, the flux density in the rotor yoke B_{ry} , the flux density in the stator yoke B_{sy} and the number of phase spires N_s [14], [15], [16], [17], [18], [19].

The beach of variation of each parameter xi $\in$ (D_i, D_e, B_g, δ, B_{ry} , B_{sy}, Ns) must respect the following constraint: $x_{imin} \leq x_i \leq x_{imax}$. The values of the lower limit x_{imin} and the upper limit x_{imax} are established following technological, physical and expert considerations.

The F_o model is coupled to a program of optimization by the method of the genetic algorithm. The progress of the program of optimization of the F_o with constraints is described by this organization diagram (figure 3) [14], [15], [16], [17], [18]:

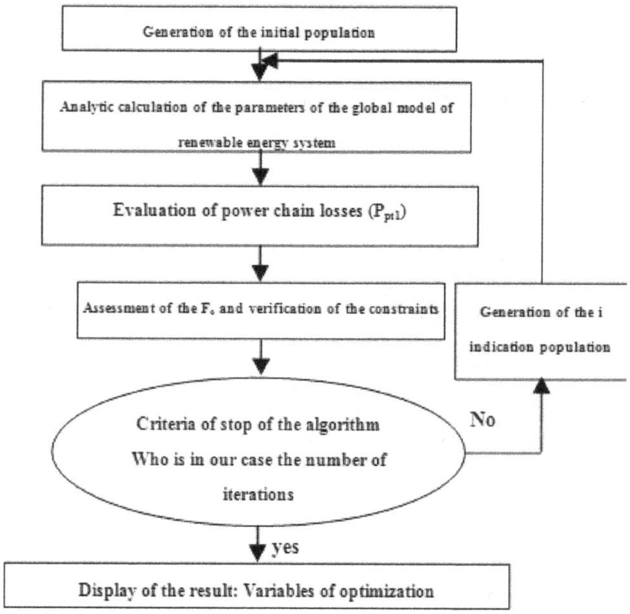

Figure 3. *Progress of the optimization program.*

8. Conclusion

An analytical model dimensioning the renewable energy generator is developed. This model is coupled to an optimization program in order to find the design prameters of the generator minimizing the power train energy losses and the generator cost. This study encourages the manufacture procedure the studied generator [9], [14], [15], [16], [17], [18], [19].

References

[1] Chaithongsuk, S., Nahid-Mobarakeh, B., Caron, J., Takorabet, N., & Meibody-Tabar, F. : Optimal design of permanent magnet motors to improve field-weakening performances in variable speed drives. Industrial Electronics, IEEE Transactions on, vol 59 no 6, p. 2484-2494, 2012.

[2] Rahman, M. A., Osheiba, A. M., Kurihara, K., Jabbar, M. A., Ping, H. W., Wang, K., & Zubayer, H. M. : Advances on single-phase line-start high efficiency interior permanent magnet motors. Industrial Electronics, IEEE Transactions on, vol 59 no 3, p. 1333-1345, 2012.

[3] C.C Hwang, J.J. Chang : Design and analysis of a high power density and high efficiency permanent magnet DC motor, Journal of Magnetism and Magnetic Materials, Volume 209, Number 1, February 2000, pp. 234-236(3)-Publisher: Elsevier.

[4] MI. Chunting CHRIS : Analytical design of permanent-magnet traction-drive motors" Magnetics, IEEE Transactions on Volume 42, Issue 7, July 2006 Page(s):1861 - 1866 Digital Object Dentifier 10.1109/TMAG.2006.874511.

[5] S.TOUNSI, R.NÉJI, F.SELLAMI : Conception d'un actionneur à aimants permanents pour véhicules électriques, Revue Internationale de Génie Électrique volume 9/6 2006 - pp.693-718.

[6] Sid Ali. RANDI : Conception systématique de chaînes de traction synchrones pour véhicule électrique à large gamme de vitesse. Thèse de Doctorat 2003, Institut National Polytechnique de Toulouse, UMRCNRS N° 5828.

[7] C. PERTUZA : Contribution à la définition de moteurs à aimants permanents pour un véhicule électrique routier. Thèse de docteur de l'Institut National Polytechnique de Toulouse, Février 1996.

[8] S. TounsI, R. NEJI and F. SELLAmI: Mathematical model of the electric vehicle autonomy. ICEM2006 (16th International Conference on Electrical Machines), 2-5 September 2006 Chania-Greece, CD: PTM4-1.

[9] R. NEJI, S. TOUNSI, F. SELLAMI: Contribution to the definition of a permanent magnet motor with reduced production cost for the electrical vehicle propulsion. Journal European Transactions on Electrical Power (ETEP), Volume 16, issue 4, 2006, pp. 437-460.

[10] P. BASTIANI : Stratégies de commande minimisant les pertes d'un ensemble convertisseur machine alternative : application à la traction électrique. Thèse INSA 01 ISAL 0007, 2001.

[11] G. Henriot : Traité théorique et pratique des engrenages : théorie et technologie 1. tome 1 Edition Dunod 1952.

[12] D-H. Cho, J-K. Kim, H-K. Jung and C-G. Lee: Optimal design of permanent-magnet motor using autotuning Niching Genetic Algorithm, IEEE Transactions on Magnetics, Vol. 39, No. 3, May 2003.

[13] Islam, M. S., Islam, R., & Sebastian, T. : Experimental verification of design techniques of permanent-magnet synchronous motors for low-torque-ripple applications. Industry Applications, IEEE Transactions on, vol 47 no 1, p. 88-95, 2011.

[14] Parasiliti, F., Villani, M., Lucidi, S., & Rinaldi, F. : Finite-element-based multiobjective design optimization procedure of interior permanent magnet synchronous motors for wide constant-power region operation. Industrial Electronics, IEEE Transactions on, vol 59 no 6, p. 2503-2514, 2012.

[15] Mahmoudi, A., Kahourzade, S., Rahim, N. A., & Ping, H. W. : Improvement to performance of solid-rotor-ringed line-start axial-flux permanent-magnet motor. Progress In Electromagnetics Research, 124, p. 383-404, 2012.

[16] Duan, Y., & Ionel, D. M. : A review of recent developments in electrical machine design optimization methods with a permanent-magnet synchronous motor benchmark study. Industry Applications, IEEE Transactions on, vol 49 no 3, p. 1268-1275, 2013.

[17] Liu, G., Yang, J., Zhao, W., Ji, J., Chen, Q., & Gong, W. : Design and analysis of a new fault-tolerant permanent-magnet vernier machine for electric vehicles. Magnetics, IEEE Transactions on, vol 48 no 11, p. 4176-4179, 2012.

[18] Lee, S., Kim, K., Cho, S., Jang, J., Lee, T., & Hong, J. : Optimal design of interior permanent magnet synchronous motor considering the manufacturing tolerances using Taguchi robust design. Electric Power Applications, IET, vol 8 no 1, 23-28, 2014.

[19] TOUNSI, R. NEJI and F. SELLAMI : Electric vehicle control maximizing the autonomy : 3rd International Conference on Systems, Signal & Devices (SSD'05), SSD-PES 102, 21-24 March 2005, Sousse, Tunisia.

Design of Voice Coil Type Linear Actuator for Hydraulic Servo Valve Operation

Baek Ju Sung

Korea Institute of Machinery&Materials, Daejeon, Korea

Email address:
sbj682@kimm.re.kr (Baek Ju Sung)

Abstract: In this study, we proposed governing equations for voice coil type linear actuator for valve operation. We draw up governing equations which are composed by combination of electromagnetic theories and empirical knowledge, and deduct the values of major design factors by use of them. We suggested the governing equations to determine the values of design parameters of linear actuator as like bobbin size, length of yoke and plunger and turn number of coil. And we also calculated the life test time of linear actuator for verification of reliability of the prototype. In addition, for reducing the life test time, the acceleration model of linear actuator is proposed and the acceleration factor is calculated considering the field operating conditions. Finally we have proven the propriety of the governing equations by accelerated life test using the valve assembly adopted the voice coil type linear actuator prototype.

Keywords: Linear Actuator, Permanent Magnet, Governing Equation, Design Factor, Voice Coil Type, Accelerated Model, Life Test, Frequency Response Test

1. Introduction

The linear actuator is a very economical motion converter due to its simple structure as electromagnetic energy converting to kinetic energy. And the linear actuator is used as key components in automobile and aircraft industry. For having higher response time and product reliability, two kinds of different techniques are needed. One is the optimal design method for linear actuator. A regarded point for design of linear actuator is flux density analysis, determination of plunger shape and mass, optimal bobbin design, selected magnetic analysis, determination of duty ratio, and calculation of coil turn number which is regarded temperature rising. For the optimal design of the linear actuator, theoretical and empirical knowledge are simultaneously needed. Theoretical knowledge governs the operational characteristics of the linear actuator, and empirical knowledge compensates for the theoretical limitation obtained from the designer's design and manufacturing experiences for various kinds of linear actuator [1]. They cannot be determined solely by calculation or simulation because the empirical knowledge is more essential than theoretical knowledge for determination of the plunger shape and value of the space factor. When designer's accumulated experiences and expertise are added to these, the most proper shape and value of them can then be obtained.

In this study, the governing equations for design of linear actuator were derived by a combination of electromagnetic knowledge and empirical knowledge. And also the no-failure test time of voice coil type linear actuator is calculated. In particular, for reducing of the no-failure test time, the acceleration model of linear actuator is proposed, and the acceleration factor is calculated with the reality. The validity of the proposed design method and deducted reliability parameters are proved by accelerated life test and performance test

2. Governing Equations

2.1. Structure of High Speed Linear Actuator

Fig.1 shows the structure of high speed linear actuator for valve operation. It is composed of an excitation coil for generation of magnetic field, yoke for flux path, plunger for creation of mechanical stroke, stationary for attraction of the plunger, bearing for guidance of movement, and centering springs [2].

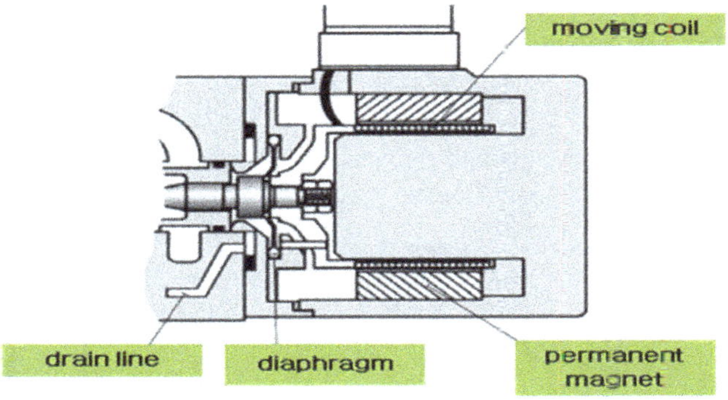

Figure 1. Structure of linear actuator.

Fig. 2 represents the simplified structure of high speed linear actuator, where the permanent magnets are excluded. Permanent magnet independently compensates the electromagnetic force of solenoid coil, and it contributes the reduction of consumption power and increasing of operational speed in comparison with the case of only used solenoid coil.

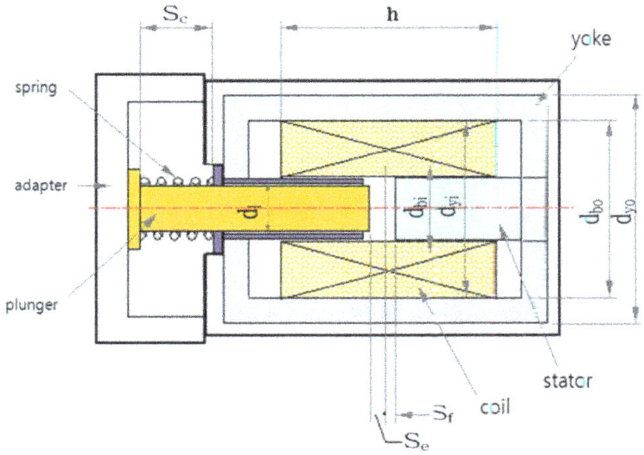

Figure 2. Simple structure of linear actuator.

2.2. Magnetic Flux Density and Magnetic Motive Force in Air Gap

The attraction force F is shown in equation (1) in the magnetic circuit of Fig. 2 [1] [3] [8].

$$F = \frac{B^2 \bullet S}{2\mu_0} [N] \tag{1}$$

Where S is cross sectional area of plunger, as it were, it is $\pi\left(\frac{d_l}{2}\right)^2$ when d_l is radius of plunger, and μ_0 is permeability in the air.

Therefore from equation (1), the magnetic flux density B needed in air gap is expressed as equation (2).

$$B = \sqrt{\frac{F \bullet 2\mu_0}{S}} \tag{2}$$

And theoretical magneto motive force U_m are shown in equation (3) and, d is maximum distance between plunger and stationary.

$$U_m = \frac{B \bullet d}{\mu_0} \tag{3}$$

Equation (4) is obtained from equations (1) and (3), and also, the design coefficient K_f can be expressed as equation (5).

$$F = \frac{K_f}{d^2} \tag{4}$$

$$K_f = \frac{\mu_0 \bullet S \bullet U_m^2}{2} \tag{5}$$

When the length of fixed air gap is S_f in Fig.2, the maximum distance d between plunger and stationary is given to equation (6) that is represented by the sum of fixed air gap S_f and plunger stroke. So, the maximum attraction force F_{max} and the minimum attraction force F_{min} become equations (7) and (8), respectively.

$$d = S_f + S_e \tag{6}$$

$$F_{max} = \frac{K_f}{S_f^2} \tag{7}$$

$$F_{min} = \frac{K_f}{d^2} \tag{8}$$

2.3. Permanent Magnet and Flux Density in Air Gap

By reference [5-6], the total magnetic flux density in the air gap generated as equation (9) [5-6].

$$B_g = \frac{B_r \bullet h_M}{\frac{A_g}{A_m} h_M + \mu_M S_f} \tag{9}$$

And the permeability of permanent magnet μ_M is like equation (10).

$$\mu_M = \frac{B_r}{H_c \mu_0} \tag{10}$$

From equation (9), we can know that the magnetic flux density of air gap approaches to the residual magnetic flux density when the length of permanent magnet is long and the length of air gap is completely short.

For decision of operating point of permanent magnet, we must consider the maximum energy area of permanent magnet and the reduced magnetic flux due to reaction of magnetic field by solenoid coil. But, in this paper, the change of characteristic of permanent magnet may not be occurred because the operating point of permanent magnet resulted from completely short length of air gap and path of magnetic flux.

2.4. Estimation of Yoke Thickness

Referring Fig.2, inner diameter d_{yi} and outer diameter d_{yo} of yoke are as equation (11) and (12).

$$d_{yi} = d_{bo} + C_g \tag{11}$$

$$d_{yo} = \sqrt{d_{yi}^2 + C_p \bullet d_l^2} \tag{12}$$

$$\text{Yoke thickness} = (d_{yo} - d_{yi}) / 2 \tag{13}$$

The empirical constant C_g in equation (11) is the length margin for smooth heat dissipation of the coil, and the empirical constant C_p in equation (12) is the length margin for smooth passing of magnetic flux [7] [9-10]. The proper values of the experience coefficients can be decided by designer's judgment depend on his experience and electromagnetic knowledge.

2.5. Temperature Rising and Bobbin Length

Heat dissipation coefficient λ is the amount of heat energy radiated form the coil surface. It can be founded in Fig. 3 [1].

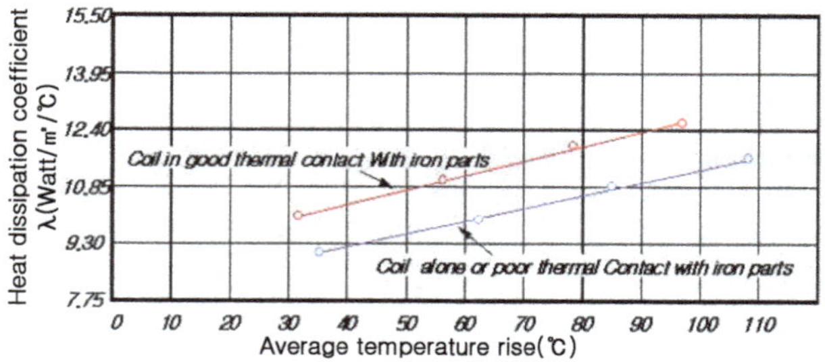

Figure 3. Heat dissipation coefficient according to temperature rising [1].

R and I passing through it produce the T_f in equation (14). By substituting equations (15) and (16) into equation (14), we can make the constructive equation of final temperature rising as equation (17). Equation (14) is usually used temperature rising equation in the coil [1].

$$T_f = \frac{W}{2 \bullet \lambda \bullet S} = \frac{I^2 \bullet R}{2 \bullet \lambda \bullet S} \tag{14}$$

$$R = \rho \frac{(l_m \bullet N^2)}{h \bullet w \bullet X_i} \tag{15}$$

$$X_i = \frac{\pi}{4} \left(\frac{d_s}{d_0} \right)^2 \tag{16}$$

$$T_f = \frac{q \bullet \rho}{d \bullet \lambda \bullet X_i \bullet w} \bullet \left(\frac{N \bullet W}{h \bullet V} \right)^2 \tag{17}$$

Where S is area of heat dissipation.

Bobbin length (coil height) used in equation (15) is calculated by equation (18).

$$h = \sqrt[3]{\frac{(q \bullet \beta \bullet \rho \bullet U^2)}{2 \bullet \lambda \bullet X_i \bullet T_f}} \tag{18}$$

That is, β is equal to $\frac{h}{w}$ referring to Fig.5, which is shown in the detailed drawing of the bobbin and yoke in Fig. 1 [7].

2.6. Number of Turns and Consumption Power Coil

Mean length of coil l_m turn is represented as equation (19).

$$l_m = \frac{\pi(d_{bo} + d_{bi})}{2} \tag{19}$$

And the relation between equivalent resistance R_t of solenoid circuit using copper wire, supply voltage V, current I, and relative resistance ρ and be expressed by equation (20).

$$R_t = \frac{V}{I} = 4\rho \left[\frac{(l_m \bullet N)}{\pi \bullet d_s^2} \right] \tag{20}$$

Diameter of bare wire, d_s is induced to equation (21) from equation (20).

$$d_s = \sqrt{\left(\frac{2 \bullet \rho \bullet (d_{bo} + d_{bi}) \bullet U}{V} \right)} \tag{21}$$

If it is assumed that insulated wire diameter is d_0 and the winding loss of a winding layer is 1 turn, the total turns number to be winded n_c in shaft direction given in equation (22). And, the total layer number m_c of coil in the radial direction is given by equation (23).

$$n_c = \left(\frac{h}{d_0} \right) - 1 \tag{22}$$

$$m_c = \frac{w}{d_0} \tag{23}$$

Therefore the total turn number N to be winded on the bobbin can be given by equation (24).

$$N = n_c \bullet m_c \tag{24}$$

By combining equations (20) and (21), the equivalent resistance R_t, which represents the total resistance of coil, is fully obtained by equation (25).

$$R_t = \frac{2 \bullet \rho \bullet (d_{bo} + d_{bi}) \bullet N}{\pi d_s^2} \tag{25}$$

According to determination of R_t, the equations of coil current I and consumption power W are determined by equations (26) and (27), respectively.

$$I = \frac{V}{R_t} \tag{26}$$

$$W = V \bullet I \tag{27}$$

2.7. Operating Frequency

The solenoid actuator can be expressed as return spring –mass system like Fig. 4. After applying power, plunger displacement is equivalent to the displacement of mass mp [8].

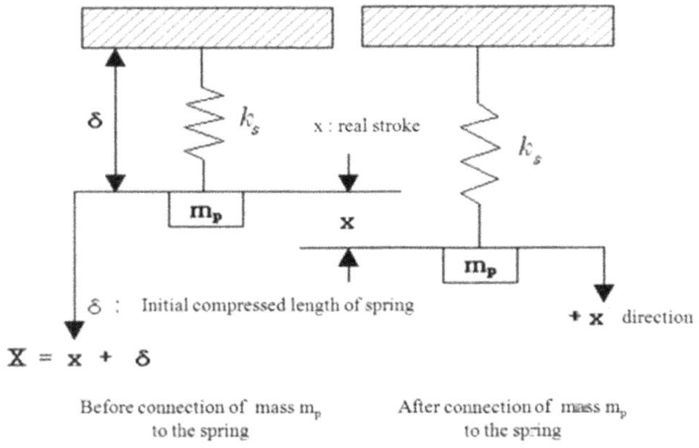

Figure 4. Mechanical model of solenoid actuator.

The state equation of mass m_p to the x-direction is equation (28). By substitution of $X = x + \delta$ to equation (28), we can achieve equation (29).

$$m_p \ddot{X} + k_s X = m_p g \tag{28}$$

$$m_p \ddot{x} + k_s x = 0 \tag{29}$$

Therefore, mathematical model about the system of Fig. 4 become to equation (34).

$$\ddot{x} + \omega^2 x = 0 \tag{30}$$

Here, the operating speed ω and operating frequency f_p of the actuator can be expressed by equation (31) and (32), respectively [8].

$$\omega = \sqrt{\frac{K_s}{m_p}} \tag{31}$$

$$f_p = \frac{\omega}{2\pi} \tag{32}$$

3. Design Program

The target specifications of prototype actuator for the hydraulic valve are as shown in Table 1.

Table 1. Material properties of SCP10.

Items	Target performance
Supply voltage	24 V
Consumption power	55 W
Operating frequency	100 Hz
Attraction force	160 N

Fig.5 shows the flow chart of the developed design program which is programmed by use of the governing equations and empirical coefficients in Chapter 2.

Designer's judgment means designer's experience which is needed to judge the fact whether the final design parameters are proper for manufacturing of target actuator or not.

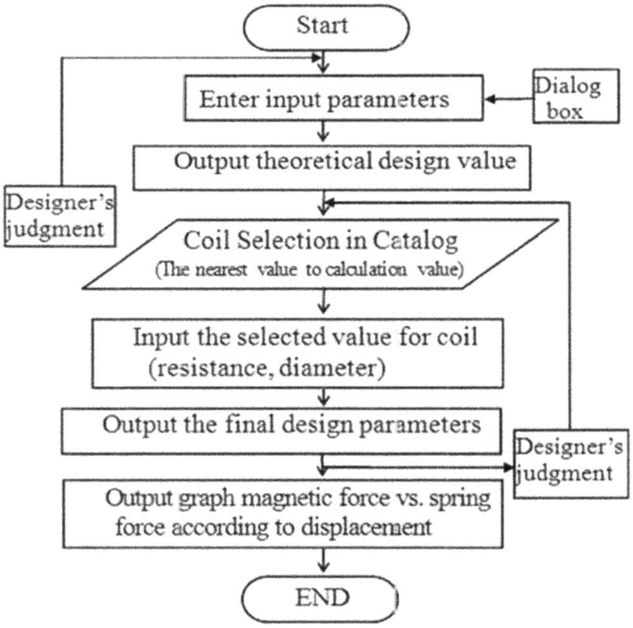

Figure 5. Flow chart of design program.

The input parameters and their values needed for design of the prototype are introduced in Table 2.

Table 2. Input parameter.

Items	Input value
F_{min} [N]	160
d_l [mm]	4
t_b [mm]	1
n_c	59
m_c	8
C_g	0.004
C_p	1.25
V [V]	24
mp [g]	9.4
Ks [kgf/mm]	4.5

important output variable	
Attraction force [N]	
F_{min}	160
Magnetic flux density [T]	
$B=2*[root(2*\mu_0*F_{min})/(d_l*root(\pi))$	2.334
bobbin inner diameter [mm]	
$d_{bi}=d_l+2(r_{air}+t_b)$	38
bobbin outer diameter [mm]	
$d_{bo}=d_{bi}+2w$	50
turn number of coil [No]	
$N=n_c*m_c$	502.775
yoke inner diameter [mm]	
$d_{yi}=d_{bo}+C_g$	50.004
yoke outer diameter [mm]	
$d_{yo}=root(d_{yi}^2+C_p*d_l^2)$	54.17
equivalent resistance [Ω]	
$R_t=(2*\rho*(d_{bo}+d_{bi})*N)/d_s^2$	10.95
coil current [A]	
$I=V/R_t$	2.19178082
magnetic motive force (at 20℃)[A•T]	
$U_{20}=I*N$	1101.9726
consumption power (at 20℃)[W]	
$W_{20}=V*I$	52.6027397
operating frequency [Hz]	
$f_p=\omega/2\pi$	110

Figure 6. *Result of design.*

By input of the values in Table 2 to the design program, we can obtain the results of Fig.6 as a numerical output.

4. Manufacturing of Prototype

The prototype is manufactured by based on the design results and FEM analysis in previous chapters. Solenoid in made by plastic bobbin and copper coil, plunger is made by processing core steel, and yoke tube is installed for offering the smooth magnetic flux path between these. And, the permanent magnets are installed in front and rear of plunger for more increasing the operating speed.

For preventing the sticking of plunger onto the core tube, non-magnetic bushings of which thickness is 0.3~0.5 t is inserted in front and rear of the core tube. The permanent magnets must put together to be confronted the same magnetic poles, at here, we put together the S poles are to be confronted by base of plunger. And, two pieces of centering springs are installed for security of neutral position of plunger in both side permanent magnets.

In addition, solenoid bearing is inserted in front of core tube. This plays a role to minimize the friction force and make the plunger locate in spatial center of inner core tube. Push rod is unified with plunger by laser welding after pressing, the rod bar transfers the attraction force of plunger to spool.

The components of manufacturing prototype shows in Fig.7, and the assembly is also shows in Fig.8.

Figure 7. *Prototype of voice coil type actuator.*

Figure 8. *Hydraulic valve assembly.*

5. Acceleration Model of Linear Actuator

Domestic industries surveyed integral servo valve operating conditions the lifetime of the field by considering the 90% confidence level B_{10} life of 1.0*10^7 cycles that were guaranteed. According to the survey of the literature, shape parameter of 1.1 Weibull distributions follows. Reliability standards for the evaluation of servo valve in the prescribed lifetime of 1.0*10^7 cycles (B_{10} life) means to guarantee the following.

-Lifetime distribution : Shape parameter(β) 1.1 Weibull distribution[10]

- Insurance life : 1.0*10^7 cycles(B_{10} Lifetime)
- Confidence level : 90 %
- Prototype : 3ea

At this point, no-failure test time was calculated equation (37) using, the result is 6.1*10^7 cycles.

$$t_n = B_{100p} \bullet \left[\frac{\ln(1-CL)}{n \bullet \ln(1-p)}\right]^{\frac{1}{\beta}} \quad (37)$$

$$t_n = 1.0\times10^7 \bullet \left[\frac{\ln(1-0.9)}{3 \bullet \ln(1-0.1)}\right]^{\frac{1}{1.1}} \cong 1.0\times10^7 cycles$$

Where,

t_n : No failure test time

B_{100p} : Assurance life

CL : Confidence level

n : Number of prototype

p : Unreliability (if B_{10}, p=0.1)

β : Shape parameter

However, because no-failure test time is too long to accelerate the model chosen, and accelerated life test of time should be calculated. Failure modes related to the pressure and flow of the servo valve. Pressure and flow are chosen to acceleration stress. Considering the pressure and flow General Log-Linear acceleration model applied to the test conditions. So the acceleration factor is calculated acceleration time fault-tolerance test. 7.0 MPa, 50 L/min and acceleration, conditions 25.2 MPa, 88 L/min was chosen as the acceleration factor calculation Thus, equation (38) 22.8096.

$$AF = \left(\frac{P_{test}}{P_{field}}\right)^m \times \left(\frac{F_{test}}{F_{field}}\right)^l \quad (38)$$

$$= \left(\frac{25.2}{7.0}\right)^2 \times \left(\frac{88}{50}\right)^1 = 22.8096$$

Where,

AF : Acceleration Factor

P_{test} P_{field} : Acceleration & field pressure (MPa)

ω_{test} ω_{field} : Acceleration & field flow (L/min)

m, l : Acceleration index (m =2, l =1)

Calculated acceleration factor equation (39) by substituting the acceleration test, time (t_{na}) is produced.

$$t_{na} = \frac{t_n}{AF} = \frac{61,000,000}{22.8096} \cong 2.7\times10^6 cycles \quad (39)$$

6. Life Test and Performance Tests

6.1. Accelerated Life Test

The propriety of the design equations has been proven equations through the accelerated life test. the 3 units the valve assembly adopted the linear actuator are used to life test. Fig. 9 shows the accelerated life test.

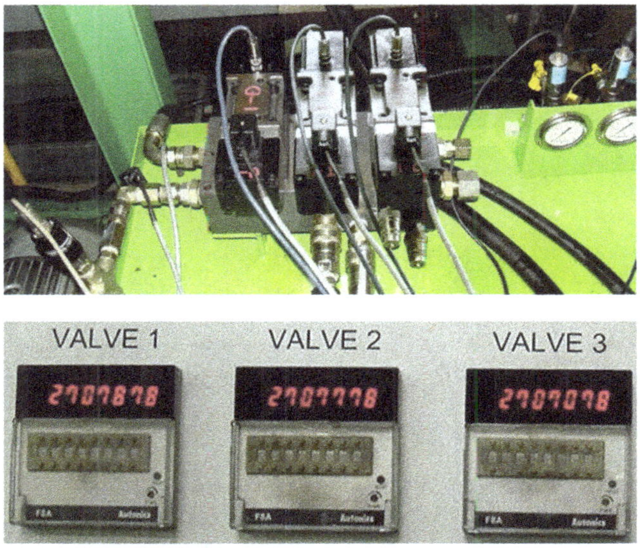

Figure 9. Accelerated life test.

6.2. Attraction Force and Linearity Test

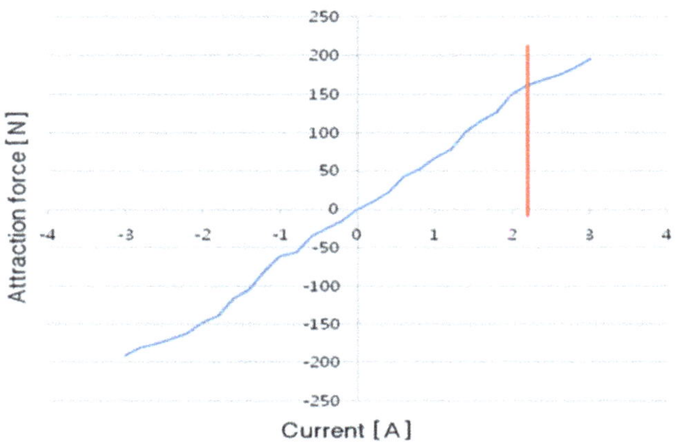

Figure 10. Result of attraction force test.

For attraction test, firstly, the prototype linear actuator is to be fastened on the attraction force test equipment, and it is connected to load cell by mechanical coupling. The attraction force should be measured changing the value of current form 0 to +3 A and form 0 to -3 A. Fig. 10 represents the measuring result of the attraction force. From Fig. 10, the attraction force is about 153 N at rated current ±2.2 A. And, the linearity is almost approaching to the first order function, f(x)=3.5x. At this time, for overall region, the error rate of linearity is 1.90 %, and the error rate of symmetrical characteristic is 3.05~-2.00 %. These mean that the test results for attraction force and linearity are generally satisfactory to the target performance in the table 1.

6.3. Test of Step Response

This test is to measure the time difference between supplying time of input step signal and reaction time of plunger. At here, the 100 % control signal(10 V) to controller is used as input step signal, and reaction of plunger is detected by output signal of LVDT. Referring to fig. 11, the step response time is 3.8 ms.

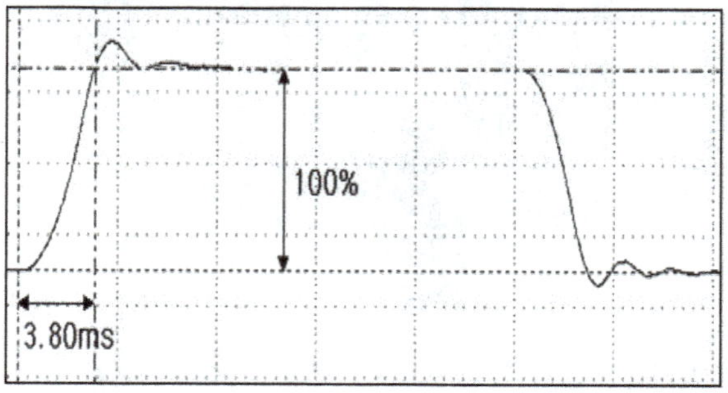

Figure 11. Result of step response test.

6.4. Test of Frequency Response Test

This test is similar to the test of step response. The input is control signal of controller and output is reaction signal of LVDT. This test performed at 25 % magnitude of input signal with 0.01 Hz ~ 500 Hz carrier frequency region.

Fig.12 is the test results for 25 % control signal of controller. It shows that the -3dB frequency is about 187 Hz in gain and 330 Hz in phase.

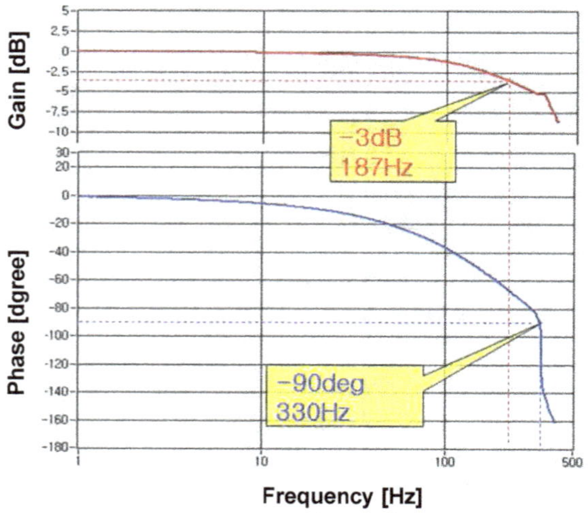

Figure 12. Result of frequency response test.

7. Conclusion

In this paper, all design courses of voice coil type high speed actuator for valve operation have been introduced. The final results are as follows:

1) The governing equations are induced for design using between electromagnetic theories and empirical knowledge. The important values of the design factors are decided as the results of optimal design through the governing equations.
2) For experiments, a prototype of the voice coil type actuator using the above design results was manufactured. As results of experiments, the attraction force test is less than target performance. But the frequency response test is better than target performance. It's difference is 230Hz. The important performance as linear actuator is high speed response for the signal.
3) These test results mean that the performance of the prototype linear actuator was satisfactory for the specifications of general high speed actuator for valve operation, and the induced governing equations are propriety for optimal design of voice coil type high speed actuator for valve operation.

Nomenclature

A_g	cross sectional area of air gap
A_m	pole area of permanent magnet
B	magnetic flux density
B_g	total magnetic flux density in the air gap
B_r	residual magnetic
C_g	empirical constant 1
C_m	empirical compensation coefficient
C_p	empirical constant 2
d	maximum distance between plunger and stationary
d_{bo}	outer diameter of bobbin
d_{bi}	inner diameter of bobbin
d_l	radius of plunger
d_o	diameter of insulated coil
d_s	diameter of bare wire
d_{yi}	inner diameter of yoke
d_{yo}	outer diameter of yoke
F	attraction force
F_{max}	maximum attraction force
F_{min}	minimum attraction force
f_p	operating frequency
h	coil height

h_M	length of permanent magnet
I	current
K_f	design coefficient
k_s	spring constant
l_m	coil mean length per single turn
m_c	total layer number
m_p	mass
N	total turn numbers
n_c	total turn number to be winded
q	duty ratio
R	resistance
R_t	equivalent resistance
S	cross sectional area of plunger
S_e	plunger stroke
S_f	length of fixed air gap
T_f	rising temperature
U	actual magneto motive force
U_m	theoretical magneto motive force
V	supply voltage
W	consumption power
w	coil layer thickness
β	ratio of bobbin height
δ	initial compressed length of spring
λ	heat dissipation coefficient
μ_0	permeability in the air
μ_M	permeability of permanent magnet
ρ	relative resistance

References

[1] C. Roters, "Electro Magnetic Device", John Wiley & Sons, Inc, 1970.

[2] B. J. Sung, E. W, Lee, H. E. Kim, "Development of Design Program for On and Off Type Solenoid Actuator: Proceedings of the KIEE Summer Annual Conference 2002(B), pp929~931, 2002.7.10.

[3] William H. Hayt, "Engineering Electromagnetics", Mc Grawhill, 1986.

[4] T. Kajima, "Dynamic Model of the Plunger Type Solenoid at deenergizing State", IEEE Transactions on Magnetics, Vol.31, No.3, pp2315~2323, May 1995.

[5] Syed A. Nasar, I. Boldea "Solenoid Electric Motors" Prentice-Hall Inc englewood Cliffs, New Jersey 1987.

[6] H.-D. Stolting, A. Beisse, "Elektrische Kleinmaschinen", B. G. Teubner Stuttgart, 1987.

[7] Hydraulic and Pneumatic Lap of KIMM, "Development of low Consumption Power Type Solenoid Valve", KIMM-CSI annual report, 2001.12.

[8] K. Ogata, "System Dynamics", Prentice Hall, 1998.1.

[9] B. J. Sung, E. W. Lee, H. E. Kim, "Characteristics of Non-magnetic Ring for High-Speed Solenoid Actuator", The eleventh Biennial IEEE Conference on Electromagnetic Field Computation, pp342, Korea, June 2004.

[10] Kanda Kunio, "Design Concept for DC Solenoid of Pneumatic Valve", KIMM research reporter, 1997.

Systemic design and modelling of a coiled rotor synchronous motor dedicated to electric traction

Aicha Khlissa, Houcine Marouani, Souhir Tounsi

School of Electronics and Telecommunications of Sfax, Sfax University, Sfax, Tunisia

Email address:
aichakhlissa@gmail.com (A. Khlissa), Houcine.marouani@isecs.rnu.tn (H. Marouani), souhir.tounsi@isecs.rnu.tn (S. Tounsi)

Abstract: In this paper, we present a methodology of design and modeling of the controlling parameters of synchronous motor with wound rotor, based on the analytical method. This methodology ensures a wide operating speed range of electric vehicles. It takes into account several physical and technological constraints. The model is highly parameterized and quickly helps to provide the dimensions and power train controlling parameters values by varying the mechanical characteristics of the vehicle. It is compatible with all brands of electric vehicle power with single motor. The analytical modeling approach is validated entirely by the finite element method.

Keywords: Coiled Rotor Motor, Analytic Design, Controlling Parameters, Systemic Control, Electric Vehicles

1. Introduction

The production of electric vehicles in large series generally suffers from relatively high costs compared to internal combustion vehicles. For this reason, our choice was directed to a structure of synchronous wound rotor motor with smooth pole (MSRB) to reduce the cost of electric vehicle, because this type of motor is with reduced cost compared to other structures of electric motors. Indeed, the engine is in a structure easy to realize, and it is with open and straight slots and concentrated winding easy to achieve. Therefore, it has a greatly reduced manufacturing cost compared to other engine structures. It has no magnets also leading to a reduction of the cost of vehicles.

In this context, this paper presents a design methodology and modeling of control parameters of the studied motor structure.

2. Motor Structure

The MSRB machine is built with the same radius for the stator and the rotor. The slots directed towards the motor's center. Three design ratios define the motor's structure [1], [2] and [3]..

The first coefficient is the ratio β of the magnet average angular width by the pole pitch ($L=\pi/p$). It adjusts the magnet width in versus the poles number chosen.

The second coefficient (α) is the ratio of the main tooth average angular width by the average angular width of a magnet. It adjusts the main tooth size and has a strong influence on the electromotive force waveform.

The last coefficient (r_{did}) fixes the inserted tooth size. It's the ratio of the main tooth average width by an inserted tooth average width.

The advantage of these coefficients is to define quickly machine shape. However, they are based on the average radius and it is necessary to compute and check higher and lower angles teeth in order to avoid any intersection.

Table 1 illustrates the values of these coefficients:

Table 1. Values of the motor parameters

Designations	β	α	r_{did}	p	N_d
Trapezoidal configuration	1	1	0.2	4	6

The MSRB structure is illustrated by figure 1:

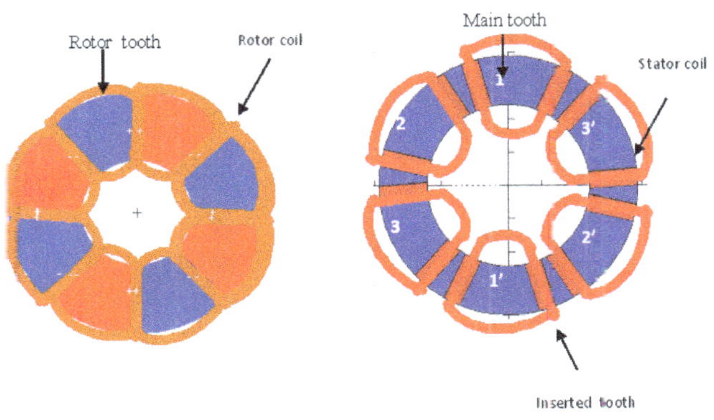

Figure 1. MSRB structure.

3. Dimensioning Torque

The sizing torque is calculated at the time of startup of the vehicle, where the current drawn by the motor is maximum. At this time, only the moment of inertia and torque of the vehicle due to gravity force are significant. The discretization of the movement equation at startup leads to the following sizing torque:

$$C_{dim} = \frac{R_r \times M_v}{r_d} \times \left(\frac{V_b}{t_d} + g \times \sin(\lambda) \right) \times \alpha_t \tag{1}$$

Where R_r is the radius of the wheel, M_v is the mass of the vehi-cle, r_d is the reduction ratio, g is the gravity force, λ is the angle between the road and the horizontal and α_t is a coefficient taking in account of the coo-ling system to integrate, it is less than 1.

The rated current can be deduced from the following relationship:

$$I_{dim} = \frac{C_{dim}}{K_e} \tag{2}$$

Where K_e is the motor constant:

$$K_e = 2 \times n \times N_s \times \frac{D_e^2 - D_i^e}{4} \times B_e \tag{3}$$

Where n is the number of motor module, D_e and D_i are respectively the external and internal motor diameters, N_s is the number of phase turn and B_e is the flux density in the air-gap.

4. Motor Sizing

The figure 2 presents the different geometric parameters of the stator:

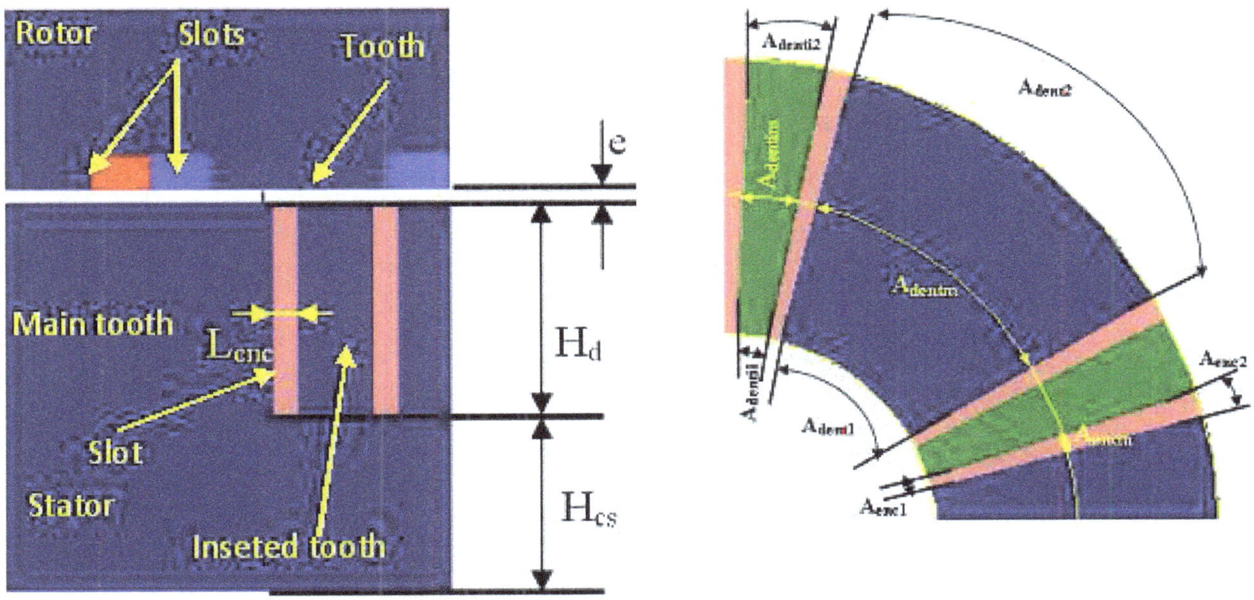

Figure 2. Geometric parameters of the stator.

The slot width of the stator is given by the following equation [4] and [5]:

$$L_{enc} = \left(\frac{D_e + D_i}{2} \right) \sin\left(\frac{1}{2} \times \left(\frac{2 \times \pi}{N_d} - \alpha \times \beta \times \frac{\pi}{p} \times (1 - r_{did}) \right) \right) \tag{4}$$

Where D_e and D_i are respectively the external and internal motor diameters, N_d is the number of main teeth and p is the number of poles pairs.

The lower angular width of stator slot is given by the following expression:

$$A_{enc1} = 2 \times A \sin\left(\frac{\frac{L_{enc}}{2}}{\frac{D_i}{2}} \right) \tag{5}$$

The superior angular width of stator slot is given by the following expression:

$$A_{enc2} = 2\times A \sin\left(\frac{\frac{L_{enc}}{2}}{\frac{D_e}{2}}\right) \quad (6)$$

The average angular width of a main tooth is expressed as follows:

$$A_{dentm} = \alpha\times\beta\times\frac{\pi}{p} \quad (7)$$

The average angular width of the inserted tooth is expressed as follows:

$$A_{dentim} = r_{did}\times A_{dentm} \quad (8)$$

The average angular width of the slot is expressed as follows:

$$A_{encm} = \frac{1}{2}\times\left(\frac{2\times\pi}{N_d} - A_{dentm} - A_{dentim}\right) \quad (9)$$

The inferior angular width of stator main tooth is given by the following expression:

$$A_{dent1} = A_{dentm} + A_{encm} - A_{dent1} \quad (10)$$

The superior angular width of stator main tooth is given by the following expression:

$$A_{dent2} = A_{dentm} + A_{encm} - A_{dent2} \quad (11)$$

For the configurations with trapezoidal waveforms the height of a tooth is given by the following equation [2]:

$$H_d = \frac{3\times2\times N_s}{2\times N_d}\times\frac{I_{dim}}{\delta}\times\frac{1}{K_f}\times\frac{1}{L_{enc}} \quad (12)$$

Where K_f is the filling factor of the slots, δ is the allowable current density in the slots, I_{dim} is the copper conductors sizing current and N_s is the number of phase spires.

The calculation method of the dimensioning current is retailed in [4].

The heights of the stator yoke are derived by applying the theorem of conservation of flux between the main tooth and the stator yoke [5]:

$$H_{cs} = \frac{B_e}{B_{cs}}\times\frac{Min(S_d, S_a)}{2\times\left(\frac{D_e - D_i}{2}\right)} \quad (13)$$

Where B_{cs} is the induction in the rotor yoke, B_e is the flux density in the air-gap, S_d is section of a stator tooth, S_a is the section of a magnet for the MSAP structure or of the rotor tooth for the MSRB structure and K_{fu} is the flux leakage coefficient

The figure 3 presents the different geometric parameters of the rotor.

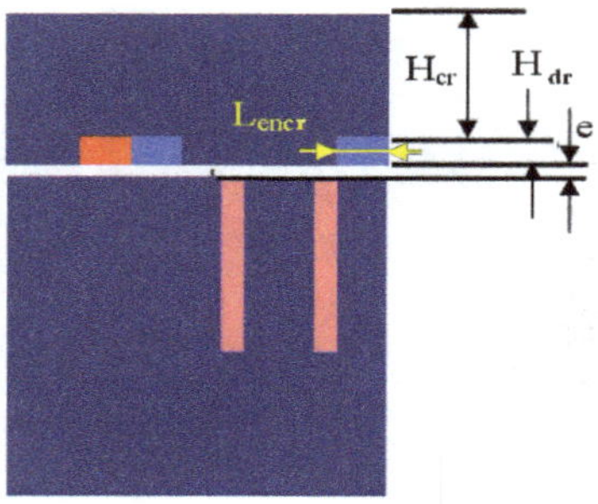

Figure 3. Different geometric parameters of the rotor.

- The middle width of a rotor slot is as:

$$A_{encmr} = \gamma\times L_a \quad (14)$$

- The slot width of these structures is given by the following equation [4] and [5]:

$$L_{encr} = \left(\frac{D_e + D_i}{2}\right)\sin\left(\frac{A_{encmr}}{2}\right) \quad (15)$$

$$L_a = \frac{\pi}{p}\times\beta \quad (16)$$

With L_a is the middle angular width of the magnet and γ is a coefficient adjusted by finite element simulations with the help of the software Maxwell 2D and can be optimized.

- The average angular width of the rotor main tooth is expressed as follows:

$$A_{dentmr} = \beta\times\frac{\pi}{p}\times(1 - 2\times\gamma) \quad (17)$$

- The lower angular width of rotor slot is given by the following expression:

$$A_{encr1} = 2\times A \sin\left(\frac{\frac{L_{encr}}{2}}{\frac{D_i}{2}}\right) \quad (18)$$

- The superior angular width of rotor slot is given by the following expression:

$$A_{encr2} = 2\times A \sin\left(\frac{\frac{L_{encr}}{2}}{\frac{D_e}{2}}\right) \quad (19)$$

- The height of a rotor tooth H_{dr} permitting to reserve the necessary space for the copper:

$$H_{dr} = \frac{n\times I_e}{\delta\times L_{encr}} \quad (20)$$

Where n is the number of rotor coil spire, I_e is the excitation current and δ is the admissible current density in the copper.

Where $K_{fu} < 1$ is the coefficient of flux leakages and e is the air-gap thickness.

Where B_c is the induction of demagnetization, B_r is the remanent induction of magnets and μ_0 is the permeability of air.

The heights of the rotor yoke is derived by applying the theorem of conservation of flux between a magnet or rotor tooth and the rotor yoke [5]:

$$H_{cr} = \frac{B_e}{B_{cr}} \times \frac{Min(S_d, S_a)}{2 \times \left(\frac{D_e - D_i}{2}\right)} \times \frac{1}{K_{fu}} \quad (21)$$

Where B_{cr} is the induction in the rotor yoke.

5. Back Electromotive Force

The figure 4 represents the distribution of the vector induction to the level of the air-gap for the functioning at no load. The level of induction reaches the value calculated analytically.

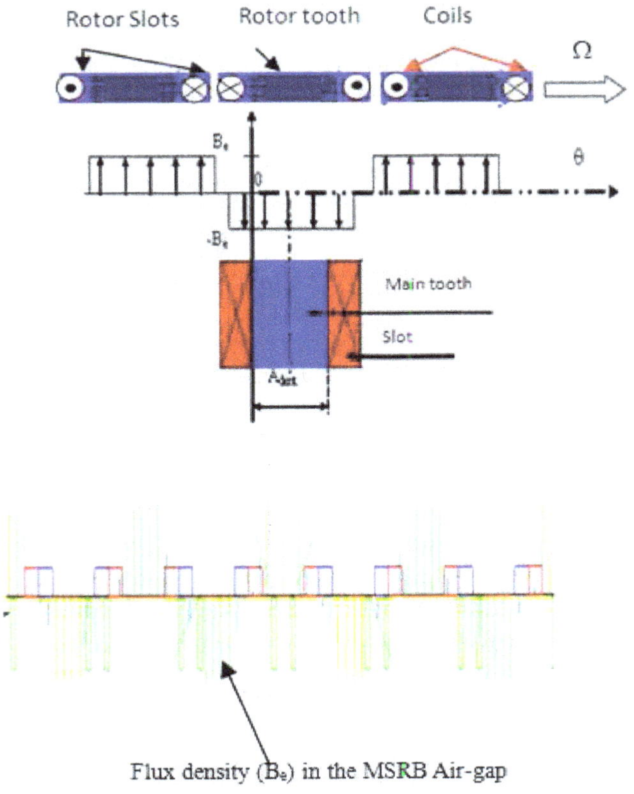

Figure 4. Initial position and induction in the air-gap.

From an initial position illustrated by figure 4, rotor moves with angular velocity ($\Omega = d\theta/dt$). Four distinct intervals appear according to magnets positions and geometrical parameters values defined previously. Table 2 illustrates these different intervals as well as flux variation. If α is equal to 1, zone 'a' disappears. In the zone 'b', the flux decreases because a part of the magnet is not in front of the tooth. In the zone 'c', a magnet of an opposite polarity overlaps also the main tooth. Consequently, the flux varies two times more quickly. Finally, the zone'd' is identical to the zone 'b'. These two zones exist only if the coefficient β is less than 1.

Table 2. Flux and electromotive force in function of motor parameters.

Zone	Position (rad)	Flux φ_b (Wb)	Emf (V)
a	$-\frac{\pi\beta}{2p}(1-\alpha) \le \theta \le \frac{\pi\beta}{2p}(1-\alpha)$	$\frac{(D_e^2 - D_i^2)}{8}\beta\frac{\pi}{p}\alpha B_e$	0
b	$\frac{\pi\beta}{2p}(1-\alpha) \le \theta \le \frac{\pi}{p}\left[1-\frac{\beta}{2}(1+\alpha)\right]$	$\frac{(D_e^2 - D_i^2)}{8}\left(\frac{\pi\beta}{2p}(1+\alpha)-\theta\right)B_e$	$N_s\Omega\ \frac{(D_e^2 - D_i^2)}{8}B_e$
c	$\frac{\pi}{p}\left[1-\frac{\beta}{2}(1+\alpha)\right] \le \theta \le \frac{\pi\beta}{2p}(1+\alpha)$	$\frac{(D_e^2 - D_i^2)}{8}(\frac{\pi}{p}-2\theta)B_e$	$2N_s\Omega\ \frac{(D_e^2 - D_i^2)}{8}B_e$
d	$\frac{\pi\beta}{2p}(1+\alpha) \le \theta \le \frac{\pi}{p}\left[1-\frac{\beta}{2}(1-\alpha)\right]$	$\frac{(D_e^2 - D_i^2)}{8}\left(\frac{\pi}{p}-\left[\frac{\pi\beta}{2p}(1+\alpha)\right]-\theta\ B_e\right.$	$N_s\Omega\ \frac{(D_e^2 - D_i^2)}{8}B_e$

The figure 5 presents the evolution of the flux and the electromotive force (e.m.f.) in function of electric angle

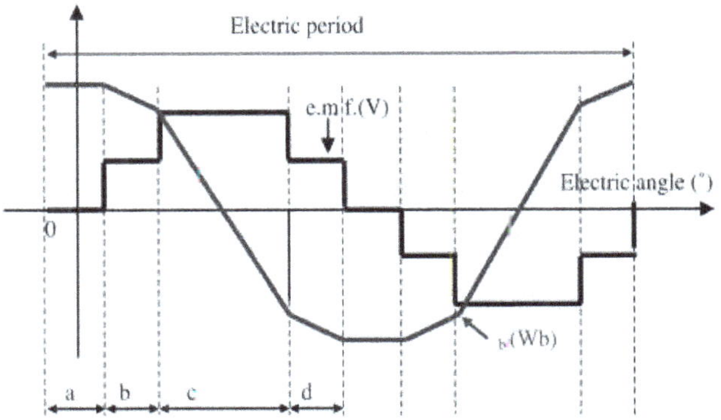

Figure 5. Flux and electromotive force in function of electric angle.

Le flux density in the air-gap is deduced from Ampere theorem:

$$B_e = \frac{\mu_0 \times n \times I_e}{e} \quad (22)$$

Where I_e is the excitation current, e is the Air-gap thickness, μ_0 is the air permeability and n is spires number of the rotor winding spires.

6. Analytical Modeling of Inductance and Mutual Inductance

6.1. Analytical Modeling of Inductance

For MSRB structures, phase inductance varies slightly in function of rotor position since the rotor slots are not deep. For these reasons, we consider that the MSRB structures is with smooth poles and the phase inductance is constant in linear regime. The figure 6 illustrates the distribution of the field lines to the level of a stator pole when the stator coil is supplied [11] and [12].

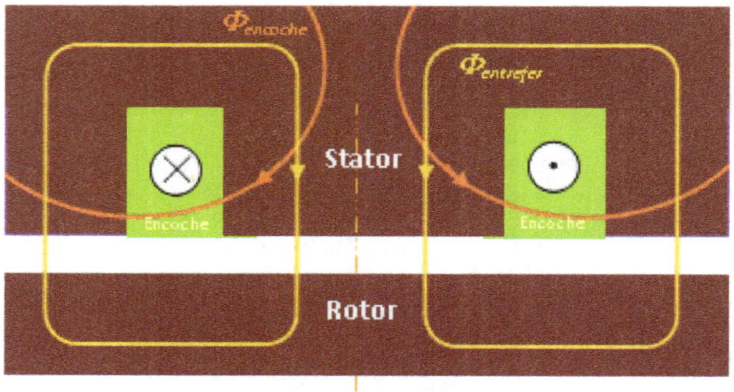

Figure 6. Distribution of the field lines for a powered coil.

This figure shows the presence of a flux leakages passing through the slot opening in a presence of leakage inductance in the slots copper, and of a main flux passing twice through the air gap and the magnet giving presence to an inductance of gap.

We recall the equations to model an inductance for a linear system:

$$L \times i_1 = N_s \times \Phi_1 \tag{23}$$

$$\Re \times \Phi_1 = N_s \times i_1 \tag{24}$$

$$L = \frac{N_s^{\ 2}}{\Re} \tag{25}$$

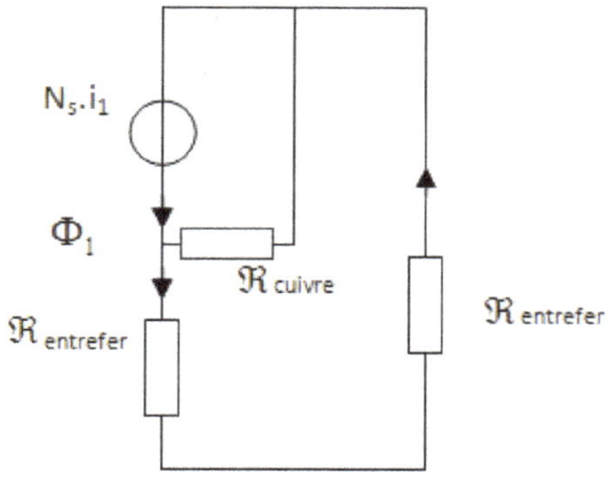

Figure 7. Network reluctance modeling inductance.

Where L is the inductance, i_1 is the current of energize, N_s is the number of spire and Φ_1 is the flux giving birth to the L inductance and $\Re$ is the réluctance of the magnetic circuit.

The figure 7 illustrates the network of réluctance modeling the inductance of total phase of the motor [11] and [12].

According to this face, we can write:

$$N_s \times i_1 = \left(2 \times \Re_{entrefer}\right) \times \Phi_{entrefer} = \Re_{cuivre} \times \Phi_{encoche} \tag{26}$$

With the reluctance of the air gap and the copper are given by the following relationship [19]:

$$\Re_{entrefer} = \frac{1}{\mu_0} \times \frac{\left(e_a\right)}{\frac{S_d}{2}} \tag{27}$$

$$\Re_{cuivre} = \frac{1}{\mu_0} \times \frac{\left(L_{enc}\right)}{\frac{D_e - D_i}{2} \times H_d} \tag{28}$$

The model of the total inductance is deduced from equations (26), (27) and (28) [7], [8] and [9]:

$$L = L_{fuite} + L_{entrefer} = \frac{N_s^2}{\Re_{cuivre}} + \frac{N_s^2}{+2 \times \Re_{entrefer}} \tag{29}$$

$$L = \mu_0 \times 2 \times \frac{N_s^2}{4} \left(\frac{\frac{S_d}{2}}{2 \times (e)} + \frac{\left(\frac{D_e - D_i}{2}\right) \times H_d}{L_{enc}} \right) \tag{30}$$

Where S_d is the surface of the main tooth, H_d is the height of the slot, H_a is the height of the magnet, L_{enc} is the width of the slot, e is the thickness of the air-gapr and $\Re$ is the reluctance.

6.2. Analytical Modeling of Mutual Inductance

The principle of the calculation of the mutual inductance rest on the supply of a coil for the calculation of the flux captured by the neighboring coil. The trajectory of the flux fixes the total reluctance of the magnetic circuit modeling this mutual inductance. The figure 8 illustrates the trajectory of the flux [8] and [9].

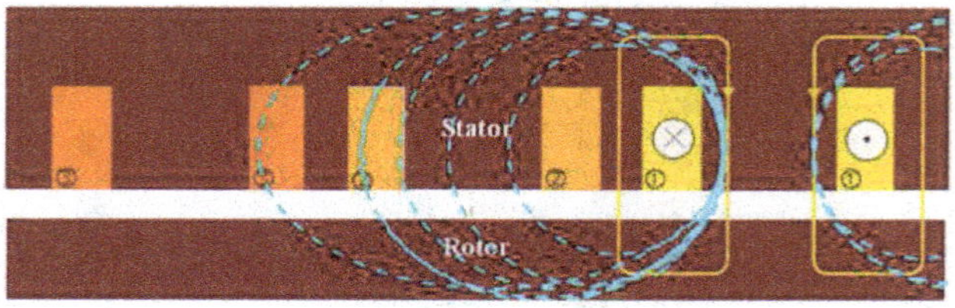

Figure 8. Distribution of the flux generated by the powered coil ① and captured by adjacent coils.

From figure 8 we deduct the network of reluctance modeling the mutual inductance (figure 9) [8], [9].

Where $\Re_1$ is the reluctance of the air-gap in front of the tooth where the coil 1 is accommodated, $\Re_2$ are the reluctance of a main tooth, $\Re_3$ is the reluctance of the stator yoke, $\Re_4$ is the reluctance of the tooth where the coil 2 is accommodated, $\Re_5$ is the reluctance of the air-gap in front of the tooth ② and $\Re_6$ is the reluctance of the rotor yoke.

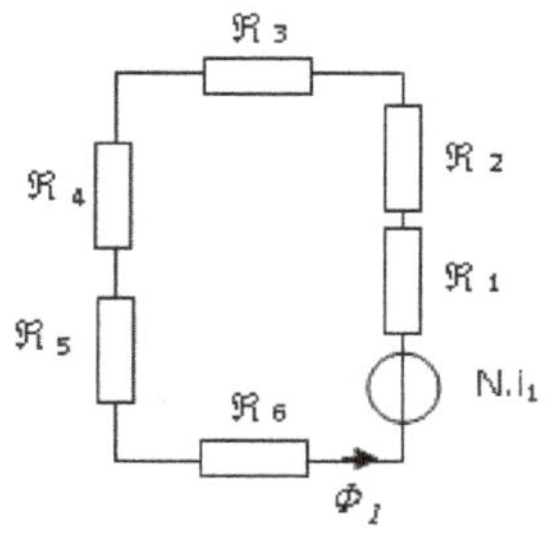

Figure 9. Network reluctance modeling mutual inductance.

The expression of the mutual inductance is given by:

$$M_{12} \times i_1 = N_s \times \Phi_1 \tag{31}$$

$$M_{12} = \frac{N_s^2}{\Re} \tag{32}$$

Where Φ_1 is the flux captured by the coil ② while energizing the coil ①, i_1 is the circulating current in the coil ① and $\Re$ is the total reluctance.

The different mutual inductances of the motor are equal since the motor is symmetrical.

It comes then [8] and [9]:

$$\Re_1 = \frac{1}{\mu_0} \times \frac{(e)}{\frac{1}{2} \times \left(\frac{D_e - D_i}{2}\right) \times \left(\frac{D_e + D_i}{4}\right) \times A_{dentm}} \tag{33}$$

$$\Re_2 = \frac{1}{\mu_0} \times \frac{2 \times H_d}{S_d} \tag{34}$$

$$\Re_3 = \frac{1}{\mu_0 \times \mu_r} \times \frac{\left(2 \times A_{encm} + \frac{1}{2} \times A_{dentm} + A_{dentim}\right) \times \frac{D_e + D_i}{4}}{H_{cs} \times \left(\frac{D_e - D_i}{2}\right)} \tag{35}$$

$$\Re_4 = \frac{1}{\mu_0} \times \frac{2 \times H_d}{S_d} \tag{36}$$

$$\Re_5 = \frac{1}{\mu_0} \times \frac{(e)}{\frac{1}{2} \times \left(\frac{D_e - D_i}{2}\right) \times \left(\frac{D_e + D_i}{4}\right) \times A_{dentm}} \tag{37}$$

$$\Re_6 = \frac{1}{\mu_0 \times \mu_r} \times \frac{\left(2 \times A_{encm} + \frac{1}{2} \times A_{dentm} + A_{dentim}\right) \times \frac{D_e + D_i}{4}}{H_{cs} \times \left(\frac{D_e - D_i}{2}\right)} \tag{38}$$

Where A_{encm} is the middle width of the slot, A_{dentm} is the middle width of the main tooth, A_{dentim} is the middle width of the inserted tooth , H_{cr} is the height of the rotor yoke, H_{cs} is the height of the stator yoke, μ_0 is the absolute permeability and μ_r is the relative permeability of the magnets.

One deducts a general expression of the mutual inductance of the motor wile neglecting the reluctance of iron:

$$M = \mu_0 \frac{\frac{S_d}{2}}{2 \times (e)} \frac{N_s^2}{4} \times 2 \tag{39}$$

7. DC Bus Voltage

The motor constant is defined by [9]:

$$K_e = 2 \times n \times N_s \times A \times B \times B_g \tag{40}$$

For the axial flux structures A and B are given by:

$$A = \frac{D_e - D_i}{2} \tag{41}$$

$$B = \frac{D_e + D_i}{2} \tag{42}$$

Where D_e and D_i are respectively the external and the internal diameter of the axial flux motor, N_s is the number of spire per phase, n is the module number and B_g is the flux density in the air-gap.

The converter's continuous voltage U_{dc} is calculated so that the vehicle can function at a maximum and stabilized speed with a weak torque undulation. The electromagnetic torque that the motor must exert at this operation point, via the mechanical power transmission system T_{Udc} (reducing + differential) is estimated by the following expression:

$$T_{Udc} = \frac{P_f}{\Omega} + T_d + \left(T_b + T_{vb} + T_{fr}\right) + \frac{T_r + T_a + T_c}{r_d} \tag{43}$$

Where T_b is the rubbing torque of the motor, T_{vb} is the viscous rubbing torque of the motor, T_{fr} is the fluid rubbing torque of the motor, T_r is the torque due to the friction rolling resistance, T_a is the torque due to the aerodynamic force, T_c is the torque due to the climbing resistance, T_d is the reducer losses torque and P_f are the iron losses and Ω is the motor angular speed.

At this operation point, the phase current is given by the following relation:

$$I_p = \frac{T_{Udc}}{K_e} \tag{44}$$

The only possibility making it possible to reach the current value I_p with a reduced undulation factor (10% for example) is to choose the converter's continuous voltage solution of the following equation [7-9]:

$$r = \frac{t_m}{t_p} = 10\% \tag{45}$$

Where t_p is the phase current maintains time at vehicle maximum speed and t_m is the boarding time of the phase current from zero to I_p [7-9]:

$$t_m = -\frac{L}{R} \times \ln\left(1 - \frac{2 \times R \times I_p}{U_{dc} - K_e \times \Omega_{max}}\right) \tag{46}$$

Where R and L are respectively the phase resistance and inductance and Ω_{max} is the maximum angular velocity of the motor.

The phase current maintains time at maximum speed of vehicle (corresponds to 120 electric degrees) is given by the following formula [7-9]:

$$t_p = \frac{1}{3} \times \frac{2 \times \pi}{p \times \Omega_{max}} \tag{47}$$

The converter's continuous voltage takes the following form [7-9]:

$$U_{dc}=\frac{2\times R\times I_p}{1-\exp\left(-\frac{2\times\pi\times r}{3\times p\times\Omega_{max}\times\frac{L}{R}}\right)}+K_e\times\Omega_{max} \tag{48}$$

8. Gear Ratio

The electric motor is controlled by a low frequency electromagnetic converter [1-6]. For this raison the insertion of a gear speed amplifier with r_d ratio is in the aims to enable the vehicle to reach the maximum speed of 80 km / h in our application. This ratio also helps ensure proper interpolation of reference voltages in order to have a good quality of electromagnetic torque.

$$r_d=\frac{2\times\pi\times R_r\times F_{ri}}{n_{qTA}\times V_{max}\times p\times n_{iTR}} \tag{49}$$

Where n_{iTR} is the reference voltages interpolation coefficient, p is the number of pair poles, n_{qTA} is the coefficient of quality of the supply voltage, F_{ri} is the switching frequency and V_{max} is the maximum speed of the vehicle.

9. Finite Element Validation

The motor is studied in 2-D by FEM finite element software with geometric provided by the analytical model. The finite element model is based on cylindrical cut plan geometric representation at the average contour. Values of back electromotive force, electromagnetic torque, inductance and mutual inductance are very close to those found by the analytical method. In conclusion, the analytical modeling approach is validated entirely by the finite element method.

10. Conclusion

This paper describes a methodology of analytical sizing and modeling of a synchronous axial flux motor with wound rotor. The model is highly parameterized. It covers thereafter a wide power range by specification data changing according to the vehicle to size. This approach is validated by the finite element method. It then presents an effective design program of these types of motors.

References

[1] S. TOUNSI « Losses modelling of the electromagnetic and IGBTs converters », International Int. J. Electric and Hybrid Vehicles (IJEHV), Vol. 5, No. 1, 2013, pp:54-68.

[2] S. TOUNSI « Comparative study of trapezoïdal and sinusoïdal control of electric vehicle power train», International Journal of Scientific & Technology Research (IJSTR), Vol. 1, Issue 10, Nov 2012.

[3] [M. HADJ KACEM, S. TOUNSI, R. NEJI «Systemic Design and Control of Electric Vehicles Power Chain », International Journal of Scientific & Technology Research (IJSTR), Vol. 1, Issue 10, Nov 2012.

[4] S. TOUNSI « Control of the Electric Vehicles Power Chain with Electromagnetic Switches Reducing the Energy Consumption», Journal of Electromagnetic Analysis and Applications (JEMAA) Vol.3 No.12, December 2011.

[5] S. TOUNSI, M. HADJ KACEM et R. NEJI « Design of Static Converter for Electric Traction », International Review on Modelling and Similations (IREMOS) Volume 3, N. 6, December 2010, pp. 1189-1195.

[6] S. TOUNSI et R. NEJI: "Design of an Axial Flux Brushless DC Motor with Concentrated Winding for Electric Vehicles", Journal of Electrical Engineering (JEE), Volume 10, 2010 - Edition: 2, pp. 134-146.

[7] S. TOUNSI, R. NEJI, and F. SELLAMI: "Design Methodology of Permanent Magnet Motors Improving Performances of Electric Vehicles", International Journal of Modelling and Simulation (IJMS), Volume 29, N° 1, 2009.

[8] A. Moalla, S. TOUNSI et R. Neji: "Determination of axial flux motor electric parameters by the analytic-finite elements method", 1nd International Conference on Electrical Systems Design & Technologies (ICEEDT'07), 4-6 Novembre, Hammamet, TUNISIA.

[9] N. Mellouli, S. TOUNSI et R. Neji: "Modelling by the finite elements method of a coiled rotor synchronous motor equivalent to a permanent magnets axial flux motor", 1nd International Conference on Electrical Systems Design & Technologies (ICEEDT'07), 4-6 Novembre, Hammamet, TUNISIA.

23

Comparative Analysis of Hysteresis and Fuzzy-Logic Hysteresis Current Control of a Single-Phase Grid-Connected Inverter

Mihail Antchev[1], **Angelina Tomova-Mitovska**[2]

[1]Power Electronics Department, Faculty of Electronic Engineering and Technologies, Technical University- Sofia, Sofia, Bulgaria
[2]Schneider Electric Slovakia, Bratislava, Slovakia

Email address:
antchev@tu-sofia.bg (M. Antchev), amtomova@gmail.com (A. Tomova-Mitovska)

Abstract: This paper presents a comparative analysis of two current control methods of single phase on-grid inverter- the hysteresis current control and the fuzzy logic hysteresis current control method. The aim of the paper is to compare the values of the main researched parameters of the on-grid inverter and to show which control method is more energy efficient. The research model is established by means of differential equations with the software MATLAB/SIMULINK. The comparison of the operation of the inverter with both control methods is analyzed in two different cases of the output power of the system - 800W and 1600W approximately.

Keywords: Hysteresis Current Control, Fuzzy Logic, On-Grid Inverter

1. Introduction

The most frequently used methods of control of on-grid inverter are the Pulse-Width Modulation (PWM) method and the hysteresis current control method. Both of them have advantages as well as disadvantages [1], [2]. One more recent and innovative method combines the mathematic concept of fuzzy logic by means of which uncertain and vague problems can be solved [3] with hysteresis current control [4], [5].

Fuzzy logic is a mathematic concept which allows the solution of not very precise problems and it is based on incertitude and vagueness [6]. This logic is used in the control of on-grid inverters [7].

The other control logic used in this paper- the hysteresis current control is based on creation of reference signal which is compared constantly with the inverter output signal. This method is applied in different power electronic devices- on-grid inverters [8], [9], active power filters [10], bi-directional converters [11].

An example of the combination of both methods- fuzzy logic and hysteresis current control for active power filters is presented in [12], [13].

This paper presents a comparative analysis of hysteresis and fuzzy logic hysteresis current control of on-grid inverter. Both models are developed in the MATLAB/SIMULINK environment and use the same on-grid inverter model realized on the basis of solving of differential equations connected to the grid via a LCL-filter. In the case of classical hysteresis current control the hysteresis value is constant and in the case of the combined fuzzy logic and hysteresis current control method, three different values of the hysteresis are used according to the current value of the difference between the transitory values of the current of the inverter and the reference current.

The operation of the inverter is studied in two cases. In the first case the power of the inverted is sufficient to supply the load and in the second case its power is lower than the desired power by the load.

2. Operation

In the first case the inverter control based on hysteresis current control uses as a feedback signal a comparison of two signals- the voltage of the point of common coupling V_{PCC} and the inverter current I_{INV}. Voltage is needed for the synchronization. The current is compared to a preliminary set reference current i_{REF}. The result of the comparison of the difference of these two currents and the hysteresis H defines which couple of power switches will be turned on [14].

In the second case when the control of the inverter is based on combined fuzzy logic and hysteresis current control, the fuzzy logic is implemented by comparison of the transitory values of the inverter output current and reference current to produce output signal based on fuzzy logic by choosing one of the three values of the hysteresis [15].

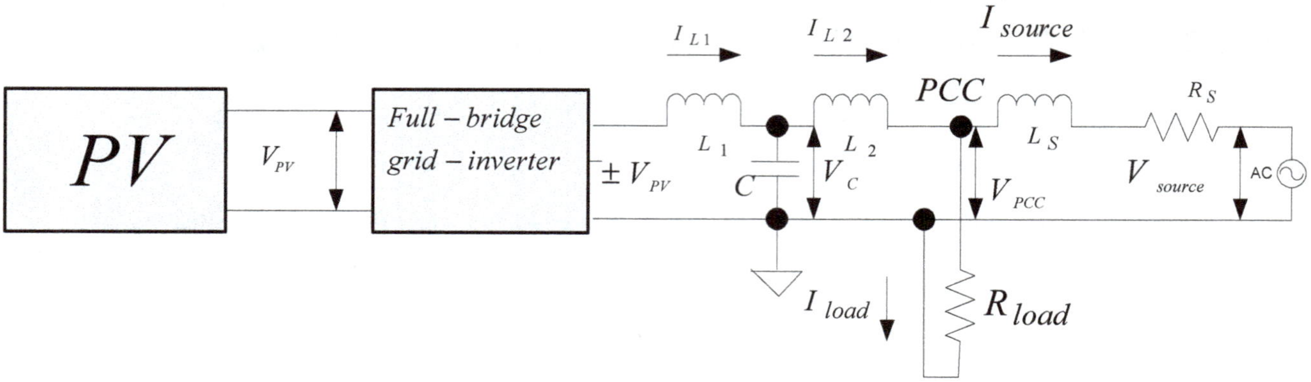

Figure 1. Block diagram of the system.

2.1. Input Data

For the simulation of the model the following data is used: $I_{ref} = 5\,A$; $V_{PV} = 400V$; $L_1 = 2.5\,mH$; $C = 2\,\mu F$; $L_2 = 40\,\mu H$; $V_S = 230V$; $L_{source} = 127\,\mu H$; $R_{source} = 12.3\,m\Omega$; $R_{load} = 68\,\Omega$, corresponding to an output power of 778W and $R_{load} = 33\,\Omega$, corresponding to an output power of 1603W.

The value of the hysteresis in the hysteresis-current control is $H = 0.25\,A$.

The three values of the hysteresis of the fuzzy-logic current control are $H_1 = 0.1\,A$, $H_2 = 0.25\,A$, $H_3 = 0.5\,A$.

For the needs of the software currents, voltages and hysteresis values are lessened by the factor of 100, in order to be operated in the range of the variables of MATLAB/SIMULINK.

2.2. Equations

The equations used to create the model of the system are:

$$i_{L_1} = \frac{v_{PV} - v_C}{s * L_1} \tag{1}$$

$$i_{L_2} = \frac{v_C - v_{PCC}}{s * L_2} \tag{2}$$

$$i_{source} = \frac{v_{PCC} - v_{source}}{s * L_S + R_S} \tag{3}$$

$$v_{PCC} = i_{load} * R_{load} \tag{4}$$

where s is the Laplace's operator.

The fuzzy-logic based equations are described here below:

$$\text{if}\,(i_{L1} \geq H_1 |\ i_{L1} \geq H_2 |\ i_{L1} \geq H_3)\ \{-1;\}\text{else if}\,(i_{L1} \leq H_1 |\ i_{L1} \leq H_2 |\ i_{L1} \leq H_3)\{1;\} \tag{5}$$

2.3. Simulation Models

2.3.1. Hysteresis-Current Control Model

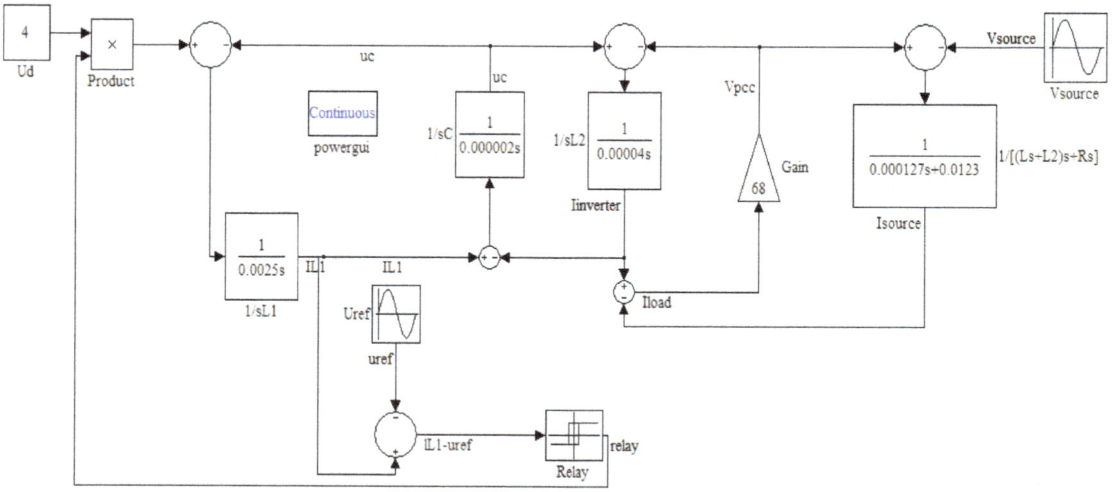

Figure 2. Simulation of hysteresis control of single-phase grid inverter.

2.3.2. Fuzzy-Logic Hysteresis-Current Control Model

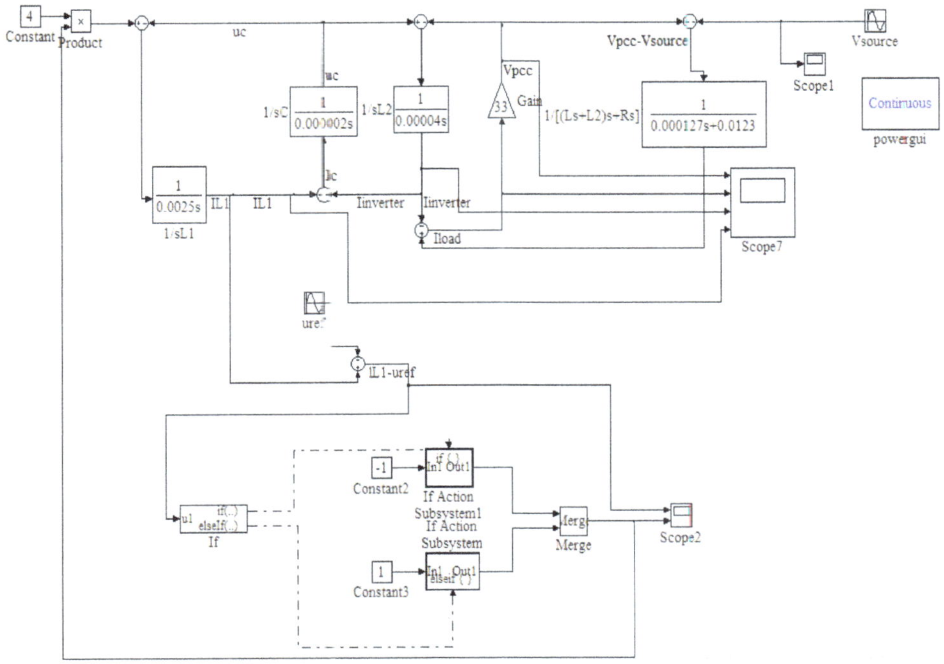

Figure 3. Simulation of fuzzy logic hysteresis control of single-phase grid inverter.

3. Results of the Computer Simulation

3.1. Resistive Load of 68Ω

3.1.1. Waveforms

In this case the power of the inverter is sufficient to supply the load.

From the two sets of waveforms on Fig.4 one can notice identical waveforms. In both cases the load current is in phase with the voltage at the point of common coupling which is normal as the load is resistive. On the other hand the inverter current is displaced in 180° to the voltage at the common point of coupling.

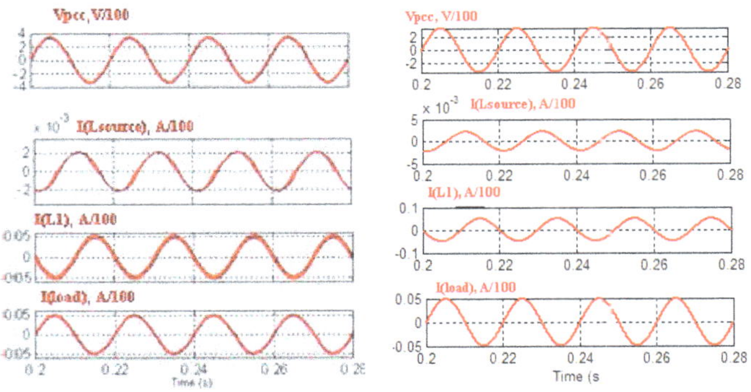

(a) (b)

Figure 4. Waveforms obtained from the simulation of the on-grid inverter model. Hysteresis current control on the left diagrams and fuzzy logic hysteresis current control on the right diagrams. From top to bottom: the voltage of the PCC, the grid current, the inverter current, the load current: a) hysteresis current control results; b) fuzzy-logic hysteresis-current control.

From the results we can notice that both methods of control have the same efficiency as the waveforms have the same shapes and values.

3.1.2. Frequency Analysis

Frequency analysis of the voltage at the point of common coupling (V_{PCC}) and inverter current (I_L) for both control methods are presented in Fig.5.

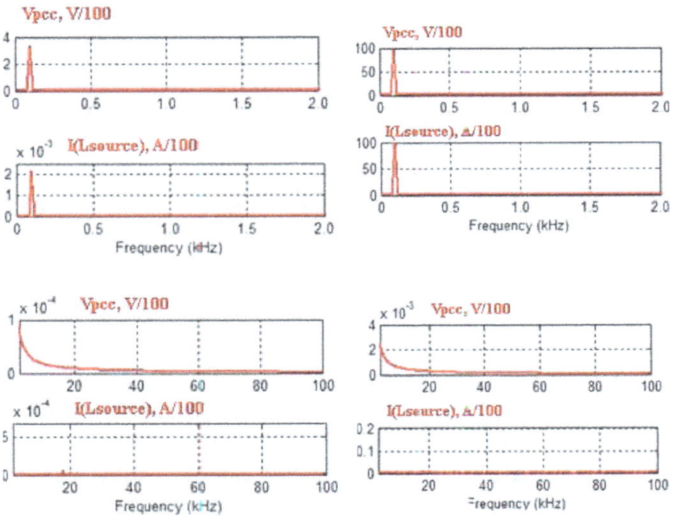

(a) (b)

Figure 5. Harmonic spectrum of the voltage of PCC- upper diagram and source current- lower diagram. Hysteresis current control on the left diagrams and fuzzy logic hysteresis current control on the right diagrams. The X-axis span from 0 to 2kHz (four upper co-ordinate systems) and from 2kHz to 100kHz (four lower co-ordinate systems).

The results are divided in two joint co-ordinate systems-from 0 to 2 kHz and from 2kHz to 100 kHz.

The shapes of the simulation results for both methods are identical. However we can notice higher values of the main harmonic for V_{PCC} and I_L for the combined fuzzy logic hysteresis control method.

3.2. Resistive Load of 33Ω

3.2.1. Waveforms

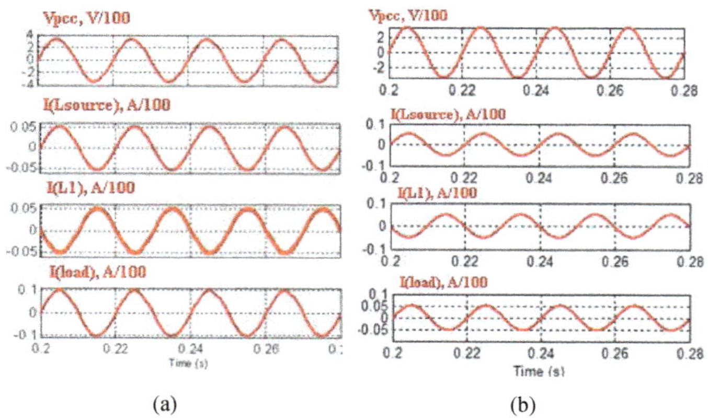

Figure 6. *Waveforms obtained from the simulation of the on-grid inverter model. Hysteresis current control on the left diagrams and fuzzy logic hysteresis current control on the right diagrams. From top to bottom: the voltage of the PCC, the grid current, the inverter current, the load current: a) hysteresis current control results; b) fuzzy-logic hysteresis-current control.*

In this case the power of the inverter is not sufficient to supply the load and the grid supplies half of the active power.

From the two sets of waveforms presented in Fig.6 one can notice identical waveforms as in the case of load of 68Ω. In both cases the load current is in phase with the voltage at the point of common coupling which is normal as the load is resistive. On the other hand the inverter current is displaced in 180° to the voltage at the common point of coupling.

As in the previous case from Fig.4 both methods of control have the same efficiency as the waveforms have the same shapes and values.

3.2.2. Frequency Analysis

Frequency analysis is done the same way as in the previous case. Results are presents in Fig.7.

In this case when power of the inverter is not sufficient to supply the load and half of the active power is supplied by the grid we can notice the same phenomena as in the case when the power of the inverter was sufficient to supply the load- same shapes, same harmonics, higher values for the combined method of control.

The one remarkable difference is the lack of harmonics for the higher harmonics in the case of the combined control, which is not the case for the hysteresis control results.

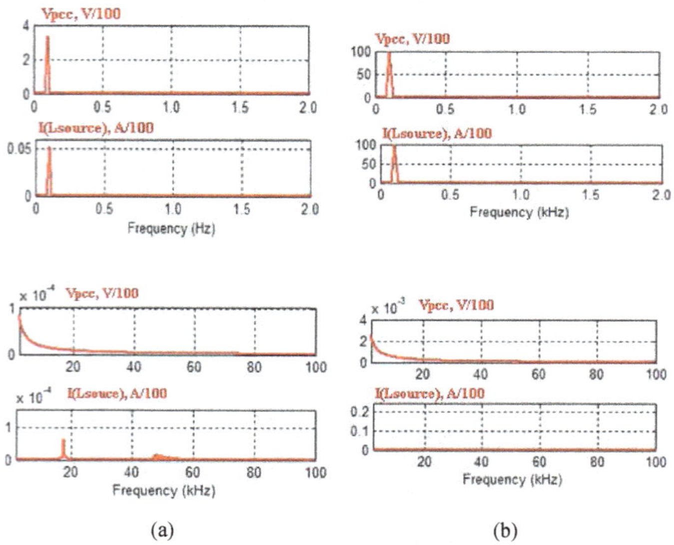

Figure 7. *Harmonic spectrum of the voltage of PCC- upper diagram and source current-lower diagram. Hysteresis current control on the left diagrams and fuzzy logic hysteresis current control on the right diagrams. The X-axis span from 0 to 2kHz (four upper co-ordinate systems) and from 2kHzto 100kHz (four lower co-ordinate systems).*

4. Conclusion

The comparative analysis of two methods of control of on-grid inverters presented in this paper proved the advantages of each of them.

From the simulation results in the time and frequency domain it is clear that both methods could provide correct control of the on-grid inverter so that it can operates in parallel with the electric grid.

The results for the shapes and values of the main parameters of the system for both control methods are identical. The only difference can be noticed in the frequency analysis simulations- where in both cases harmonic values of the study for the combined method are higher.

This paper presents only computer simulation results but they are significant enough to show that in case of a study on system or grid which is more sensitive to harmonics or which have some special requirements for the high harmonics we should use the hysteresis current control to do the study.

In case of no special requirements regarding the high harmonics and physical realization of the system we can use the combined method as it is easier to be implemented.

References

[1] H. Mao, X. Yang, Z. Chen, Z. Wang, "A hysteresis current controller for single-phase three-level voltage source inverters," IEEE Transactions on Power Electronics, vol. 27, no. 7, pp. 3330-3339, 2012.

[2] M. Kazmierkowski, L. Malesani, "Current control techniques for three-phase voltage-source PWM converters," IEEE Transactions on Industrial Electronics, vol. 45, no. 5, pp. 691-703, 1998.

[3] C. C. Lee, "Fuzzy logic control systems: Fuzzy logic controller, Part II," IEEE Transactions on Systems, Man and Cybernetics, vol. 20, no. 2, pp. 419-435, 1990.

[4] Z. Yao,L. Xiao, "Two-switch dual-buck grid-connected inverter with hysteresis current control," IEEE Transactions on Power Electronics, vol. 27, no. 7, pp. 3310-3318, 2012

[5] Y. Ounejjar, K. Al-Haddad, L. A. Dessaint, "A novel six-band hysteresis control for the packed U cells seven-level converter: Experimental validation," IEEE Transactions on Industrial Electronics, vol. 50, no. 10, pp. 3808-3816, 2012

[6] C. C. Lee, "Fuzzy logic control systems: Fuzzy logic controller, Part II," IEEE Transactions on Systems, Man and Cybernetics, vol. 20, no. 2, pp. 419-435, 1990.

[7] I. Sefa,N. Altin, "Simulation of fuzzy logic controlled grid interactive inverter," Univeristy of Pitesti- Electronics and Computer Science, Scientific Bulletin, vol. 2, no. 8, pp. 30-35, 2008.

[8] Z. Yao,L. Xiao, "Two-switch dual-buck grid-connected inverter with hysteresis current control," IEEE Transactions on Power Electronics, vol. 27, no. 7, pp. 3310-3318, 2012.

[9] J. Bauer,J. Lettl, "Solar power station output inverter control design," Radioengineering, vol. 20, no. 1, pp. 258-262, 2011.

[10] M. H. Antchev, M. P. Petkova, A. Kostov, "Hysteresis current control of single-phase shunt active power filter using frequency limitation," in Ninth IASTED Conference Power and Energy Systems, 2007, pp. 234-238.

[11] M. Makhlouf, F. Messai, H. Benalla, "Modeling and control of a single-phase grid connected photovoltaic system," Journal of Theoretical and Applied Information Technology, vol. 37, no. 2, pp. 289-296, 2012.

[12] Petkova M., M. Antchev, A. Tomova, "Computer investigation and verification of the operation of single-phase on-grid inverter with hysteresis current control" in Proc. EUROCON 2013, 1-4 July 2013, Zagreb, Croatia, pp.976-983, (978-1-4673-2232-4/13)

[13] Tomova A., M. Antchev, M. Petkova, H. Antchev, "Fuzzy Logic Hysteresis Control of A Single-Phase on-Grid Inverter: Computer Investigation", International Journal of Power Electronics and Drive System (IJPEDS), Vol. 3, No. 2, June 2013, pp. 1~8, ISSN: 2088-8694

Design and modeling of a synchronous renewable energy generation system

Wiem Nhidi[1], Souhir Tounsi[2], Mohamed Salim Bouhlel[3]

[1]National School of Engineers of Gabes (ENIG), Sfax University, SETIT Research Unit, Gabès, Tunisia
[2]Ational School of Electronics and Telecommunications of Sfax, Sfax University, SETIT Research Unit, Sfax, Tunisia
[3]Institut Supérieur de Biotechnologie de Sfax (ISBS), Sfax University, SETIT Research Unit, Sfax, Tunisia

Email address:
nhidiwiem@gmail.com (W. Nhidi), souhir.tounsi@isecs.rnu.tn (S. Tounsi), medsalim.bouhlel@enis.rnu.tn (M. S. Bouhlel)

Abstract: In this paper we describe a design and modeling methodology of synchronous generation system for renewable energy. Our choice fell on a synchronous generator structure with permanent magnet and axial flux simple to manufacturing to reduce the production cost of the energy generation system. The modeling approach presented leads to a scientific results of high level and opens the line of research to the study of the optimization of the energy recovered by the energy accumulator.

Keywords: Design, Modeling, Synchronous Generator, Renewable Energy, AC-DC Converter, Matlab-Simulink, Simulations

1. Introduction

In light of strong oil crises and undeniable air pollution problem, the generation of renewable energy from wind has become a current project. In this context several research addresses this problem. Our work is situated in this context too. The general problem of wind turbine systems is in generally the low efficiency due to total losses in energy conversion chain. This study illustrates a parameterized model of the energy generation system that can be a problem of recovered energy optimizing.

In this direction, this paper is organized as follows:

- The first part concern the presentation of the conversion chain structure.
- The second part presents the different models of the conversion chain.
- And the third part presents the results of simulations.

2. Conversion Chain Structure

The conversion chain (figure 1) has a propeller to recover the energy generated by wind. This mechanical energy is converted into an alternating electrical energy via the synchronous generator. The electrical energy developed is also converted into a continuous power through a three-phase rectifier. Finally, the continuous energy is elevated via a DC-DC converter to optimize the energy recovered by the battery [1-5].

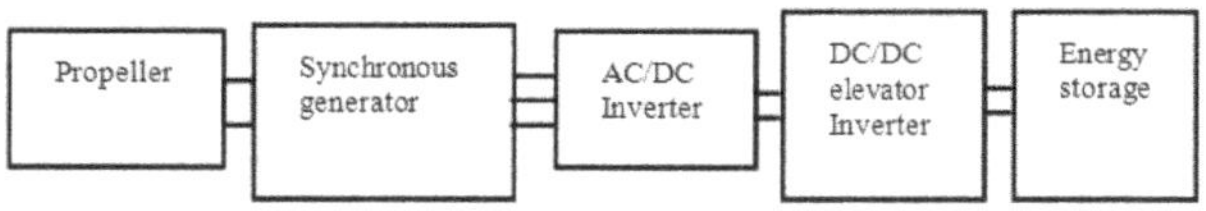

Figure 1. *Structure of the renewable enrgy system.*

3. Movement Equation

The equation of motion of the shaft of the synchronous generator is given by the following relationship:

$$J\times\frac{d\Omega}{dt}=T_{em}-\frac{P_{iron}}{\Omega}-\frac{P_m}{\Omega} \tag{1}$$

Where J is the moment of inertia of the rotating parts, P_{iron} are the iron losses, Ω is the angular velocity of the propeller, P_m are the mechanical losses and T_{em} is the torque on the shaft of the synchronous generator.

$$T_{em}=\frac{1.918\times r_p^{\,2}V^3}{\Omega} \tag{2}$$

Where r_p is the pale length and V is the wind speed.

Mechanical losses are expressed by the following relationship [1-5]:

$$P_m = s + k \times \Omega + \gamma \times \Omega^2 \tag{3}$$

Where s is a dry friction coefficient, k is a viscous friction coefficient, and γ is a coefficient of fluid friction.

The iron losses are expressed by the following relationship [1-5]:

$$P_{iron} = 1.1 \times \left(\frac{f}{50}\right)^{1.5} \left(M_{cs} \times B_{cs}^2 + M_{ds} \times B_d^2\right) \tag{4}$$

Where f is the frequency of the currents delivered by the synchronous generator, M_{cs} is the stator yoke mass, M_{ds} is the mass of stator teeth, B_{cs} is the stator yoke flux density and B_d is the stator teeth flux density.

The equation of motion is implanted within the Matlab-Simulink environment according to Figure 2.

4. AC-DC Conversion

The torque recovered on the generator shaft allows the drive the latter to the angular velocity Ω. This movement allows the induction of three-phase trapezoidal electromotive forces. These three electromotive forces are converted into a DC voltage filtered by a capacitance of high value by a three-phase rectifier.

The Simulink model of the conversion system of mechanical energy into a continuous electrical energy is illustrated by the following figure [1-3]:

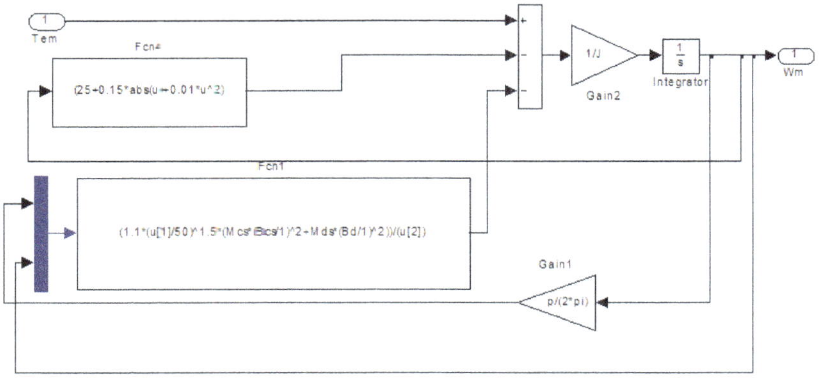

Figure 2. Simulink model of the movement equation.

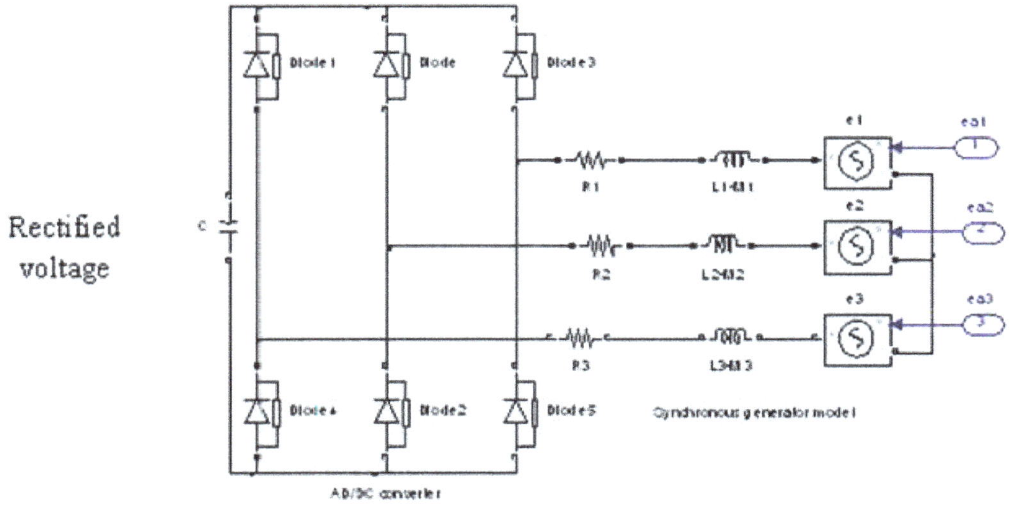

Figure 3. Simulink model of AC-DC conversion system.

The three electromotive forces are estimated from the following three equations:

$$a = \cos\left(p \times \Omega \times t + \frac{\pi}{2}\right) \tag{5}$$

$$b = \cos\left(p \times \Omega \times t - \frac{2 \times \pi}{3} + \frac{\pi}{2}\right) \tag{6}$$

$$c = \cos\left(p \times \Omega \times t - \frac{4 \times \pi}{3} + \frac{\pi}{2}\right) \tag{7}$$

The models of the electromotive forces (e_1, e_2, e_3) are estimated from the following algorithm:

{Begin
if
a>1/2;

a1=1/2.K_e*.*Ω*:*
else
a1=0;
if
a<-1/2;
a2=-1/2.K_e*.*Ω*;*
else
a2=0;
e1=a1+a2;
if
b>1/2;
b1=1/2.K_e*.*Ω*:*
else
b1=0;
if
b<-1/2;
*b2=-1/2*K_e*.*Ω*;*
else
b2=0;
e_2*=b1+b2;*
if
c>1/2;
c1=1/2.K_e*.*Ω*:*
else
c1=0;
if
c<-1/2;
c2=-1/2.K_e*.*Ω*;*
else
c2=0;
e_3*=c1+c2;*
end}.

With K_e is the electromotive forces constant, Ω is the motor angular speed and p is the number of pole pairs.

The Simulink model of the electromotive forces is illustrated by the figure 4.

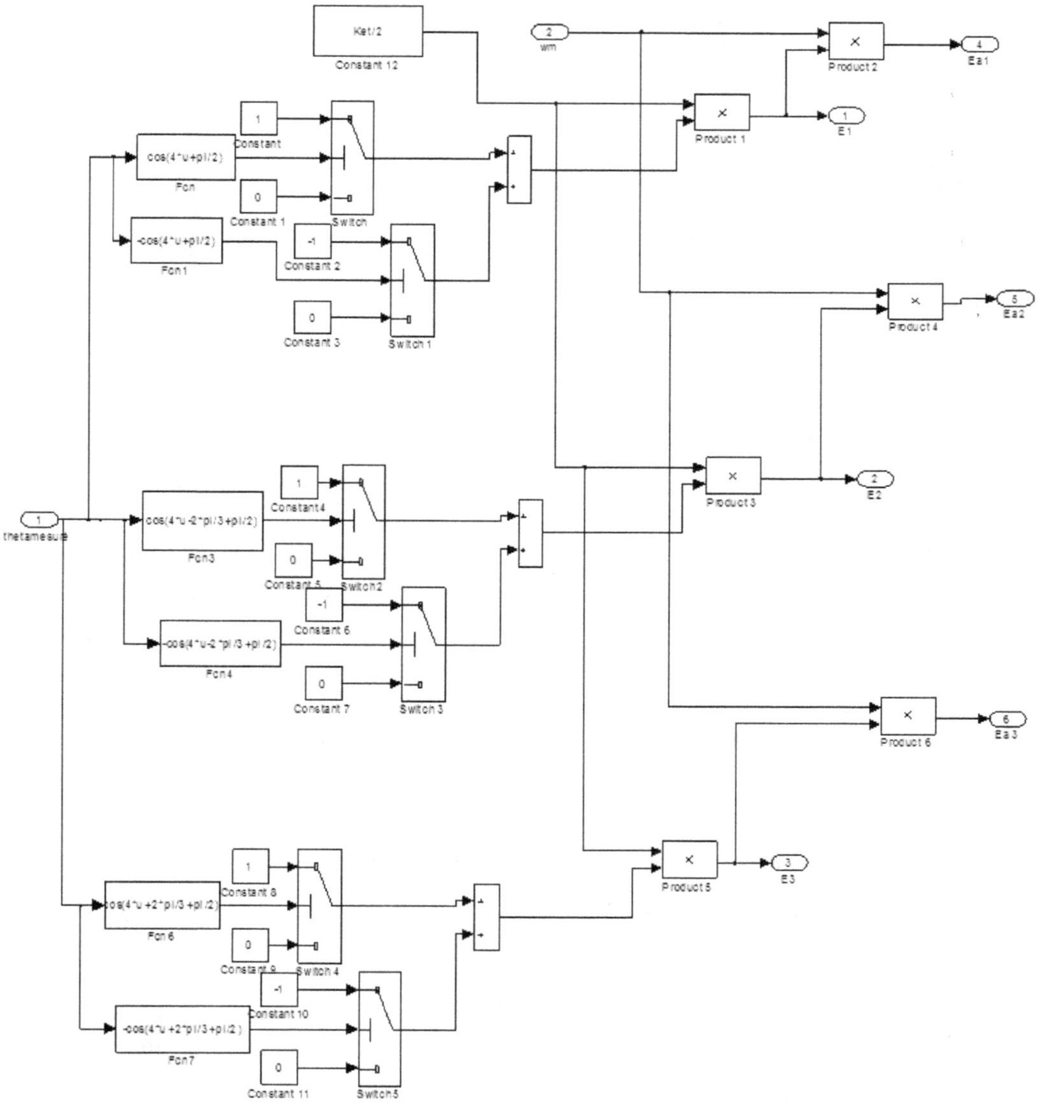

Figure 4. *Simulink model of the electromotive forces.*

5. DC-DC Conversion

A DC-DC boost converter (figure 5) capable of converting the rectified and filtered voltage into to a higher voltage amplitude is used in order to optimize the recovered energy. The cyclic ratio of the converter control signal is optimized following several simulations in order to maximize the energy recovered by the energy accumulator.

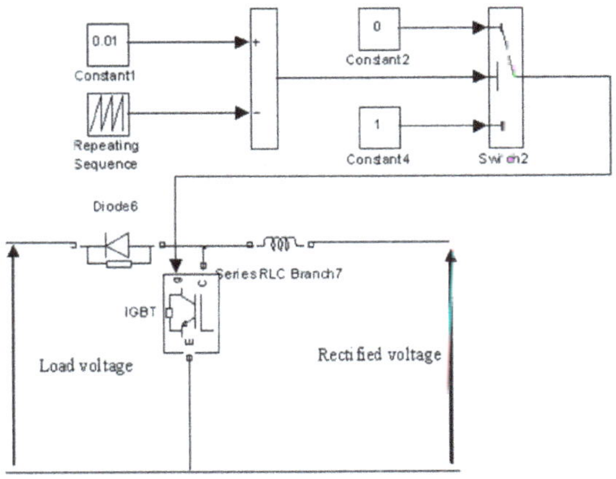

Figure 5. DC-DC elevator inverter.

6. Energy Accumulator

The energy accumulator comprises batteries in parallel with supper capacitor to increase the storage capacity. The Simulink model of the battery is illustrated by figure 6 [5].

C2 is a capacitor for holding into account the transitional regim and R_1 is the internal resistance of the battery.

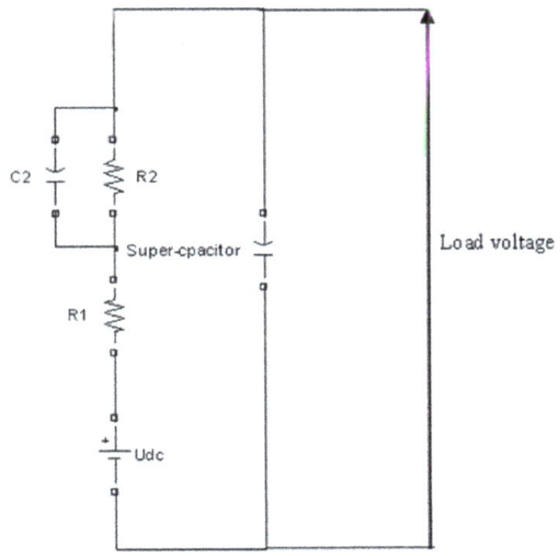

Figure 6. Simulink model of the energy accumulator.

7. Global Model of Renewable Energy Generation System

The global model of the energy generation system is based on the connection of the different Simulink models of the components making up this chain (Figure 7).

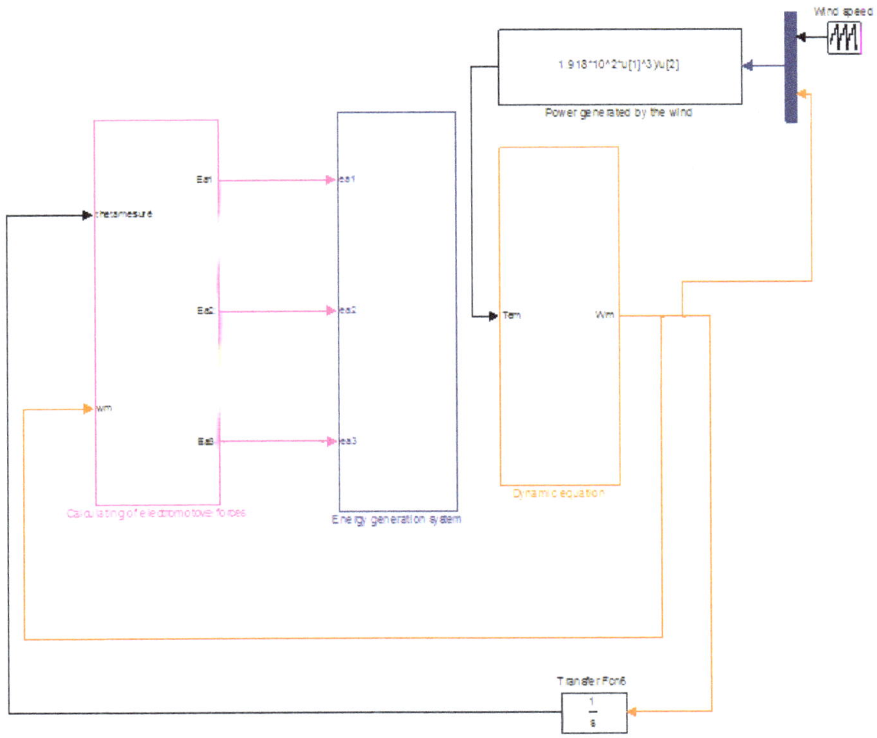

Figure 7. Simulink model of the global system.

8. Simulations Results

The evolution of the wind speed is illustrated in Figure 8.

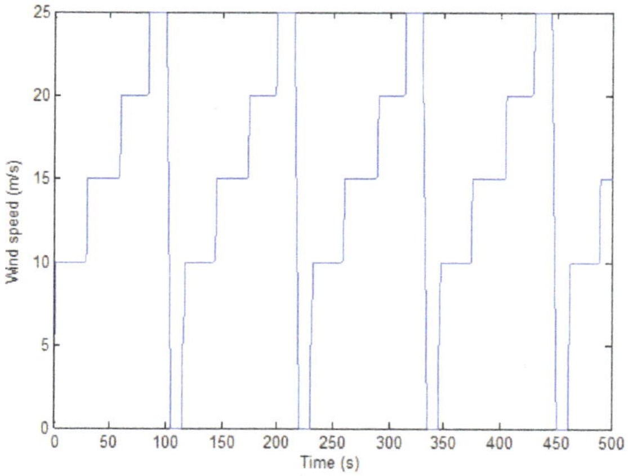

Figure 8. *Evolution of the wind speed.*

The evolution of the angular speed of the generator shaft is illustrated in Figure 9.

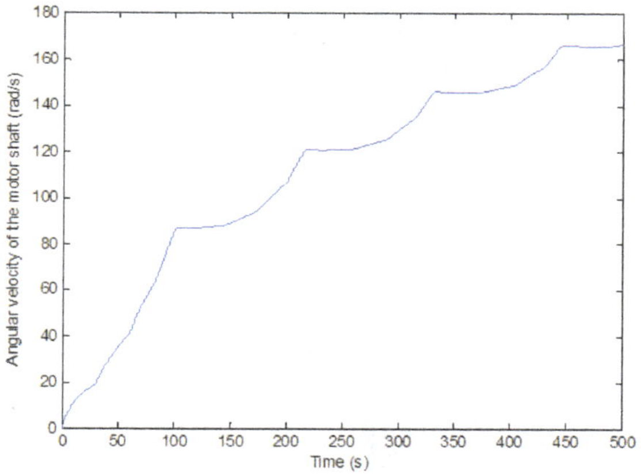

Figure 9. *Evolution of the angular speed of the generator shaft.*

The three-phase voltage system generated by the synchronous generator is shown in Figure 10.

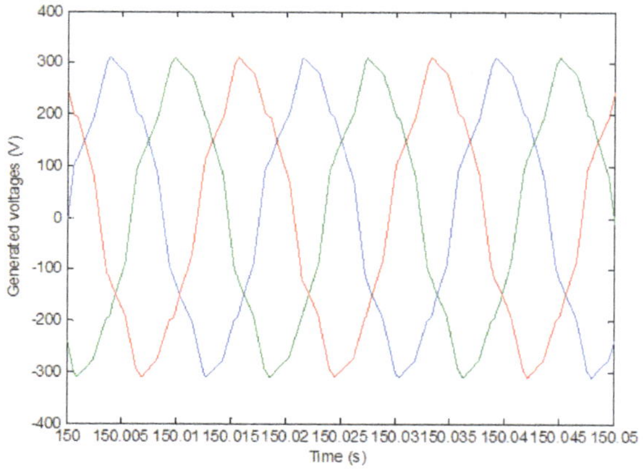

Figure 10. *Phase voltages generated by the synchronous generator.*

The phase currents generated by the synchronous generator are illustrated by Figure 11.

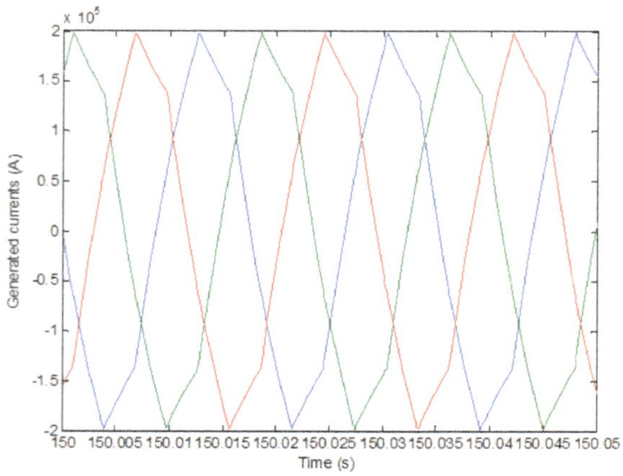

Figure 11. *Phase currents generated by the synchronous generator.*

The slight deformation in the shapes of voltages and currents is mainly due to the presence of a strong inertia of the rotating parts and the strong phase inductance.

The evolution of the load voltage is illustrated by the figure 12.

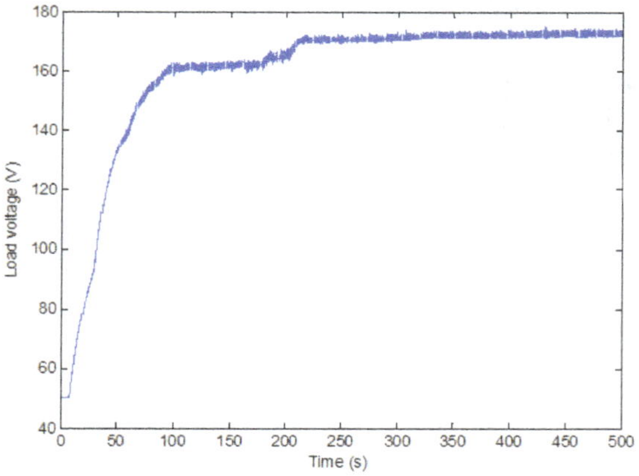

Figure 12. *Evolution of the load voltage.*

The evolution of the load current is illustrated by the figure 13.

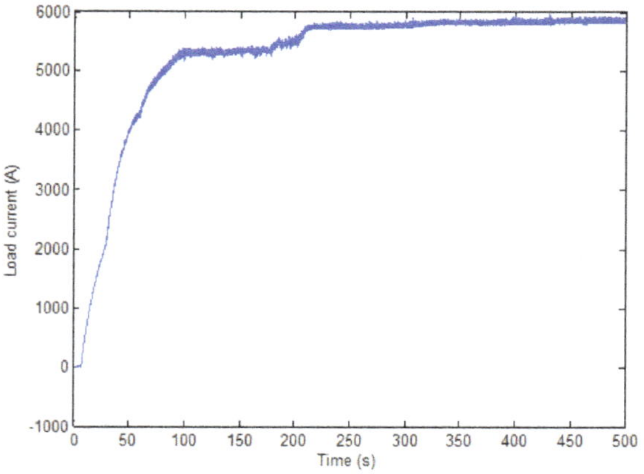

Figure 13. *Evolution of the load current.*

The generated and recovered powers are illustrated by figure 14.

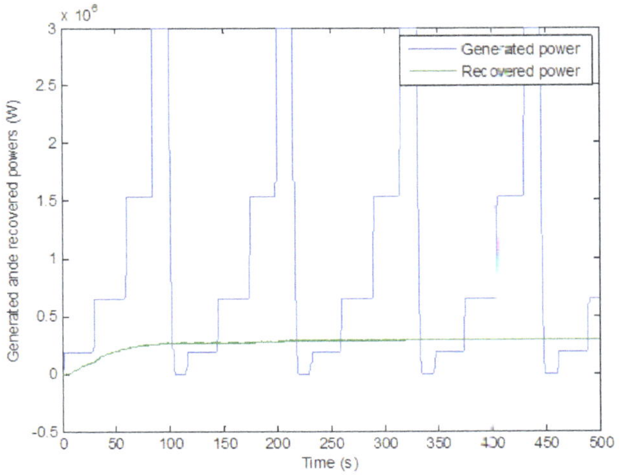

Figure 14. Generated and recovered powers.

The recovered energy is illustrated by figure 15.

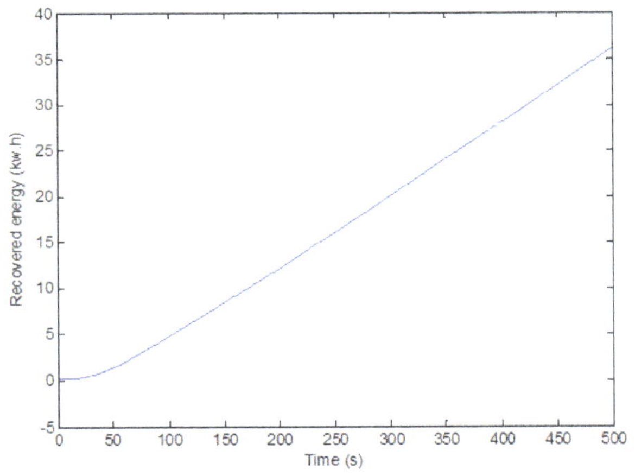

Figure 15. Recovered energy.

9. Conclusion

In this paper we present a methodology for the design and modeling of a parameterized system for generating renewable energy. The implementation of the energy generation chain global model under the Matlab-Simulink environment leads to the scientific results of very good level. Being parametrized, the developed model poses an optimization problem of the recovered energy for future research work.

References

[1] S. TOUNSI « Comparative study of trapezoïdal and sinusoïdal control of electric vehicle power train», International Journal of Scientific & Technology Research (IJSTR), Vol. 1, Issue 10, Nov 2012.

[2] M. HADJ KACEM, S. TOUNSI, R. NEJI «Control of an Actuator DC Energy-saving Dedicated to the Electric Traction», International Journal of Computer Applications: IJCA (0975 – 8887) Volume 54– No.10, pp. 20-25, September 2012 .

[3] S. TOUNSI « Control of the Electric Vehicles Power Chain with Electromagnetic Switches Reducing the Energy Consumption», Journal of Electromagnetic Analysis and Applications (JEMAA) Vol.3 No.12, Deember 2011.

[4] S. TOUNSI, M. HADJ KACEM et R. NEJI « Design of Static Converter for Electric Traction », International Review on Modelling and Similations (IREMOS) Volume 3, N. 6, December 2010, pp. 1189-1195. [8] S. TOUNSI et R. NEJI: "Design of an Axial Flux Brushless DC Motor with Concentrated Winding for Electric Vehicles", Journal of Electrical Engineering (JEE), Volume 10, 2010 - Edition: 2, pp. 134-146.

[5] S. TOUNSI, R. NEJI, and F. SELLAMI: "Design Methodology of Permanent Magnet Motors Improving Performances of Electric Vehicles", International Journal of Modelling and Simulation (IJMS), Volume 29, N° 1, 2009.

Energy Saving Lamps as Sources of Harmonic Currents

Zahari Ivanov[1], Hristo Antchev[2]

[1]Department of Electrical Supply, Electrical Equipment and Electrical Transport, Faculty of Electrical Engineering, Technical University - Sofia, Sofia, Bulgaria
[2]Power Electronics Department, Faculty of Electronic Engineering and Technologies, Technical University - Sofia, Sofia, Bulgaria

Email address:
zai@tu-sofia.bg (Z. Ivanov), hristo_antchev@tu-sofia.bg (H. Antchev)

Abstract: This paper presents the results from the research of the indexes in regard to the power supply network, when a change is applied to the power supply tension of the Compact Fluorescent Lamps and LED Lamps with power up to 30W of different manufacturers. Comparision data is presented for the power factor, the displacement factor and the total harmonic distortion of the current, and the relevant conclusions have been made.

Keywords: Compact Fluoroscent Lamps, Light-Emitting Diodes, Harmonic Distortion

1. Introduction

In the past years, on a worldwide scale is being conducted the replacement of lamps with incandescent light bulbs with new, energy saving lamps - Compact Fluorescent Lamps and LED Lamps. These new light sources require an electrical converter, connected to the power supply network. It influences the indexes of the light source in regard to this network - power factor, the displacement factor and the total harmonic distortion of the current. These indexes are subject to research at different power levels of the lamps by different authors, such as [1, 2, 3, 4]. Fot his purpose some of them have developed special generators [5]. The matter with the influence of the powersaving lamps on the power supply network becomes even more topical in relation to the creation of smart grids and the use of renewable energy sources [6].

Based on their research, some authors offer shunt active power filters for compensating the harmonic distortions of the Compact Fluorescent Lamps [7]. It must be noted, that the more common approach is the use of active correction of the power factor in the electrical converter itself [8, 9]. Usually this is done when the power input is more than 25W.

The goal of this work is to conduct a research of the indexes in relation to the power supply network of Compact Fluorescent Lamps and LED Lamps with different power inputs - less than 25W and more than 25W. Based on this, conclusions will be made about their influence over the power supply network in conjunction with the international standards. In Part 2 are presented the main indexes against the network, as well and the standard requirements. In Part 3 are given the results of the experimental research. In Part 4 are presented the relevant conclusions based on the conducted research.

2. Main Indexes and Standards

Main index of the influence of the connected to the network consumers (in the case of light sources) is the power factor PF, representing the relation between active $P$and total Spower:

$$\mathrm{PF} = \frac{\mathrm{P}}{S} = cos\varphi_1 . v \tag{1}$$

It has to be noted, that for power factor in literature different designations are used, such as K_p, and often in relation to light sources – λ.

The power factor is the product of the displacement factor $\cos\varphi_1$, defined by the phase shift between the tension of the power supply network and the first harmonic of its current - I_1, and distortion factor - v. This last one represents the relation between the efective value of the first harmonic of the current from the power supply network I_1 and the effective value of its current I:

$$\mathrm{v} = \frac{I_1}{I} = \frac{I_1}{\sqrt{\sum_{k=1}^{n} I_k^2}} \tag{2}$$

Except with the distortion factor , the distortion of the current is characterized as well with total harmonic distortion – THD:

$$THD = \frac{\sqrt{\sum_{k=2}^{n} I_k^2}}{I_1} \quad (3)$$

The connection between ν and THD is:

$$\nu = \frac{1}{\sqrt{1+THD^2}} \quad (4)$$

The requirements fo the harmonic constitution of the current in relation to the power supply network with current <16A on phase are defined by: IEC 6100-3-2 [10, 11]. In relation to this the lihgting equipment falls into class C equipment, in regards to which the requirements for harmonic constitution of the current are presented in table 1 [11]. It has to be noted that the standard enforces these requirements on powers greater than 25 W.

On some converters with uncontrollable rectifier at the entry of the Compact Fluorescent Lamps and LED Lamps, the form of the current could be related to the one, presented in the standard for lighting equipment class D. The requirements for this class, though, are for powers greater than 75 W.

Due to the abovementioned peculiarities of the standard, the requirements for the harmonic constitution of the current are violated by some of the manufacturers on power outputs lesser than 25W.

In regards to the form of the current, the crest factor K, is also defined, representing the relation between the maximal I_M and effective I_{RMS} values:

$$K = \frac{I_M}{I_{RMS}} \quad (5)$$

Table 1. *Limits for Clas C Equipment.*

Harmonic order, n	Maximum permessible harmonic current, (% of fundamental)
2	2
3	30xcircuit power factor
5	10
7	7
9	5
$11 \leq n \leq 39$	3

Sinusoidal current form: $K = \sqrt{2}$.

In regards to the displacement factor $\cos\varphi_1$ another standard exists, in which are enforced restrictions on the lesser power lamps as well. In table 2 are presented the requirements regarding the displacement factor $\cos\varphi_1$ for different powers, in accordance with [12].

Table 2. *Recommendet values for displacement factor.*

Metric	Limit			
Kdisplacement ($\cos\varphi_1$)	P≤2W	2W<P≤5W	5W<P≤25W	P>25W
	No limit	≥0.4	≥0.7	≥0.9

3. Results of the Investigations

For the experimental research, 10 lamps with different powers have been used, taken randomly. In order to avoid eventual misunderstandings with the companies-manufacturers, in the below presented data, the lamps are given names from 1 to 10 and only the type and power of the lamp are indicated – Compact Fluorescent Lamp (CFL) or LED Lamp. Lamp 7 is induction, non-electrode. Research has been made using three different values of the power supply network voltage – normal 230V, 230V-10% and 230V+10%. The results for the main indexes - $PF, \cos\varphi_1, THD, K$ are presented in Table 3. Fluke Power Quality Analyzer 434 has been used.

Table 3. *Indexes in accordance with the power supply network for the researched lamps.*

Lamp№	Type, Power W	Supply Voltage, Urms, V	PF	$\cos\varphi_1$	THD,%	K
		207	0.51	-	80.8	152.4
1	CFL, 15	230	0.51	-	80.8	140.2
		253	0.51	-	81.6	104.8
		207	0.54	0.89	79.5	54.5
2	CFL, 30	230	0.54	0.89	80.9	60.5
		253	0.51	0.91	82.6	55.6
		207	-	-	49.3	29.7
3	CFL, 15	230	-	-	46.6	24.1
		253	-	-	44.3	20.6
		207	0.52	0.89	82.8	80.5
4	CFL, 15	230	0.51	0.88	80.4	92.9
		253	0.48	0.89	84.2	85.4
		207	0.47	0.89	84.4	79.1
5	CFL, 20	230	0.45	0.9	86	95.5
		253	0.45	0.91	86.5	85.6
		207	0.48	-	80.6	116.3
6	CFL, 8	230	0.47	-	81.4	119.3
		253	0.47	-	82.2	91.4
		207	0.49	0.91	84.6	64.3
7	Induction L, 23	230	0.47	0.91	84.7	58.1
		253	0.49	0.92	84.7	
		207	0.54	-	76.7	110.6
8	LED lamp,15	230	0.52	-	77.5	134.1
		253	0.53	-	79.1	120.4
		207	0.71	0.99	67.1	11.8
9	LED lamp,30	230	0.69	0.98	70.1	12.8
		253	0.67	0.98	70.5	14
		207	-	0.98	22.5	10.9
10	LED lamp,15	230	-	0.98	24.5	12.5
		253	-	0.98	27.3	15.5

No strict dependence exists between THD and the value of the power supply network voltage. In most cases THD grows with the growth of the voltage. Table 3 and fig. 10 show the significantly better indexes of lamp 10, despite its power of 15W. On the other hand, from fig.9 we see that lamp 9, with power of 30W, also does not cover the requirement of the standard in regards to the harmonic constitution of the current.

The waveforms from the experiments for each of the lamps and with tension of the power supply network are presented in fig. 1 to 10. Oscilloscope Tektronix 2012 has been used.

CH1 is a waveform of the tension of the power supply network, while CH2- its current. The waveform of the current is with ratio 100mV/A.

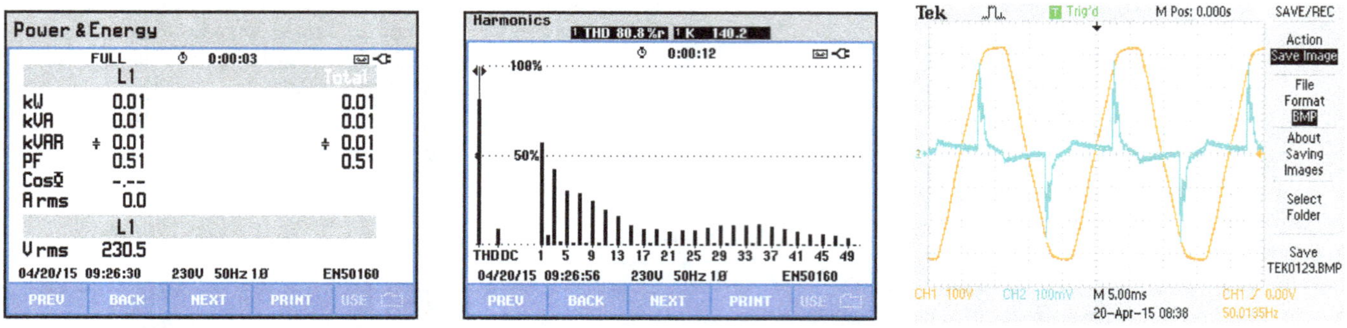

Figure 1. *Wavwforms for lamp 1.*

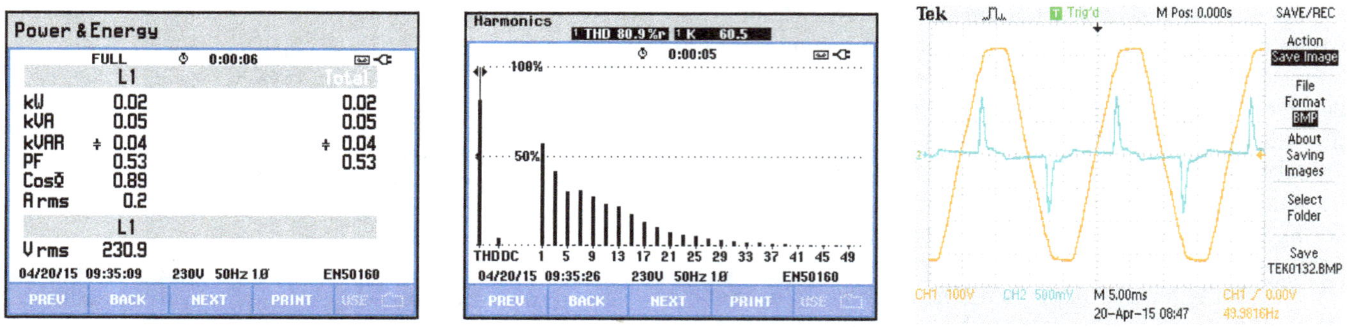

Figure 2. *Wavwforms for lamp 2.*

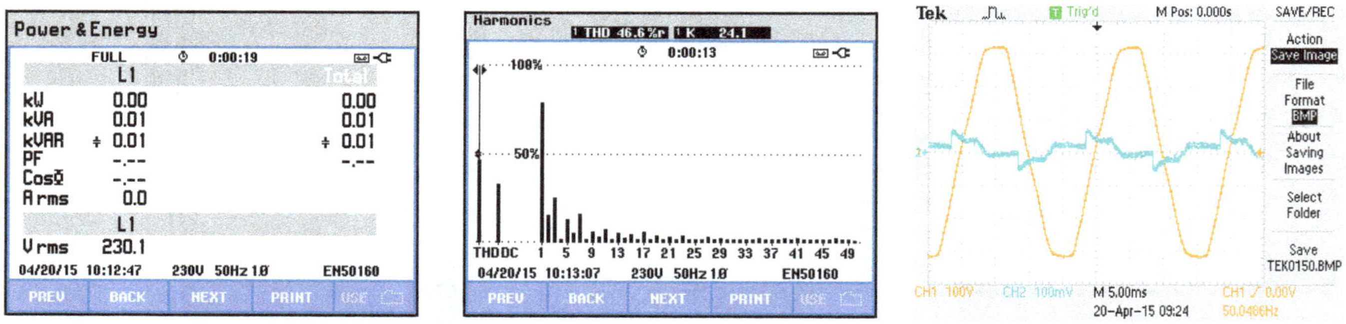

Figure 3. *Wavwforms for lamp 3.*

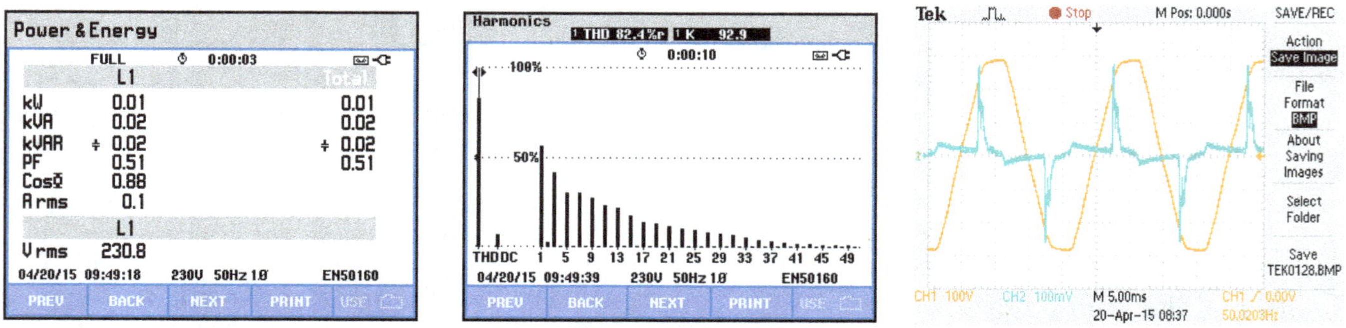

Figure 4. *Wavwforms for lamp 4.*

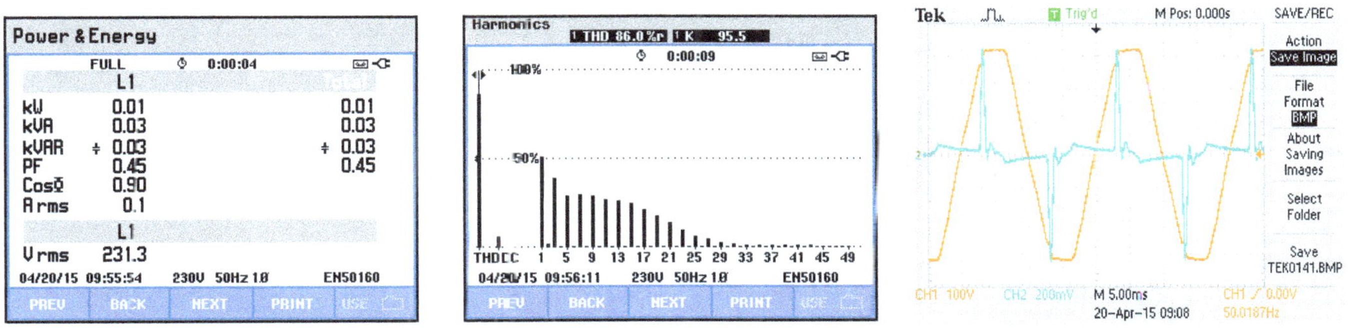

Figure 5. *Wavwforms for lamp 5.*

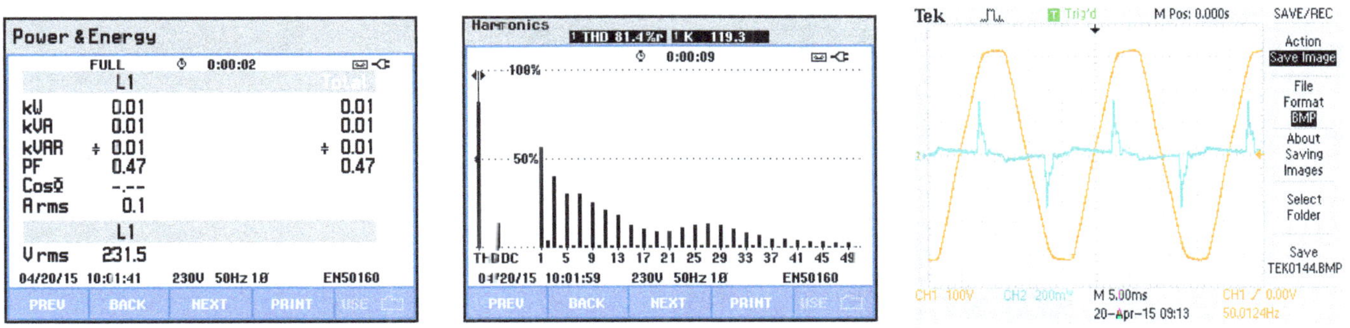

Figure 6. *Wavwforms for lamp 6.*

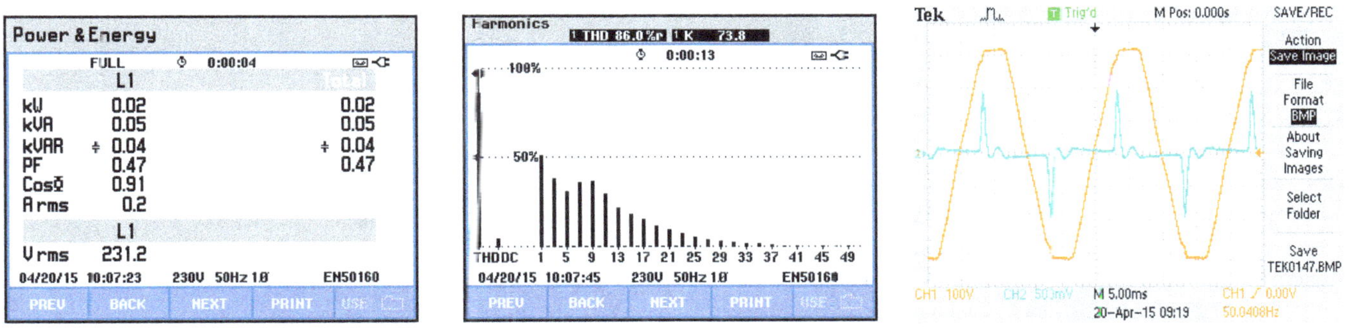

Figure 7. *Wavwforms for lamp 7.*

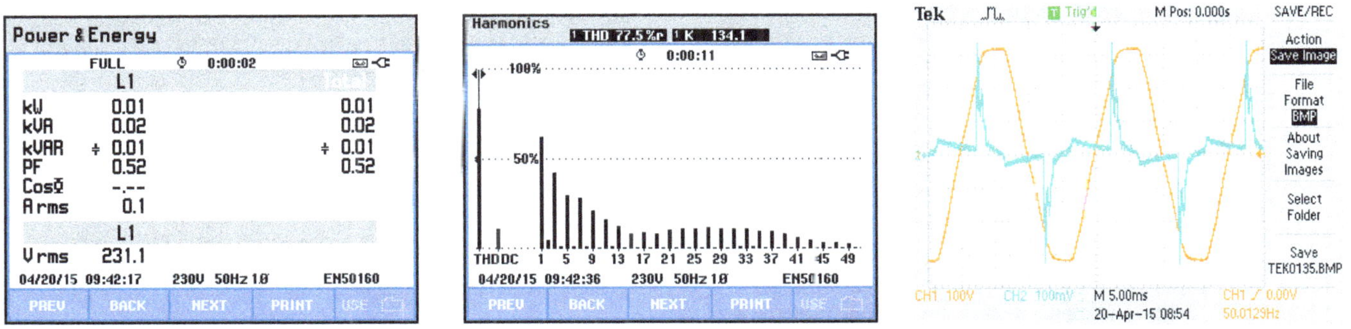

Figure 8. *Wavwforms for lamp 8*

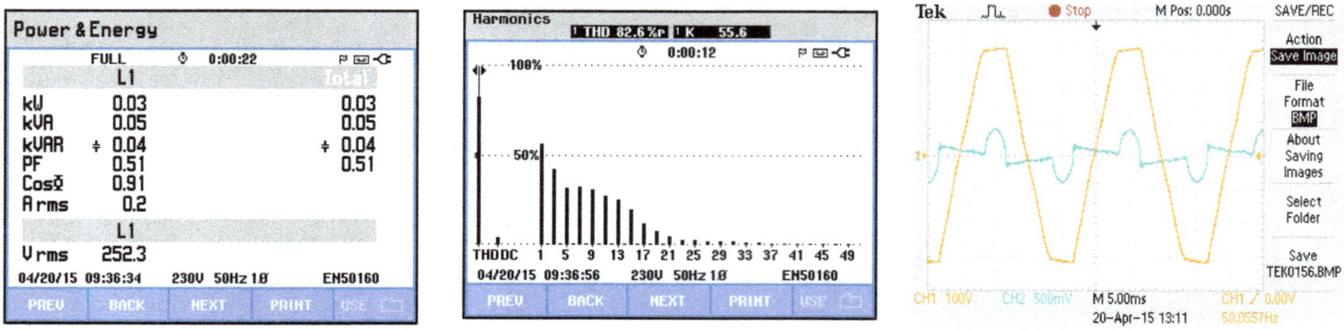

Figure 9. *Wavwforms for lamp 9.*

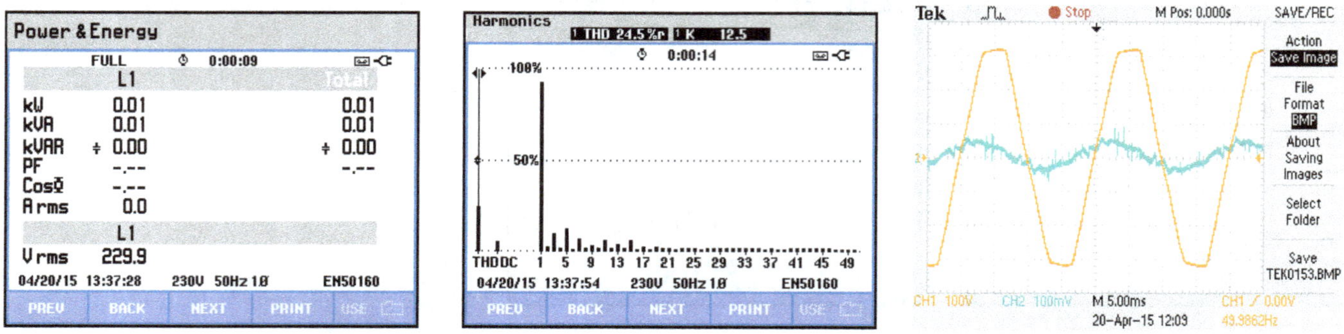

Figure 10. *Wavwforms for lamp10.*

4. Conclusion

The results of the conducted experimental research show a bad harmonic constitution in most lamps with a power of less than 25 W – high THD and K. For the resolution of this matter, there are enough technical means at hand – active or passive correction of the power factor. A comprehensive evaluation might be necessary and eventually correction in the standard with restrictions in respect to the harmonic constitution of the current and with power under 25 W. This matter is also connected to a particular economic evaluation, with which to be registered the increased costs of a separate lamp when entering a means of correction from one side and the cumulative negative effect on the supply network at bad harmonic constitution.

References

[1] Nassar A., M.Mednik, "Introductory physics of harmonic distortion in fluorescent lamps", American Journal of Physic., Vol.71, No6, pp.577-579, June, 2003.

[2] George V., Bagaria A., Rampattiwar S. R., "Comparision of CFL and LED lamps – harmonic disturbances, economics (cost and power quality), and maximum possible loading in a power system", International Conference and Utility Exhibition, 2011, pp.1-5.

[3] Dolara A., S.Leva, "Power Quality and Harmonic Analysis of End User Devices", Energies, No 5, 2012, pp.5453-5466, ISSN 1996-1073.

[4] F. V. Topalis F. V., I. F Gonos, M. B. Kostic, "Effects of changing line voltage on the harmonic current of compact fluorescent lamps", Proceedings of the IASTED International Conference Power and Energy Systems, November 8-10, Las Vegas, USA, pp.1-4.,1999.

[5] F. V. Topalis F. V., I. F Gonos, M. G.A.Vokas," Arbitrary wavwform generator for harmonic distortion tests on compact fluorescent lamps, Measurement, Elsevier, no.30, pp.257-267, 2001.

[6] Islam M. S., Chowdhury N. A. Sakil A. K., Khandakar A., "Power Quality Effect of using Incandescent, Fluorescent CFL and LED Lamps", First Workshop on Smart Grid and Renewable Energy, Qatar, 2015, pp.1-5.

[7] Moulahoum S., Houassine H., N. Kabahe, "Shunt Active Power Filter to Mitigate Harmonics Generated by Compact fluorescent lights ", 18th International Conference of Methods and Models in Automation and Robotics, 2013, pp.496-501.

[8] Chun C., H.L.Cheng, T.Y. Chung, "A Novel Single-State High-Power-Factor LED Street-Lighting Driver With Coupled Inductors" IEEE Transactions on Industry Applications, Vol.50, Issue 5, 2014, pp.3037-3045.

[9] Lam J. C. W., Shangzhi P., Jain P. K., "A Single-Switch Valley-Fill Power –Factor-Corrected Electronic Ballast for Compact Fluorescent Lightings With Improved Lap Current Crest Factor", IEEE Transactions on Industrial Electronics, Vol.61, Issue 9, 2014, pp.4654-4664.

[10] IEC 6100-3-2, Electromagnetic compatibility (EMC), Part3: Limits-Section 2: Limits for harmonic current emissions (equipment current < 16A per phase).

[11] European Power Supply Manufacturers Association, Harmonic Current Emissions, Guidelines to the Standart EN 6100-3-2, Revision data 2010-11-08.

[12] IEC 60969, Self-ballasted compact fluorescent lamps for general lighting services – performance requirements.

Modelling and control of electric vehicle power train

Souhir Tounsi

National School of Electronics and Telecommunications of Sfax-(SETIT): Research Unit, Sfax University, Sfax, Tunisia

Email address:
souhir.tounsi@isecs.rnu.tn

Abstract: This paper describes the choice and the design of electric vehicles power train structure reducing considerably the energy consumption. Indeed The converter feeding the motor is naturally with IGBTs leading on the one hand to important losses and on the other hand to many control problems. This structure is replaced by another with electromagnetic switch leading to a strong reduction of the losses and to an increase of the electric motor control reliability. The power train contains an energy recuperation system during the deceleration phases, where the motor functions in generator. The motor is controlled by vector control method maintaining the current Id equal to zero, leading to the maintain of the current in phase with electromotive force, what also leads to the reduction of the energy consumption. A supper-capacity is added in parallel with the energy accumulator leads to an increase of the storage energy capacity. All these factors lead to the increase of the autonomy for a known stocked energy.

Keywords: Power Chain, Design, Battery, Converter, Thermal Model, Simulation

1. Introduction

Currently and in look of the strong petroleum crises, during these last decades and the problems of atmospheric pollution, the electrification of the vehicles project became a project of actuality. In this context, several works of research are thrown on this thematic [1], [2], [3], [4] and [5].

Following several works of research a single motor configuration provided with a differential is kept. The motor is with permanent magnet and sinusoidal wave-form, having an axial structure. Naturally the power converter is with IGBTs, leading to an important energy losses [6], and to many control problems: such as the floating voltage and the tail current at the commutation time and the problems of the static and dynamic latch-up requiring a complicating control system. In our case, we choices a static converter structure with electromagnetic switch leading practically to the annulment of losses and to the increase of the reliability of the control.

The power train includes an energy recuperation system during decelerations phases leading to the reduction of the energy consumption and thereafter to the increase of the autonomy.

We choices a vector control strategy fixing the electromotive force in phase with the current (Strategy Id =0) also leading to the economy of energy and thereafter the increase of the autonomy. The design of this power train takes in account of most technological constraints and others attached to reliability [6].

This paper describes the choice, the design methodology and the control strategy of this power train.

2. Electric Vehicle Power Train Structure

The synoptical schema of the electric vehicle power train is illustrated by figure1:

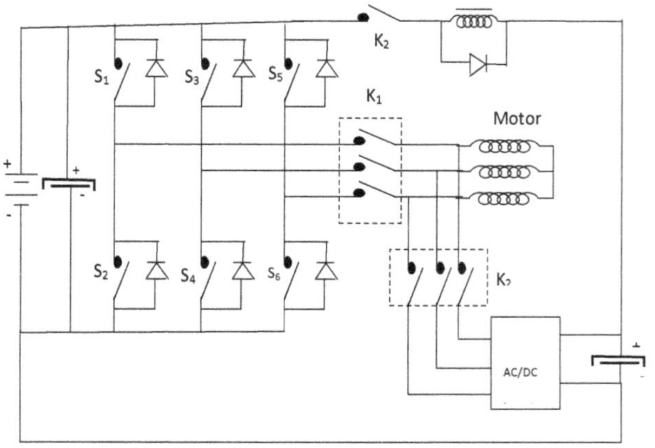

***Figure 1.** Electric vehicle power train structure.*

Naturally, DC/AC converter powering the motor is with IGBTs. In our case, we have chosen a structure with electromagnetic switches leading to a reduction of the energy losses. The control of these switches is assured by six generating windings. Being powered by a sufficient current, these windings attract their ferro-magnetic cores, leading to the closings of these switches according to the vector control strategy fixing the current in phase with the electromotive force (Id = 0). At the time of their dice-power these windings free the energy stocked through a free wheel diode. Two working phases are possible:

- Working in motor phase: in this phase the K1 switches are closed and generating windings assures the opening and the closing of the converter's switches according to a vector control strategy reducing the energy consumption. This phase is possible for working either in accelerated phase or in constant speed. During this phase the K2 switches are open.
- Working in generating phase: this phase is possible for a working in decelerated phase. In this phase, the motor function in generator. The control system opens the switches of the static converter and the K1 switches. At this moment the energy recuperation system functions. In this phase, the K2 switches close itself to convert the three electromotive forces of the motor that absorbed himself according to the speed decreasing during the time on DC voltage. The recovered DC voltage is filtered by a capacitor. This voltage source is converted in a current source permitting the injection of electrons in the battery. This last is in charge thereafter. This phase is named energy recuperation phase. The duration of this phase is until the stability of the speed or the acceleration of the vehicle.

3. Dimensioning Torque Energy Accumulator

The energy accumulator is a coupling of several elementary batteries, whose structure is given by the figure 2:

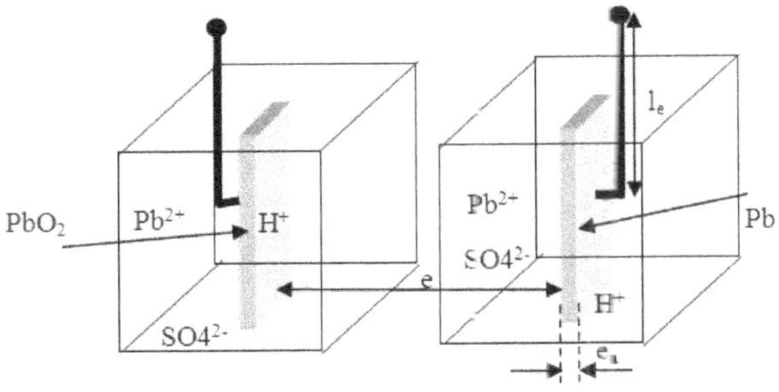

Figure 2. Elementary structure of battery.

Every element is composed of reservoirs with different electric polarity, in contact with two armatures. This structure is equivalent to a capacity temporarily charged by the quantity of charge recovered by the armature. Two electrodes permit the recuperation of the voltage generated by the battery. The whole armature more electrodes present a resistance in series with the capacity. A capacity in parallel with this resistance exists. The existence of this capacity can be explained by the polarity difference between electrode-armature and electrode output voltage contacts, what leads to the equivalent diagram of a battery element (figure3):

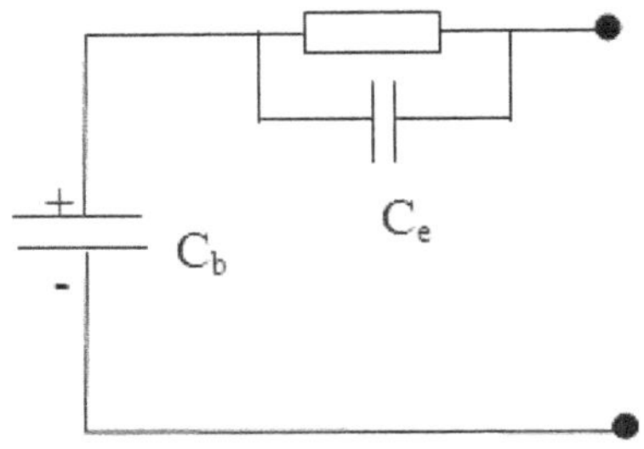

Figure 3. Model of the battery elementary structure

$$C_b = \frac{\varepsilon e}{S_a} \tag{1}$$

Where ε and e are respectively the permitivity and the distance of the place separating the armatures and Sa is the surface of the armature.

$$C_e = \frac{\varepsilon l_e}{S_e} \tag{2}$$

l_e and S_e are respectively the length and the section of one electrode.

$$R_b = 2\frac{\rho_a(t)\, e_a}{S} + 2\frac{\rho_e(t) l_e}{S_e} \tag{3}$$

Where ρ_a and ρ_e are respectively the materials resistivity of the armature and the electrode, S is the section of armature and ea is the armature thickness.

The number of elements in series and in parallel depends on the stocked energy and on the battery delivered voltage. We expect the coupling of a super-capacitor in parallel with the battery to increase the storage capacity of the accumulator of energy. If we disregard R_b and Ce, the energy stocked in the accumulator is deduced by the following formula:

$$W_b = \frac{1}{2}(C_{bt} + C_s)U_b^2 \tag{4}$$

Where C_{bt} and C_s are respectively the equivalent capacities of the battery and the super-capacity and U_b is the battery voltage.

The equation of oxydo-reduction is as follow:

- Left reservoir : $P_b^{2+} + 2H_2o \rightleftarrows P_bO_2 + 4\,e^- + 4\,H^+$
- Right reservoir : $P_b^{2+} + 2\,e^- \rightleftarrows P_b$

$[H^+]$ and $[P_b^{2+}]$ are the molar concentrations of hydrogen and lead.

The voltage to the level of the two electrodes left and right are given by the following NRENST formula:

$$E_G = E_G^0 + \frac{R \times T}{2 \times A} \times Ln\left(\left[P_b^{2+}\right]\right) \tag{5}$$

$$E_D = E_D^0 + \frac{R \times T}{2 \times A} \times Ln\left(\frac{[H^+]^4}{[P_b^{2+}]}\right) \quad (6)$$

Where EG0 and ED0 are the standard voltages of the two red-ox couples, T is the temperature and R, A are constants.

From where the difference of the two electrodes voltages is expressed by the following formula:

$$U_e = E_D^0 - E_G^0 + \frac{R \times T}{2 \times A} \times Ln\left(\frac{[H^+]^4}{[P_b^{2+}]}\right) - \frac{R \times T}{2 \times A} \times Ln([P_b{}^{2+}]) \quad (7)$$

At 25°C the voltage of one element of battery is expressed as follow:

$$U_e = E_D^0 - E_G^0 + \frac{0.06}{2} \times Ln\left(\frac{[H^+]^4}{[P_b^{2+}]^2}\right) \quad (8)$$

The number of elements to couple in series to get the wished battery voltage is deducted therefore from the following formula:

$$n_{es} = \frac{U_b}{E_D^0 - E_G^0 + \frac{0.06}{2} \times Ln\left(\frac{[H^+]^4}{[P_b^{2+}]^2}\right)} \quad (9)$$

The number of elements to couple in parallel to complete the stocked energy reserves is expressed as follow:

$$n_{ep} = \frac{W_b - \left(\frac{1}{2} \times n_{es} \times C_e \times U_e^2 + \frac{1}{2} \times C_s \times U_b^2\right)}{\frac{1}{2} \times C_e \times U_e^2} \quad (10)$$

4. Design Methodology of Electric Motor

The chosen of electric motor structures is oriented to permanent magnets, axial flux and modular structure with more combination and its radial flux equivalent motor [7].

Figure 4 illustrates one configuration with axial flux and its radial flux equivalent structure:

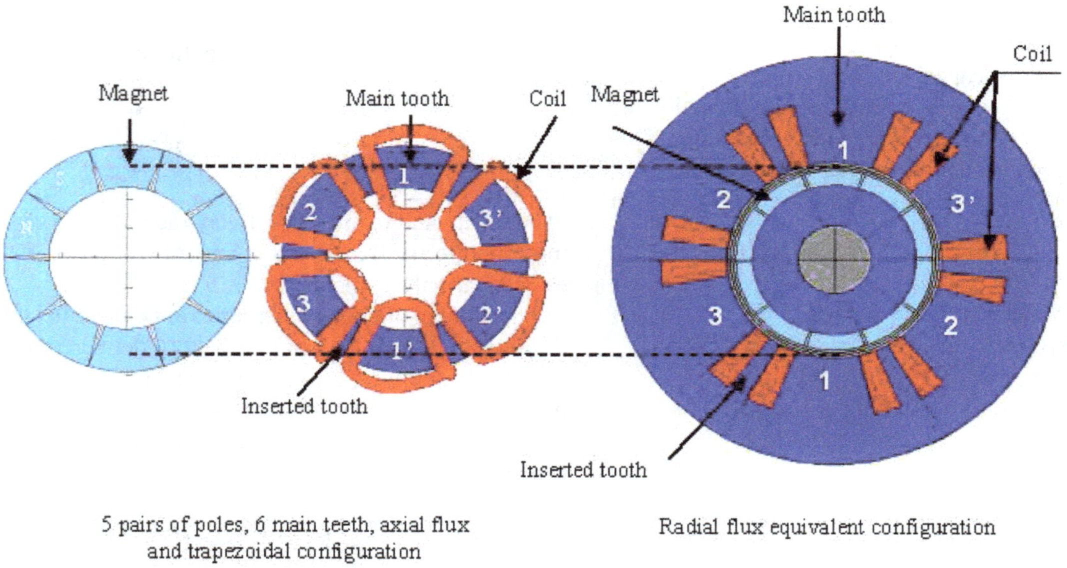

Figure 4. *Example of motor configuration.*

The design methodology consists of the determination of geometrical and control parameters of the motor-converter improving Autonomy. The motor must function on a broad beach of speed and without demagnetization. This methodology requires the development of an analytical and parameterized model of the all motor-converter. This latter last makes it possible to establish the relation between data, such as: the data of schedule conditions, the constant characterizing materials, the expert data, the motor configuration and the outputs such as: geometrical and electromagnetic motor magnitudes [8].

This model is validated by finite elements method [8]. Indeed, the motor is drawn according to its geometrical magnitudes extracted from analytical model with the software Maxwell-2d, and is simulated in dynamic and static in order to compare the results obtained with those found by the analytical method.

The coupling of this model to a model evaluating the autonomy poses an optimization problem with several variables and constraints. This latter is solved by the genetic algorithms (GAs) method [9].

The global architecture of the design methodology is illustrated in figure 5.

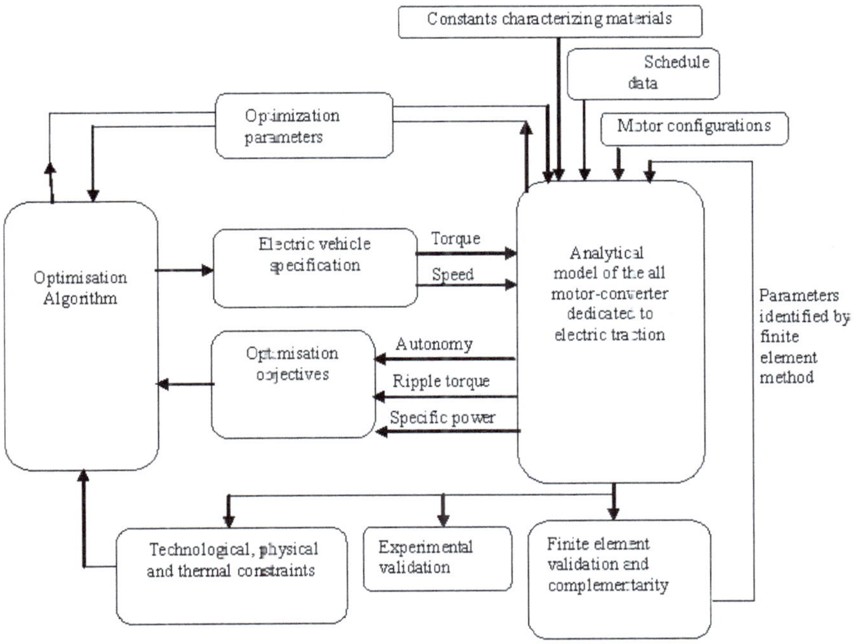

Figure 5. Design methodology of traction motor.

The structure of the motor is modular that is to say multi-stages. The design analytic model of the motor is found on [7].

A thermal nodal model of the electric motor is developed to respect thermals constraints [8]. This thermal model of the motor modular structure is developed while considering that the flux of heat propagates itself axially. The figure 6 illustrates this property:

The nodal model of the motor structure is illustrated by the figure 7:

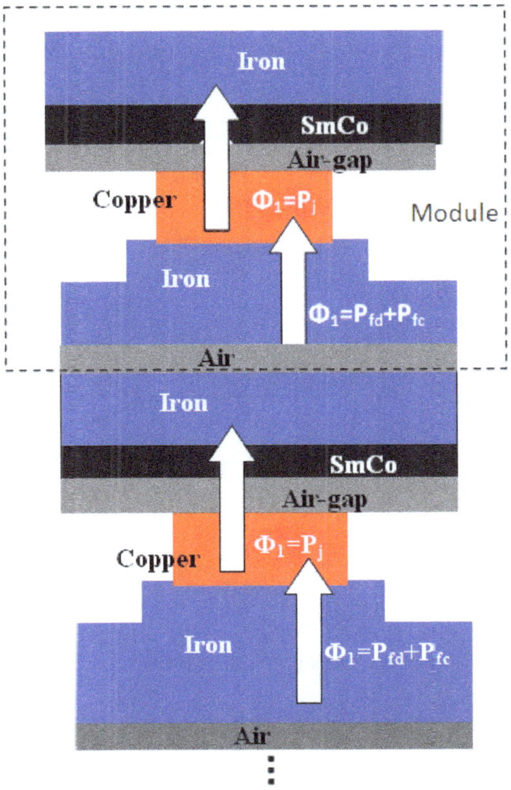

Figure 6. Thermal flux propagation.

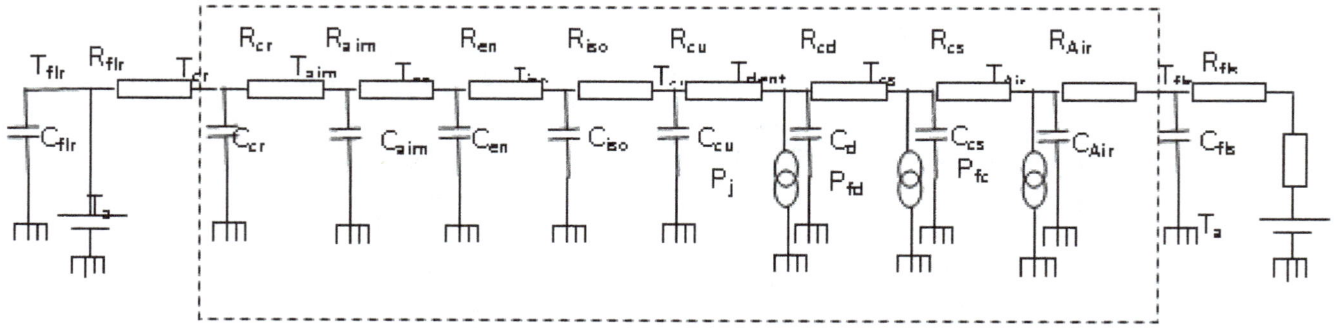

Figure 7. Thermal model of the motor structure.

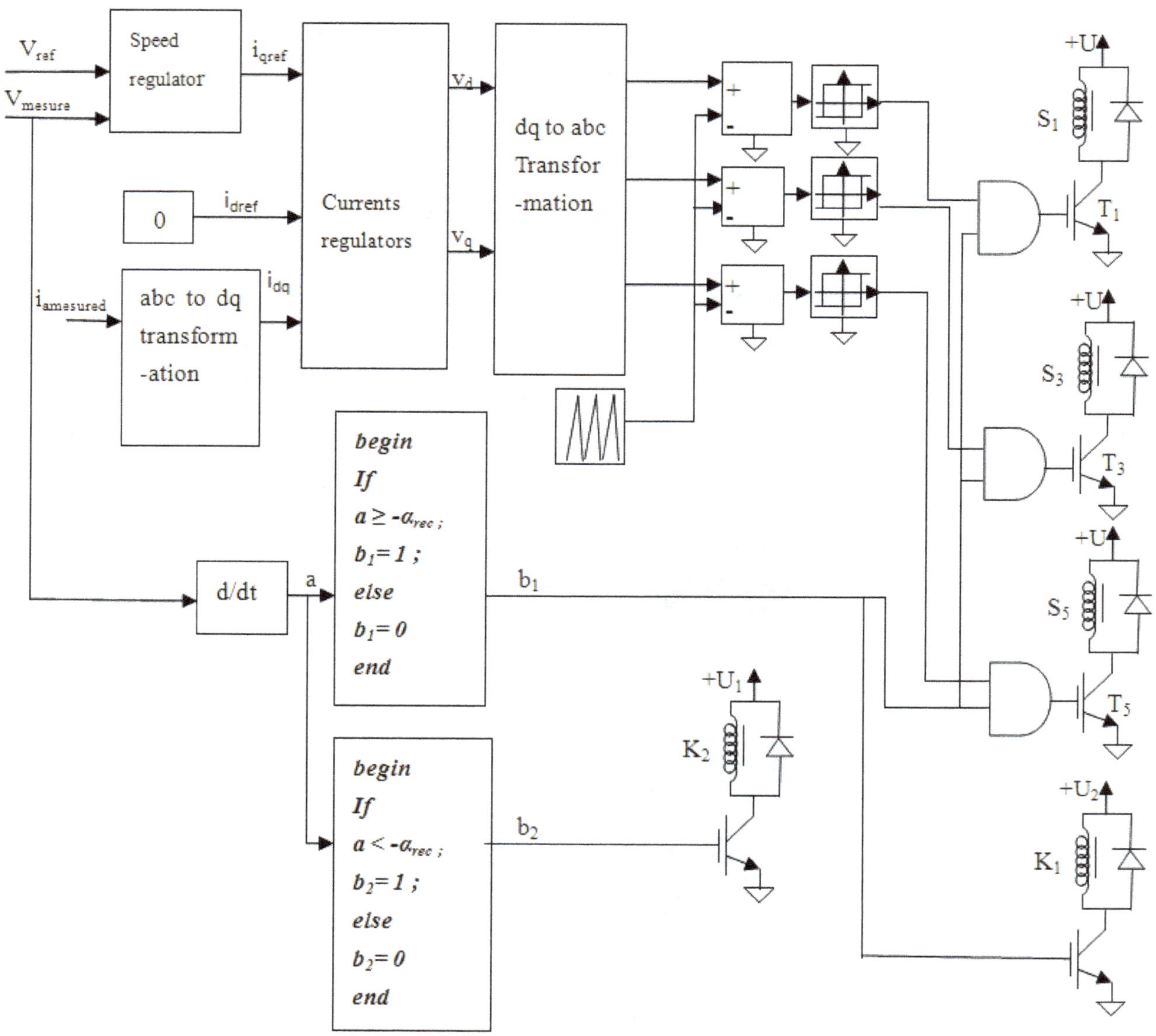

Figure 8. Control generator structure.

Table 1. *illustrates the nomenclature of the nodal model diagram*

h_l	Free convection coefficient
S_r	Active section of the rotor
h_f	Forced convection coefficient
S_s	Active section of the stator
R_{cr}	Rotor yoke conduction resistance
R_c	Magnets conduction resistance
R_{ce}	Air-gap conduction resistance
R_{cu}	Copper conduction resistance
R_{cd}	Teeth conduction resistance
R_{cs}	Stator yoke conduction resistance
R_{Air}	Air conduction resistance
P_j	Copper losses
P_{fd}	Teeth iron losses
P_{fc}	Stator yoke iron losses
T_a	Ambient temperature
T_{cr}	Rotor yoke temperature
T_{ai}	Magnets temperature
T_{en}	Air-gap temperature
T_{cu}	Copper temperature
T_d	Teeth temperature
T_{cs}	Stator yoke temperature
C_{cr}	Thermal capacity of the rotor yoke
C_{aim}	Thermal capacity of magnets
C_{en}	Thermal capacity of the air-gap
C_{cu}	Thermal capacity of the copper
C_d	Thermal capacity of the teeth
C_{cs}	Thermal capacity of the stator yoke
C_{AIR}	Thermal capacity of the air
C_{iso}	Thermal capacity of insulation

5. Control Generator Structure

The control generator structure is illustrated by figure 8.

The control generator assures the transmission of the control signals to the T1 transistors, T3 and T5 dragging the excitation respectively of the three generating windings S1, S3 and S5 according to the vector control law (Id=0 strategy) during the phases of working to constant speed or in accelerated phase. These three windings attract their cores according to the nature of control signals. Three other windings S2, S4 and S5, no schematized in the control circuit will be powered by signals complementary respectively to the control signals: S1, S3 and S5.

During the energy recuperation phase, b1 signals drags the opening of the DC/AC converter switches and the opening of the K1 switches. b2 signal drags the closing of the K2 switches following the excitation of the K2 winding. This phase is only possible for decelerations lower to a doorstep where the recuperation is not negligible.

The model of the control generator is implanted under Matlab/Simulink environment (figure 9).

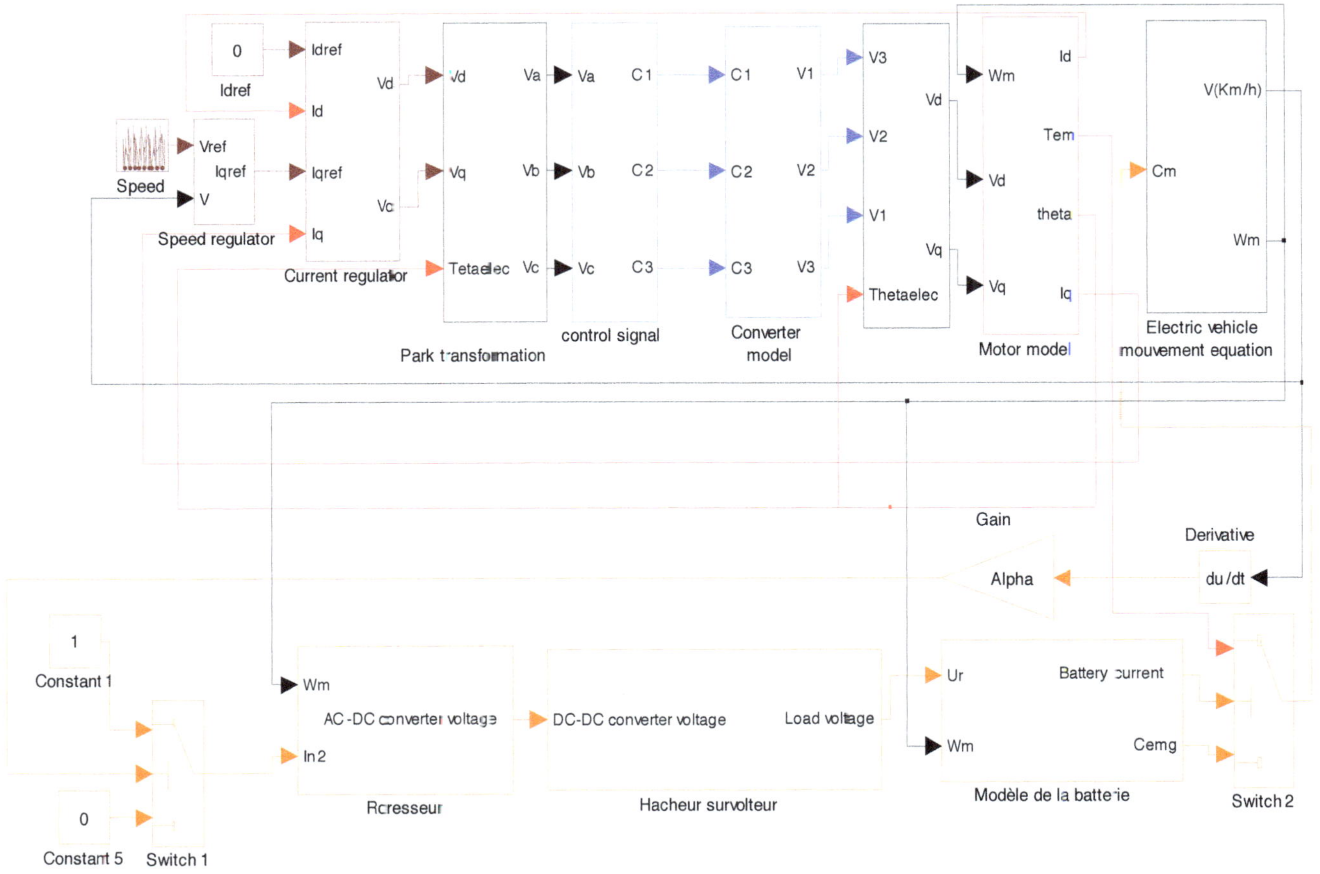

Figure 9. Simulink model of the control generator.

6. Generating Winding

The generating windings assure the closing and the opening of DC/AC converter electromagnetic switches, according to the chosen of control law. When a winding is powered by a sufficient current, it attracts its core and drags the closing of one or switches attached to its Ferro-magnetic core thereafter. The current powering this windings, must be sufficient to defeat the opposite strength generated by the recall spring.

The structure of the generating winding is illustrated by the figure 10:

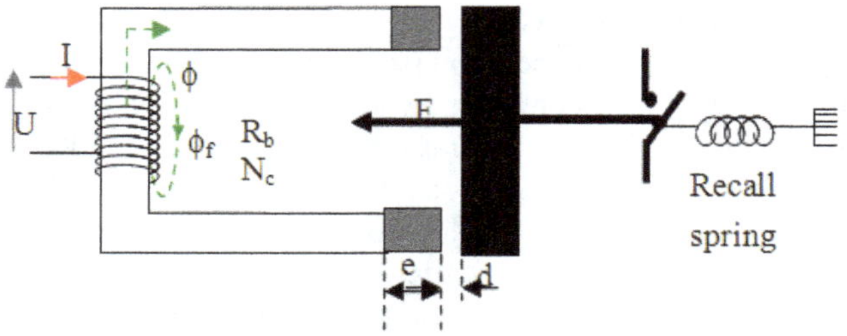

Figure 10. *Generating winding.*

The generating winding inductance depends on the displacement of the mobile iron core:

$$L_b = \frac{N_c^2}{\Re_t} = \frac{N_c^2}{\frac{2\times e}{\mu_0 \times S} + \frac{2\times(d-x)}{\mu_0 \times S}} = \frac{\mu_0 \times S \times N_c^2}{2\times(e+d-x)} \quad (11)$$

Where N_c is the spires number of winding, μ_0 is the air permeability, S is the section of the iron core, x is the displacement of the iron core.

The energy stocked in the winding is given by the following equation:

$$W = \frac{1}{2}\times L_b \times I^2 \quad (12)$$

The power drifting of this energy is given by the following equation:

$$P = \frac{dW}{dt} = \frac{I^2 \times \mu_0 \times S \times N_c^2}{4} \times \frac{d}{dt}\left(\frac{1}{e+d-x}\right) \quad (13)$$

This electric power turns into a mechanical power to the level of the mobile iron core:

$$P = F \times V \quad (14)$$

Where F and V are respectively the attraction strength and the speed of the mobile iron core.

We deduct from the two equations (13) and (14) the expression of the attraction strength depending on the displacement x:

$$F = \frac{\mu_0 \times U^2 \times S \times N_c^2}{R_b^2} \times \frac{d}{dx}\left(\frac{1}{e+d-x}\right) = \frac{\mu_0 \times U^2 \times S \times N_c^2}{R_b^2} \times \left(\frac{1}{(e+d-x)^2}\right) \quad (15)$$

The equation that describes the working of the mobile iron core and switch, drift of the dynamics fundamental relation:

$$M_n \times \frac{dV}{dt} = F - m \times K \times x \quad (16)$$

Where Mn is the mass of the mobile iron core, K is recall spring constant and m is the number of attracted switches.

While replacing F by its expression, (16) becomes:

$$M_n \times \frac{dV}{dt} = \frac{\mu_0 \times U^2 \times S \times N_c^2}{R_b^2} \times \left(\frac{1}{(e+d-x)^2}\right) - m \times K \times x \quad (17)$$

At the balance we have:

V = 0 and d = x, from where we deduct the expression of the powering voltage:

$$U = \sqrt{\frac{4 \times m \times K \times e^2 \times R_b^2}{\mu_0 \times S \times N_c^2}} \quad (18)$$

The active section of the copper depends on the admissible current density in the copper:

$$S_c = \frac{I}{\delta} = \frac{U}{R_b \times \delta} \quad (19)$$

From the equations (18) and (19) we deduct the expression of the active section of winding copper thread:

$$S_c = \sqrt{\frac{4 \times m \times K \times e^2}{\mu_0 \times \delta^2 \times S \times N_c^2}} \quad (20)$$

The generating winding resistance is given by the following equation:

$$R_b = \frac{\rho \times L_e}{S_c} \quad (21)$$

Where ρ is the copper resistivity and Le is the winding length:

$$L_e = 2 \times N_{c/c} \sum_{n=1}^{N_{cc}} \left(a + b + 2 \times n \times \sqrt{\frac{S_c}{\pi}}\right) \quad (22)$$

Where Nc is the total number of winding spires, Nc/c is the number of thread layer rolled up, a and b are respectively the iron core width and thickness.

$$N_{c/c} = \frac{E_B}{2\times\sqrt{\frac{S_c}{\pi}}} \quad (23)$$

Where EB is the thickness of the copper thread rolled up.

The number of thread by layer is given by the following equation:

$$N_{cc} = \frac{N_c}{N_{c/c}} \quad (24)$$

From where the winding resistance is deduced by the following equation:

$$R_b = \frac{\rho\times\left(2\times N_{c/c}\sum_{n=1}^{N_{cc}}\left(a+b+2\times n\times\sqrt{\frac{\sqrt{\frac{4\times m\times K\times e^2}{\mu_0\times\delta^2\times S\times N_c^2}}}{\pi}}\right)\right)}{\sqrt{\frac{4\times m\times K\times e^2}{\mu_0\times\delta^2\times S\times N_c^2}}} \quad (25)$$

7. Simulation Results

The model of the losses is implanted under the environment of Matlab/Simulink. The thermal fluxes are calculated while leaning on the inverse gait of modelling of the power chain.

The simulation of the thermal model with a natural ventilation (Coefficient of convection equal to 30 W/m²K) and for a working to speed consolidated equal to 80 km/h, give the evolution of the temperatures in the different active parts of the motor (figure11).

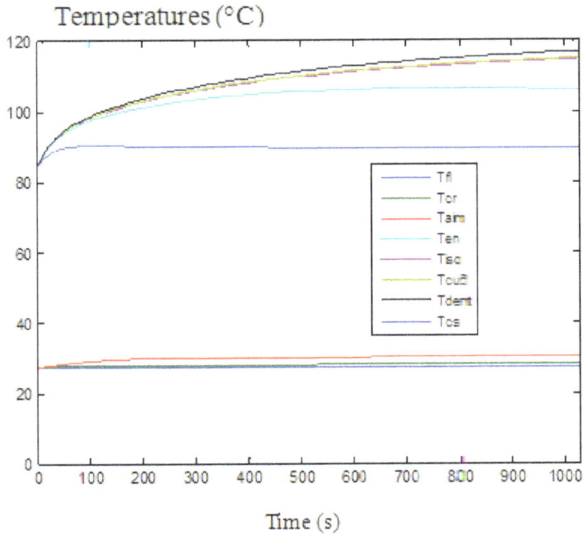

Figure 11. *Evolution of the temperatures in the different active parts of the motor for a speed consolidated of 80 km/h (h=30 W/m²K).*

Where Tfl is the average temperature of flabby, Tcr is the average temperature of the rotor yoke, Taim is the average temperature of magnet, Ten is the average temperature of air-gap, Tiso is the average temperature of isolating, Tcu is the average temperature of copper, Tdent the average temperature of stator teeth and Tcs the average temperature of the stator yoke.

This face shows that there is an overtaking of 47 °C for the copper and the resin, what proves the necessity of a cooling system. Several simulations are thrown for several values of forced convection coefficient to a system of cooling by forced ventilation, led to the fixing of this coefficient to 300 W/m²K.

The evolution of the temperatures in the different active parts of the motor for a working with a system of cooling to forced ventilation with a convection coefficient equal to 300 W/m²K is illustrated by the figure 12.

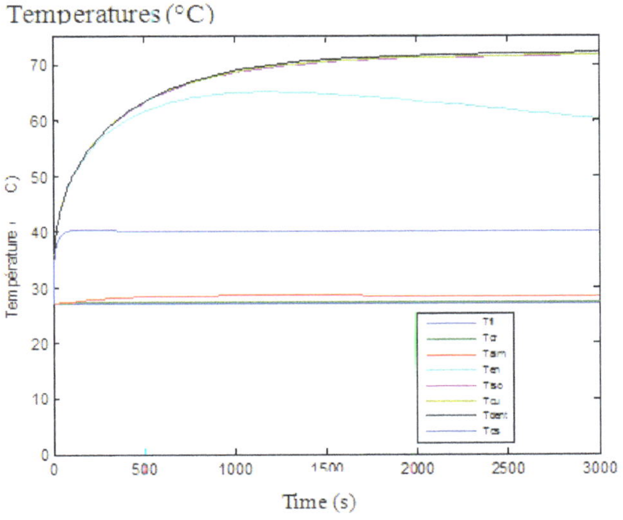

Figure 12. *Evolution of the temperatures in the different active parts of the motor for a speed consolidated to 80 km/h (h=300 W/m²K).*

The energy recovered (figure 13) believes at the time of the phases of strong decelerations and remain constant during the phases of working in motor.

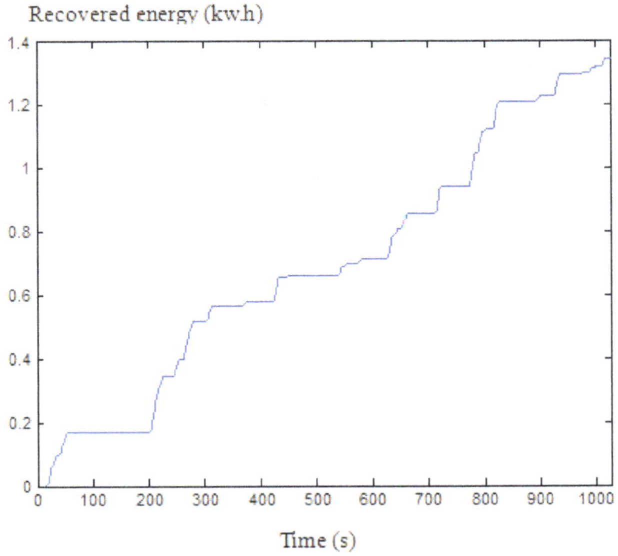

Figure 13. *Recovered energy.*

This energy is equal in middle value to 0.6942 kw.h on a

duration of 1027 s. This value is important, what shows the efficiency of the system of energy recuperation.

8. Conclusion

The choice done, the design methodology and control of this power train, increases the autonomy, reliability, considerably. This power train structure presents an attractive solution to solve the problem of the electric vehicles weak autonomies. It's interesting to study the problem of the excessive cost and the problem of battery load infrastructure in future.

List of Symbols

Cb	Elementary battery capacity
Ce	Electrode armature capacity
ε	Permitivity of the place separating the armatures
e	Length of the place separating armatures
le	Length of electrode
Sa	Armature section
Se	Electrode section
ρa	Resistivity of the armature material
ρe	Resistivity of the electrode material
Wb	Energy stocked in the accumulator
Rb	Armature-electrode resistance
Cbt	Equivalent capacity of the battery
Cs	Super-capacity
Ue	Voltage of elementary battery
nes	Number of elementary battery coupled in series
nep	Number of elementary battery coupled in parallel
Pj	Copper losses
Pfd	Iron losses in teeth
Pfc	Iron losses in stator yoke
Ta	Ambient temperature
Lb	Winding inductance
Nc	Spires number
$\mu 0$	Air permeability
S	Section of iron core
x	Displacement of iron core
W	Energy stocked in winding
I	Current feeding the winding
P	Electric power of the winding
F	Attraction iron core strength
U	Voltage feeding the winding
Mn	Mass of the mobile iron core
K	Recall spring constant
V	Velocity of the mobile iron core
m	Number of attracted switches
Sc	Active section of the copper
δ	Current density
Nc	Number of winding spires
Nc/c	Number of thread layer rolled up
Ncc	Number of thread by layer
EB	Thickness of the copper thread rolled up

References

[1] Naomitsu Urasaki, Tomonobu Senjyu and Katsumi Uezato: "A novel calculation method for iron loss resistance suitable in modelling permanent-magnet motors", IEEE TRANSACTION ON ENERGY CONVERSION, VOL. 18. NO 1, MARCH 2003.

[2] B. Ben Salah, A. Moalla, S. Tounsi, R. Neji, F. Sellami: "Analytic design of Permanent Magnet Synchronous motor Dedicated to EV Traction with a Wide Range of Speed Operation", Interneational Review of Electrical Engineering (I.R.E.E), VOL 3, NO 1 January-February 2008"

[3] Sid Ali. RANDI : Conception systématique de chaînes de traction synchrones pour véhicule électrique à large gamme de vitesse. Thèse de Doctorat 2003, Institut National Polytechnique de Toulouse, UMRCNRS N° 5828.

[4] C. C. Chan and K. T. Chau: "An Overview of power Electronics in Electric Vehicles", IEEE Trans. On Industrial Electronics, Vol, 44, No 1, February 1997, pp.3-13.

[5] C. PERTUZA: "Contribution à la définition de moteurs à aimants permanents pour un véhicule électrique routier". Thèse de docteur de l'Institut National Polytechnique de Toulouse, Février 1996.

[6] S. TOUNSI, R. NEJI, F. SELLAMI: "Contribution à la conception d'un actionneur à aimants permanents pour véhicules électriques en vue d'optimiser l'autonomie". Revue Internationale de Génie Electrique, Volume 9/6-2006, pp. 693-718. Edition Lavoisier.

[7] S. Tounsi : "Modélisation et Optimisation de la Motorisation et de l'Autonomie d'un Véhicule Electrique".Thèse de docteur de l'Ecole National d'Ingénieur de Sfax Tunisie, February 2006.

[8] Sid Ali. RANDI: Conception systématique de chaînes de traction synchrones pour véhicule électrique à large gamme de vitesse. Thèse de Doctorat 2003, Institut National Polytechnique de Toulouse, UMRCNRS N° 5828.

[9] S. TOUNSI, R. NEJI and F. SELLAMI : Electric vehicle control maximizing the autonomy: 3rd International Conference on Systems, Signal & Devices (SSD'05), SSD-PES 102, 21-24 March 2005, Sousse, Tunisia.

Micro Network Protection by Synchronous Generators by the Use of Fault Current Limiter

J. Beiza, H. Mohebalizadeh, A. Kh. Hamidi

Department of Electrical Engineering, Shabestar Branch, Islamic Azad University, Shabestar, Iran

Email address:
Jamalbeiza@gmail.com (J. Beiza), Hmpeed@gmail.com (H. Mohebalizadeh), Abzhamidi@yahoo.com (A. Kh. Hamidi)

Abstract: Micro protection is one of the challenges ahead of micro network expansion. Micro network functions in two states of connection to network and island. The short connection level of micro network is different at two functional levels. Therefore, the protection system which needs to diagnose the faults at two states would function improperly. In the preset article fault current limiter and oriented over-current relays optimal adjustment are implemented for micro network protection under the study. Additive particles optimal algorithm is used to adjust relays and fault current limiter impedance. The results show that by this method the micro networks can be secured at both two modes of functions.

Keywords: Micro Network, Protection Coordination, Over-Current Relays, Particles Optimal Algorithm

1. Introduction

Increasing problems related to air pollution the end of fossil fuels, depletion and congestion of transference system has caused the scattered products to be attended more than before. Micro-network is a new concept for a better utilizing of scattered product units and solving their technical problems. Micro-network is a complex of resources and loads which act under a single control unit from the network perspective. Micro-network has two types of functions: related to the network and island. In connected format the Micro-network invokes economic exchanges with the major network, anti-peak and offering complementary services. In times of turbulence in the main network the micro network can be separated from the main network and provide its loads in an island mode. The existence of scattered product units at distribution networks can lead to loss of coordination among these networks. Distribution networks mostly have radius structure where the fault current from the upper network is injected to the lower network by installing the scattered products the fault current level is increased and in addition to that we will observe the bidirectional fault currents in the distribution network. The current rate of short-circuit synchronous generators is considerable. The fault current of general electronic transformers is at 2-1 pre-unit level and has less impact on short-circuit level [3-1].

Several methods have been implemented to establish protection coordination within distribution networks by the presence of scattered product units. These methods can be divided into two parts of optimizing and geometric parts. In optimizing methods, the objective function is the performance of all relays and under the optimizing conditions minimizing the function is pursued [9-4]. In the methods of the second group the attempt is to make use of the graphic theory to distinguish the major relays and sensitive ones and then the arrangement of other relays are done [11-10]. References [12-13] protection coordination is made for the configuration of different power systems by considering the inlet and outlet of the generators References [14-15] have done the general analysis of the distribution systems protection impact on the scattered products. Reference [16] has implemented fault current in series format by scattered generation units and optimal arrangement of over current relays. In this research only the connected mode to the scattered products network; and the island mode has not been studied. To preserve the Micro-networks various methods are suggested. Two main protection challenges ahead of the Micro-network consist of changes of the short-circuit current for two functional modes and the low cooperation of units with power electronic transformer in the fault current. Various methods have been suggested for the protection of the Micro-network so far and the researches in this respect are proceeding. At (2) digital relays with communication links on the basis of differential protection are implemented for the protection of the Micro-networks and voltage protection is introduced as the supportive protection. A similar method has been implemented in (3) on the basis of signal transference by the use of power lines and differential protection to keep the feeders have been presented. Micro-network protection system on the basis of central protection and communicational relations among the relays and protection unit has been presented. The central protection unit by determining the delay intervals and performance indexes and considering the mode of the Micro-network provides the apposite micro-network protection system (5-6-7-13). At (14) over current relays which are capable of identifying the performance mode for the protection of Micro-network protection system. Most of the suggested methods for the Micro-network protection system require communicational link between relays.

In the present article by the use of fault current limiter and optimal regulation of relays the Micro-network protection system will be dealt with. This article, is an attempt to solve, the extreme differences of short-circuit levels at two functional modes which makes use of additive particles algorithm for optimizing. The second part states the protection coordination issue and the third part describes the methodology and the simulated results are presented in part four.

2. Protection Coordination

The protection coordination of protective devices within a network can be achieved by the use of optimal optimizing and regulation. As mentioned earlier the function of the problem objective is as the following equation.

$$MIN \ \sum_{m=1}^{2} \sum_{f=1}^{f_{max}} \left(\sum_{i} \left(t_i + \sum_{j} t_{ij}\right)\right) \qquad (1)$$

Where i stands for main relay and j stand for protective relays t_i is the time of main relay performance and t_{ij} show the protective relays performance time equal to f error. m shows island or micro network performance mode with the network. The time of the protective relays performance need to be behind the main relay performance time to the amount of time coordination feedback. This condition is displayed at (2).

$$t_{ij}{}^{f} - t_i{}^{f} \leq CTI \qquad (2)$$

In this article standard time reverse over current relays has been implemented. The features of the performance of these relays have been displayed at (3).

$$t = \frac{0.14 \times TDM}{\left(\frac{I_f}{I_p}\right)^{0.02} - 1} \qquad (3)$$

Where I_p is relay threshold current TDM is the relay regulation index and I_f is observed error current by relay

Relays current values are regulated at around 1.1-1.2 equal to relays currents. Relay regulation index changes in a disconnected way in practice but for the ease of action it has been used in an integral format. Since micro-networks function at two modes therefore, the load of relays will be different for both performance modes. In this study the maximum current load in both modes has been considered as the normal current load.

3. Fault Current Limiter and the Recommended Method

In the connected mode to the network because of the cooperation of the main network in providing the micro-network current faults we would have a higher level of short-circuit fault compared to island format. As can be seen in the following fig. 1 if the relays at island form get regulated properly the observed fault at connected mode would increase and therefore, the coordination of relays would be lost. By the installation of a fault current limiter the provided current fault of the network can be reduced and thereby transfer the protection coordination point towards island mode. This transference can be continued as far as protection coordination conditions are established. Therefore, by the use of fault current limiter we can compensate for the different short-circuit level at functional mode.

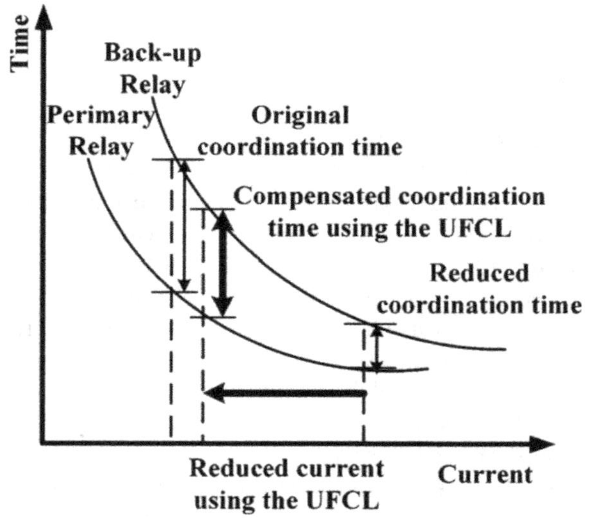

Fig. 1. Regulation relays at island form.

In order to optimize the time of relays we need to regulate the amount of fault current limiter and relays so that the total performance time gets reduced. To this end, additive particles algorithm has been implemented to determine the optimal value of fault current and relay regulations.

Additive particles algorithm is an ultra-explorative optimizing which has imitated the group fly of the birds. In this algorithm each answer of the question is modeled as a particle with n dimension features in the problem atmosphere. Particles or the answer candidates to the question move under the influence of three factors of current position, best position so far and the best position of the complex to reach the answer to the question. The main relation of particles optimizing algorithm is presented at (4).

$V(t)$ and $X(t)$ respectively speed and position of the particle at frequency t, X_{pbest} the best position of the particle so farX_{gbest} and the best experienced position by the total of the particles. ω inertia index and c_1 and c_2 are the related ratios.

In this article first, the problem solution candidates which consist of fault current limiter data and the relays regulations are generated. Then, by the use of impedance matrix the connected mode fault current to the network by considering the size of the fault current limiter is calculated. Since the fault current limiter connects the network at the connection place it would not affect the island mode short-circuit current level. By having the faults values at two functional modes algorithm calculates the sum of objective function. It needs to be mentioned that lack of observing the coordination conditions is considered as a heavy punishment at the objective function. At the end from the relation (4) answer candidates are updated at each repetition. Fig. 2 displays the implemented flowchart.

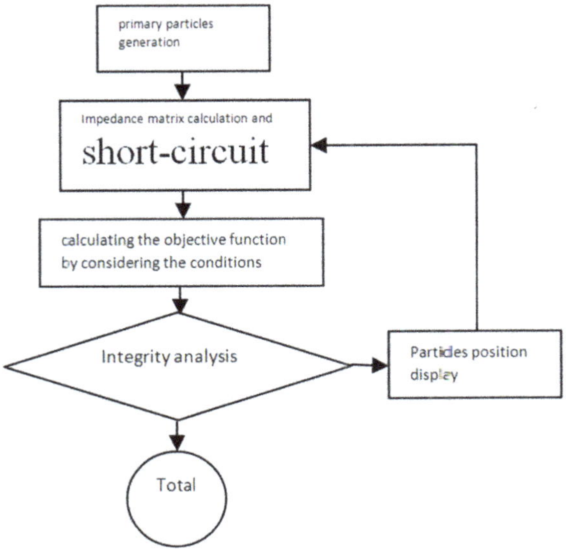

Fig. 2. Implemented flowchart.

4. Simulation Results

The micro network under the study is drawn in Fig. 3. This micro-network consists of three units of synchronous scattered generation which is linked to the main network by bass. The micro-network data has been presented at table 1. Over current relays have been used for the protection of the micro-network.

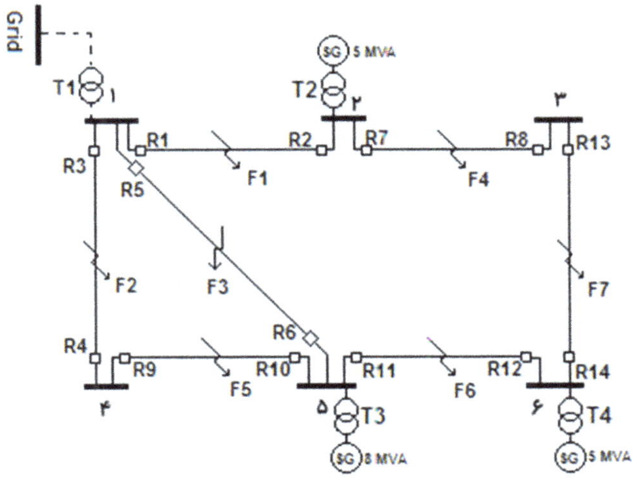

Fig. 3. The micro network under the study.

Table 1. The micro-network data.

12K W20, MVA	Micro-network basis
0.04+0.04j	Impedance line
0.1j	Transient Impedance of generator (on the basis of generator)
0.1j	Impedance of transformator connected to generators' micro networks
200MVA	Short circuit level of main network
0.08j	Impedance of the transformator connected to the micro network to the main network

Table 2 shows relay load currents for two island modes and connected to the network. The assumption is that when in island mode 50% of load depletion would happen. Given the shortness of the fault lines in the middle of the analysis line in the present study three phase faults are considered.

Table 2. Relay load current.

End of the line basbar	Beginning of the line basbar	Load current (per-unit-island mode)	Load current (connected mode to the network-per unit)
1	2	0.056	0.16
1	4	0.0357	0.16
1	5	0.062	0.107
2	3	0.0547	0.126
4	5	0.092	0.055
5	6	0.03	0.109
3	6	0.0626	0.073
Bass 2 unit		0.22	0.18
Bass 5 unit		0.27	0.292
Bass 6 unit		0.168	0.18
Network share			0.65

The sum of the short-circuit currents is calculated by means of impedance matrix method. To this end some virtual bass are use in the middle of the lines. It is supposed that the relay threshold current can be considered at 11-2 times the nominal current. By the use of particles optimizing method for the network under the study optimal protection coordination has been done. The position of the fault limiter is stable and in the place of connection to the network. After the optimizing fault limiter impedance was obtained at around 0.67 per unit. Table 3 shows the relays optimal regulation sums for the micro network under the study. The protection system total performance time equals 60.9496 seconds. Main relays time performance matrix and support for different faults are shown at table 3.

Table 3*. Relays time performance matrix.*

Relay No	Regulation ratio	Threshold current
R1	0.2571	0.2064
R2	0.2522	0.176
R3	0.1691	0.176
R4	0.181	0.1984
R5	0.2139	0.1681
R6	0.3089	0.1468
R7	0.2	0.1831
R8	0.2691	0.1671
R9	0.1392	0.1221
R10	0.2854	0.1199
R11	0.2516	0.2028
R12	0.2582	0.123
R13	0.2879	0.1030
R14	0.3433	0.805
R15	0.1896	0.371
R16	0.2286	0.3562
R17	0.2307	0.2393
R18	0.0322	0.7247

Table 4. *Relays performance time.*

Supporting relays performance time	Main relay performance time	Fault
0.8406, 0.866, 0.8141, 0.8658	0.7017, 0.6258	Island -F1
0.9572, 0.7186, 0.8502	0.6957, 0.566	Island -F2
0.7584, 0.709, 0.974	0.4057, 0.5538	Island -F3
0.5757, 0.7956, 0.8303	0.5567, 0.3995	Island -F4
0.789, 0.8122, 1.3818, 1.3803	0.6387, 0.6215	Island -F5
0.8181, 0.8222, 2.30	0.667, 0.5869	Island -F6
0.842, 0.8184, 0.8084	0.6205, 0.6782	Island -F7
0.8428, 0.7814, 0.7811, 0.8923	0.6311, 0.6205	Connected -F1
0.7024, 0.8221, 0.8644	0.6658, 0.5491	Connected -F2
0.6927, 0.7757, 0.8519, 1.04	0.3774, 0.5383	Connected -F3
0.513, 0.7862, 0.8399, 1.5889	0.5424, 0.362	Connected -F4
0.878, 0.7375, 0.7929, 0.8257, 0.8872	0.6262, 0.5875	Connected -F5
0.7887, 0.822, 0.8108	0.6567, 0.52	Connected -F6
0.7988, 0.7584, 0.8161	0.6079, 0.6488	Connected -F7

Results of the simulation confirm that by the use of fault current limiter we can improve short-circuit level difference in two modes and protect the micro network. Fault current limiter along with optimal regulation as the protect system independent from the micro network performance can be implemented.

5. Conclusion

In the present article fault current limiter was implemented to protect micro network. By the use of this element the short-circuit level difference in two modes of operation would not differ significantly. By the use of particle optimizing algorithm for a suggested micro network over current relays optimal regulation and the fault current limiter impedance was determined. The results show that protection system is able to protect micro network in both modes of function.

References

[1] A. Chowdhury and D. Koval, Power Distribution System Reliability: Practical Methods and Applications. Hoboken, NJ: Wiley-IEEE, Mar. 2009.

[2] B. Hussain, S. Sharkh, and S. Hussain, "Impact studies of distributed generationon power quality and protection setup of an existing distribution network," in Proc. Int. SPEEDAM, 2010, pp. 1243–1246.

[3] N. Nimpitiwan, G. T. Heydt, R. Ayyanar, and S. Suryanarayanan, "Fault current contribution from synchronous machine and inverter based distributed generators," IEEE Trans. Power Del., vol. 22, no. 1, pp. 634–641, Jan. 2007.

[4] P. Bedekar, S. Bhide, and V. Kale, "Optimum coordination of over current relays in distribution system using dual simplex method," in Proc. 2nd ICETET, Dec. 2009, pp. 555–559.

[5] M. Mansour, S. Mekhamer, and N.-S. El-Kharbawe, "A modified particle swarm optimizer for the coordination of directional overcurrent relays," IEEE Trans. Power Del., vol. 22, no. 3, pp. 1400–1410, Jul. 2007.

[6] P. Bedekar, S. Bhide, and V. Kale, "Optimum coordination of overcurrent relays in distribution system using genetic algorithm," in Proc. ICPS, 2009, pp. 1–6.

[7] P. P. Bedekar and S. R. Bhide, "Optimum coordination of directional overcurrent relays using the hybrid GA-NLP approach," IEEE Trans. Power Del., vol. 26, no. 1, pp. 109–119, Jan. 2011.

[8] C. So and K. Li, "Time coordination method for power system protection by evolutionary algorithm," IEEE Trans. Ind. Appl., vol. 36, no. 5,pp. 1235–1240, Sep./Oct. 2000.

[9] M. Barzegari, S. Bathaee, and M. Alizadeh, "Optimal coordination of directional overcurrent relays using harmony search algorithm," in Proc. 9th Int. Conf. EEEIC, May 2010, pp. 321–324.

[10] H. Sharifian, H. Askarian Abyaneh, S. Salman, R. Mohammadi, and F. Razavi, "Determination of the minimum break point set using expert system and genetic algorithm," IEEE Trans. Power Del., vol. 25, no. 3, pp. 1284–1295, Jul. 2010.

[11] Q. Yue, F. Lu, W. Yu, and J. Wang, "A novel algorithm to determine minimum break point set for optimum cooperation of directional protection relays in multiloop networks," IEEE Trans. Power Del., vol. 21, no. 3,pp. 1114–1119, Jul. 2006.

[12] A. Noghabi, J. Sadeh, and H. Mashhadi, "Considering different network topologies in optimal overcurrent relay coordination using a hybrid GA," IEEE Trans. Power Del., vol. 24, no. 4, pp. 1857–1863, Oct. 2009.

[13] A. Noghabi, H. Mashhadi, and J. Sadeh, "Optimal coordination of directionalovercurrent relays considering different network topologies using interval linear programming," IEEE Trans. Power Del., vol. 25, no. 3, pp. 1348–1354, Jul. 2010.

[14] P. Barker and R. De Mello, "Determining the impact of distributed generationon power systems. I. Radial distribution systems," in Proc. IEEE Power Eng. Soc. Summer Meeting, 2000, pp. 1645–1656.

[15] A. Girgis and S. Brahma, "Effect of distributed generation on protective device coordination in distribution system," in Proc. LESCOPE, 2001, pp. 115–119.

[16] W. El-Khattam and T. Sidhu, "Restoration of directional overcurrent relay coordination in distributed generation systems utilizing faultcurrent limiter," IEEE Trans. Power Del., vol. 23, no. 2, pp. 576–585, Apr. 2008.

Design and Investigation of High Voltage Isolated Silicon-Controlled Rectifier (SCR) Gate Driver

Kostadin Milanov[1, 2], Mintcho Mintchev[2], Hristo Antchev[3]

[1]Department of Electrical Apparatus, Faculty of Electrical Engineering, Technical University - Sofia, Sofia, Bulgaria
[2]Department of Electrical Apparatus, Faculty of Electrical Engineering, Technical University - Sofia, Sofia, Bulgaria
[3]Power Electronics Department, Faculty of Electronic Engineering and Technologies, Technical University - Sofia, Sofia, Bulgaria

Email address:
k.milanow@tu-sofia.bg (K. Milanov), mintchev@tu-sofia.bg (M. Mintchev), hristo_antchev@tu-sofia.bg (H. Antchev)

Abstract: The present article examines the design and implementation specifics of High Voltage Isolated Driver for control of serially connected SCR. The driver features the use of current transformers for the impulses applied to the control electrodes of simple-connected primary coils, as well as control pulses transmission through optical fibre. Results from the experimental study are provided.

Keywords: Silicon-Controlled Rectifier, Driver, High Voltage Isolation

1. Introduction

The correct and reliable operation of the controlled rectifiers depends to a great extent on their management system operation, part of which are the thyristor control drivers [1]. The latter must provide control pulses with the required waveform and parameters at minimum power loss on its elements. The available literature describes generally drivers when using pulse voltage transformers [2]. A summary of the possible solutions for different devices has been made in [3]. In practice high supply voltages require serially connected thyristors, which imposes also the corresponding requirements to their control drivers. A more systematic consideration of the possible circuit solutions is presented in [4]. The control of serially connected thyristors is described in [5]. A driver for serially connected thyristors, using current transformers, is proposed in [6]. The proposed connection of primary coils is complicated and creates implementation difficulties.

This article describes the specificities of driver design and implementation for serially connected thyristors, based on current transformers with common primary coils and simplified connection. Each of them consists of a wire, passing through the transformers' cores in different direction. Part 2 of the article proposes a complete circuit diagram and driver operation description, while part 3 is dedicated to current transformers' design and implementation. Part 4 shows the results from the experimental study of the driver.

2. Operation Description

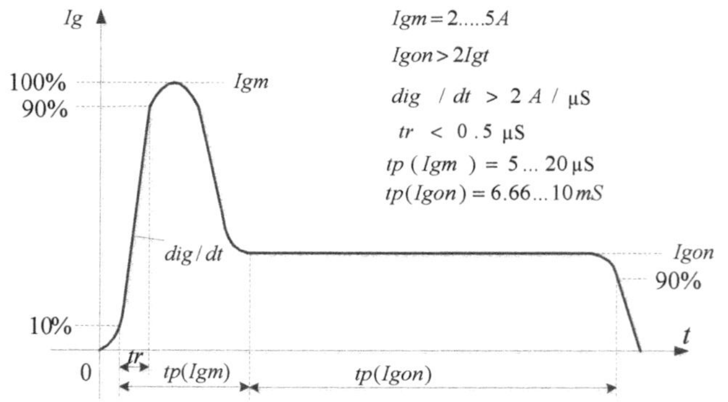

Figure 1. *Requirements to the control pulse waveform.*

The requirements to the control pulse waveform for the particular thyristor type T123-160 are shown on fig.1 [7]. They are typical for all rectifying thyristors. Fig.2 presents a schematics of the driver. For the control of each thyristor are used two current transformers - CTr1 and CTr2. Both wires forming the primary coils pass through the cores of the transformers in different direction, as shown on fig.2. Therefore, when the current flows through one of the primary coils, the voltages of the secondary coils have different polarity. Thyristor enabling pulse of an approximate level 0V is obtained at the output of the optical receiver HFBR-2524 [8], inverted by transistor VT1 and applied at the input of a driver integrated circuit TC4422 [9]. Thus, for the control pulse duration the transistor VT3 is enabled and the current I_{PULSE} flows through the current transformers primary coil connected with $R10$. The initial peak of the control current I_{gm} for time $tp(I_{gm})$ is ensured by current transformer 1 (CTr1), while the current I_{gon} for time $tp(I_{gon})$ is provided by current transformer 2(CTr2).

During the thyristor control pulses pause, the transistor VT2 is enabled also via driver integrated circuit TC4422. The current I_{PAUSE} flows through the primary coil, connected with $R9$ and demagnetizes the current transformer coils. Blocks 2, 3 and 4, shown on fig. 2, contain the secondary coils and the elements thereof for thyristors VS2, VS3 and VS4 respectively.

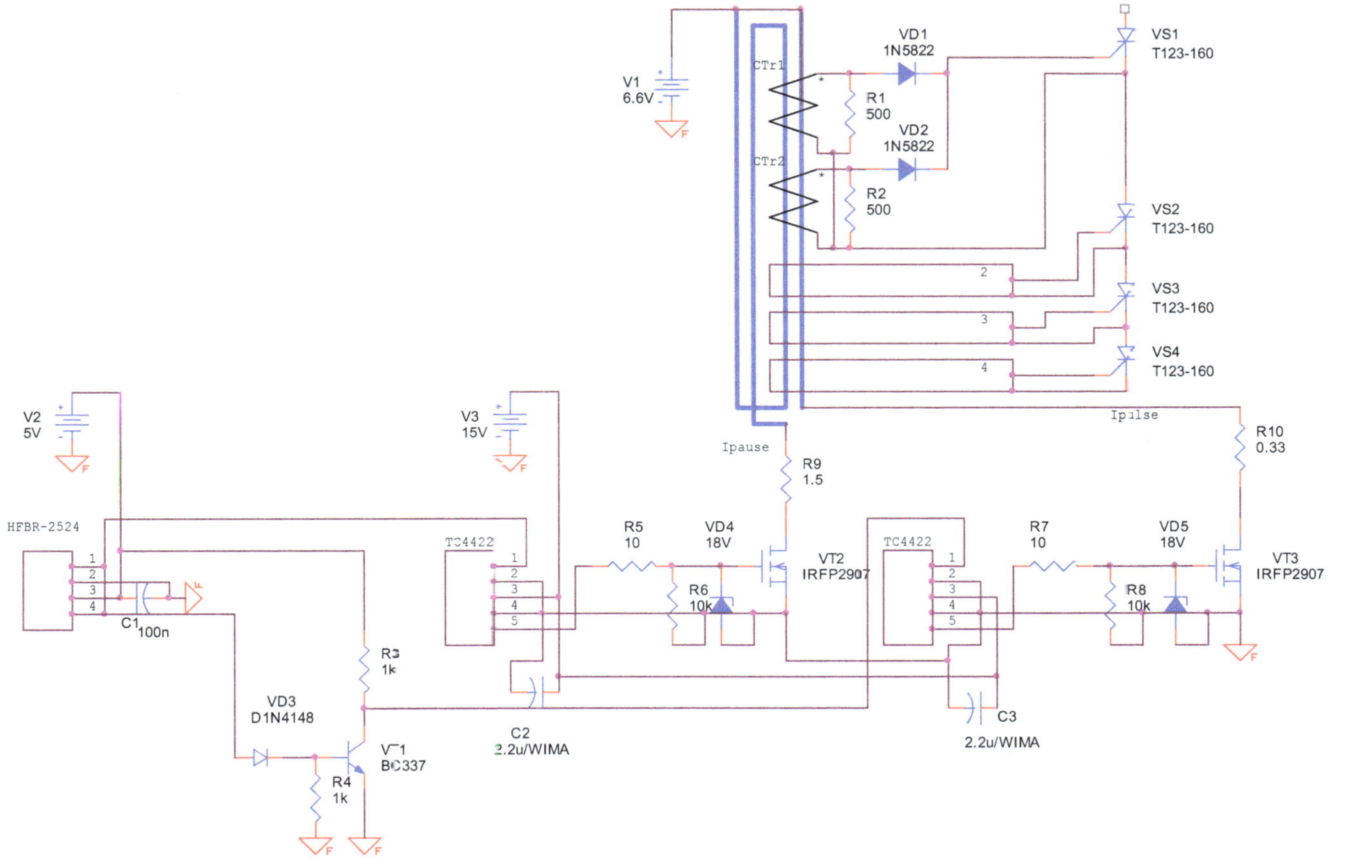

Figure 2. *Driver circuit diagram.*

3. Design and Implementation Specifics of Current Transformers

The output data for transformers design result from the thyristors input characteristics and the default values of the amplitude I_{gm} and the plateau I_{gon} of the thyristor control current and their durations, according to fig.1. To these currents correspond certain voltages on the control electrode of the thyristors, indicated as U_{gm}, respectively U_{gon}.

The amplitude value of the control current is produced by the current transformer, conditionally designated as CTr1, and the plateau – by the current transformer - CTr2.

The current transformer CTr1 has a more important function - it has to provide not only big amplitude of the control current, but also a steep rising edge. This can be effectively achieved by using high quality magnetic core, for example from an amorphous alloy or even better from nanocrystalline soft magnetic alloy. Ferrite magnetic cores can also be used, but they are not effective enough.

The current transformer CTr2 has the tough job to provide long-duration control current, which in this case it is assumed to exceed the normal duration of thyristors on-state – 6,66mS. It is appropriate to be implemented with magnetic core made of high quality electrical steel.

The shape of the magnetic core and that of both current transformers is chosen to be toroidal in order to achieve the best performance. The size of the internal diameter should be such that after laying their secondary coils, there should remain sufficient space for the primary coil and the insulating barrier between both coils.

In order to synthesize the construction of the considered transformers, the material of their magnetic cores need to be selected and in particular the magnetic induction at which they can operate – B. The value of the current through the primary coil of the current transformers I_{1m} should be also selected.

The number of the transformers secondary coils is determined by the expression:

$$\text{For CTr1 - } w_{21} = \frac{I_{1m}}{I_{gm}} \tag{1}$$

and for CTr2 respectively

$$w_{22} = \frac{I_{1m}}{I_{gon}} \tag{2}$$

The section of the transformers' A_e magnetic cores is determined by an expression, giving the relationship between the voltage on the control electrode at the respective current, the current flow duration and the magnetic induction with which the respective magnetic core can operate.

$$A_e = \frac{U_g t_p}{2 w_2 B} \tag{3}$$

In the case of the discussed sample, transformer CTr1 has a secondary coil of 7 windings and a magnetic core of nanocrystalline alloy with section $0{,}1.10^{-4}m^2$, while transformer CTr2 has a secondary coil with 40 windings and electrical steel magnetic core section of $4{,}5.10^{-4}m^2$. The current through the primary coils of both transformers, while generating control pulse, is equal to 20A. The demagnetizing current does not exceed 5A.

4. Experimental Investigation

The driver thus presented is used for the implementation of a rectifier with output voltage 6.5 kV and current up to 30 A. The supply voltage of the rectifier is 3x 4.5 kV. A three-phase bridge circuit has been used, each thyristor consisting of 4 serially connected thyristors as shown on fig. 2. All four thyristors form a driver controlled thyristor block.

Fig. 3 shows a photo of one of the thyristor blocks together with the current transformers and the elements in their secondary coils.

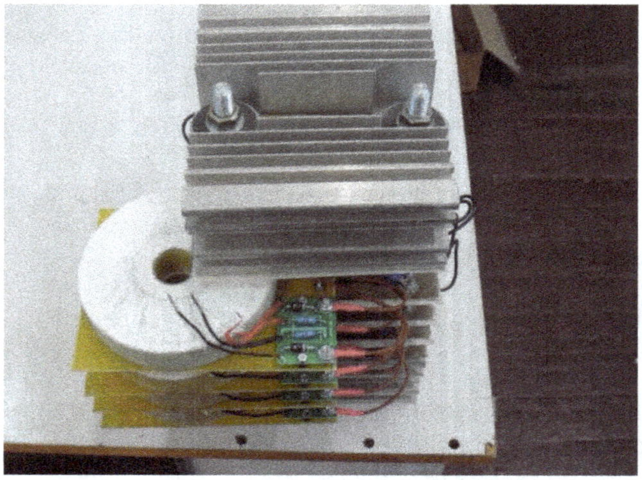

Figure 3. *General view of a thyristor block with the current transformers and the elements thereto.*

Fig. 4 shows 3 drivers (for one of the rectifier thyristor groups - anode or cathode), together with the power supplies elements. Fig.5 shows the experimentally captured control pulse waveform for one of the serially connected thyristors. The waveform of the current is with ratio 100mV/A.The measured parameters of the controlling pulse are as follows:

$$I_{gm} = 2.5A;\ I_{gon} = 400mA;\ {}^{di_g}/_{dt} = 2.5\,{}^{A}/_{\mu S};\ t_r = 0.4\mu S;$$

$$t_p(I_{gm}) = 20\mu S;\ t_p(I_{gon} = 6.7mS)$$

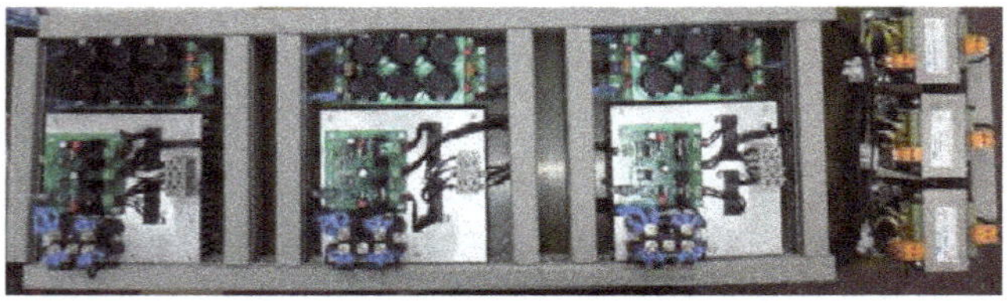

Figure 4. *General view of the drivers for one thyristor group together with the power supplies.*

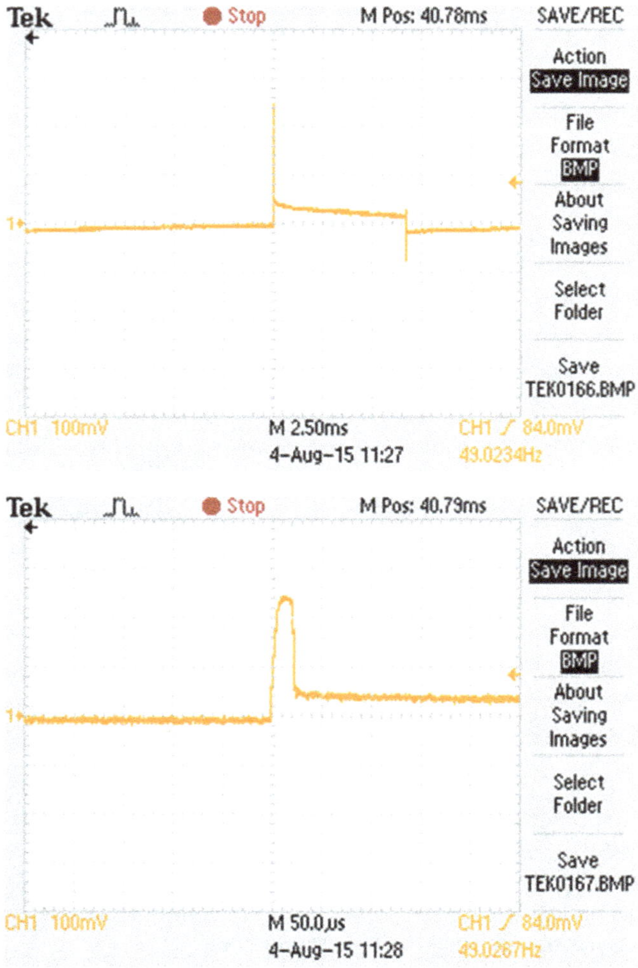

Figure 5. *Experimental control pulse waveform for one of the thyristors.*

5. Conclusion

The present article describes a high voltage isolated driver for serially connected thyristors, using current transformers. A special feature of the transformers is the simplified construction of the primary coils, consisting of a single wire each. Each thyristor uses two current transformers, one of which transmits the initial peak of the controlling short duration pulse, and the other – the set value of the longer duration controlling pulse. The experimental investigation and the use of the proposed driver in a controlled thyristor rectifier with 6.5kV output voltage and 30A current, show that the required control pulses waveform and reliable operation have been achieved.

References

[1] SARI Energy, HVDC Control System-Overview, derived from www.sari-energy.org 25 July 2015.

[2] Mohan N, Undeland T., Robbins W., Power electronics:Converters, Applications and Design: 3rd Edition, 2003, John Wiley and Sons.

[3] Afsharian J., B. Wu, Zargari N., Self-powered Supplies for SCR, IGBT, GTO and IGCT Devices: A Review of the State of the Art, Canadian Conference of Electrical and Computer Engineering, CCECE'09, 2009, pp.920-925.

[4] Wahl F. P., Firing Series SCRs at Medium Voltage: Understanding the Topologies Ensures the Optimum Gate DriveSelection, derived from www.researchgate.net 15 July 2015.

[5] Geun-Hie Rim, Shenderey S., Series Connection of Thyristors with Only One Active Driver for Pulsed Power Generation, proceedings of 29th Annual Conference of the IEEE Industrial Elecronics Society, IECON'03, Vol.1, 2003, pp.107-116.

[6] Applied Power Systems, BAP1289 High Voltage Isolated SCR Gate Driver, Derived from www.appliedps.com 02 July 2015.

[7] Elecroviprymitel, Phase Control Thyristor T123-160, Derived from www.elvpr.ru 29 July 2015.

[8] Avago Technologies, HFBR0501-Series Versatile Link The Versatile Fiber Optic Connection, Derived from www.avagotech.com 10 July 2015.

[9] Microchip, TC4421/TC4422 9A High-Speed MOSFET Drivers, Derived from www.microchip.com 28 July 2015.

Design and optimization of axial flux brushless DC motor dedicated to electric traction

Mariem Ben Amor[1], Souhir Tounsi[2], Mohamed Salim Bouhlel[3]

[1]National School of Engineers of Sfax (ENIS), Sfax University, SETIT Research Unit, Sfax, Tunisia
[2]National School of Electronics and Telecommunications of Sfax, Sfax University, SETIT Research Unit, Sfax, Tunisia
[3]Higher Institute of Biotechnology of Sfax (ISBS), Sfax University, SETIT Research Unit, Sfax, Tunisia

Email address:
souhir.tounsi@isecs.rnu.tn (S. Tounsi), medsalim.bouhlel@enis.rnu.tn (M. S. Bouhlel)

Abstract: In this paper, we present an analytic model of the whole motor converter taking in account of several systemic and physical constraints. Being couple to a model of the losses of the power chain and to a model of the mass of the motor, this analytic model puts a problem of conjoined optimization of the consumption and the cost of the motor. This problem is solved by genetic algorithms method.

Keywords: Electric Vehicle, Motor, Electromagnetic Converter, Design, Optimization

1. Introduction

Currently and in look of the strong petroleum crises, during these last decades and the problems of atmospheric pollution, the electrification of vehicles project became a project of actuality. In this context, several works of research are thrown on this thematic [1], [2], [3], [4] and [5].

The electric vehicle production in big series is braked by their elevated cost as well as their weak autonomy. In this context, a modular axial motor structure and with permanent magnet reducing the cost of manufacture is chosen.

We choice the analytic method to conceive the whole motor converter seen its compatibility to optimization approaches. Indeed, it's fast and product results quickly and without iterations.

The coupling of power train losses model and the model of the motor mass to the program dimensioning the motor - converter, pose an optimization problem. This last is solved by the software of optimization based on the Genetic Algorithm method.

2. Structures of the Traction Motor

2.1. Manufacturing Cost Reduction

The motor structure is modular i.e. it can be with several stages. This technology allows the reduction of the production cost of these motor types. The slots are right and open what facilitates the coils insertion and reduces the motor manufacturing cost. The concentrated winding is used because of its advantages:

- Reduction of the manufacturing time of this motor (insertion of coils in one block).
- Reduction of the end-windings.
- Reduction of the motor bulk.

Figure 1 illustrates the first trapezoidal configuration (n=1) with axial flux only one stage [6].

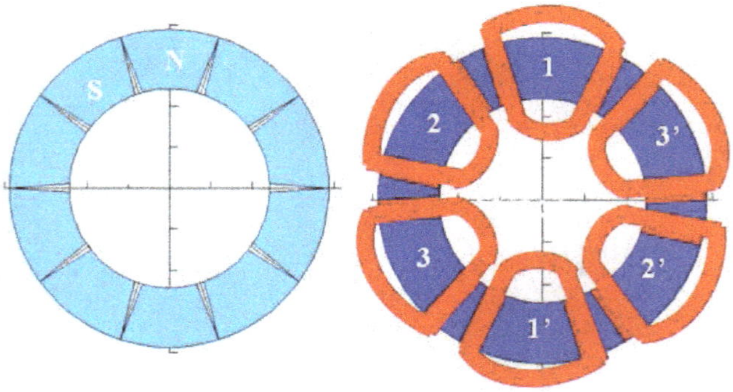

Figure 1. *5 pairs of poles, 6 main teeth, axial flux and trapezoidal configuration.*

Five configurations with a trapezoidal wave-form are found while being based on optimization rules of the ripple torque and cost. Each configuration is characterized by a variation law of the pole pairs number (p) according to an integer number n varying from one to infinity, the ratio (r) of the number of main teeth (N_t) by the number of pole pairs, the ratio (ν) between the angular width between two main teeth and that of a principal tooth, the ratio (α) between the angular width of a principal tooth and that of a magnet and the ratio (β) between the angular width of a magnet and the polar step. Table .1 gives these ratios for these configurations [6], [7], [8], [9].

Table 1. Found configurations

Trapezoidal configurations	p	r	ν	α	β
1	2.n	1.5	1/3	1	1
2	5.n	1.2	2/3	1	1
3	7.n	6/7	4/3	1	1
4	4.n	0.75	5/3	1	1
5	5.n	0.6	7/3	1	1

2.2. Converter Continuous Voltage

The motor constant is defined by [9]:

$$K_e = 2\times n\times N_s\times A\times B\times B_g \quad (1)$$

For the axial flux structures A and B are given by:

$$A = \frac{D_e - D_i}{2} \quad (2)$$

$$B = \frac{D_e + D_i}{2} \quad (3)$$

Where D_e and D_i are respectively the external and the internal diameter of the axial flux motor, N_s is the number of spire per phase, n is the module number and B_g is the flux density in the air-gap.

The converter's continuous voltage U_{dc} is calculated so that the vehicle can function at a maximum and stabilized speed with a weak torque undulation. The electromagnetic torque that the motor must exert at this operation point, via the mechanical power transmission system T_{Udc} (reducing + differential) is estimated by the following expression:

$$T_{Udc} = \frac{P_f}{\Omega} + T_d + \left(T_b + T_{vb} + T_{fr}\right) + \frac{T_r + T_a + T_c}{r_d} \quad (4)$$

Where T_b is the rubbing torque of the motor, T_{vb} is the viscous rubbing torque of the motor, T_{fr} is the fluid rubbing torque of the motor, T_r is the torque due to the friction rolling resistance, T_a is the torque due to the aerodynamic force, T_c is the torque due to the climbing resistance, T_d is the reducer losses torque and P_f are the iron losses and Ω is the motor angular speed.

At this operation point, the phase current is given by the following relation:

$$I_p = \frac{T_{Udc}}{K_e} \quad (5)$$

The only possibility making it possible to reach the current value I_p with a reduced undulation factor (10% for example) is to choose the converter's continuous voltage solution of the following equation [7]:

$$r = \frac{t_m}{t_p} = 10\% \quad (6)$$

Where t_p is the phase current maintains time at vehicle maximum speed and t_m is the boarding time of the phase current from zero to I_p [9]:

$$t_m = -\frac{L}{R}\times \ln\left(1 - \frac{2\times R\times I_p}{U_{dc} - K_e\times\Omega_{max}}\right) \quad (7)$$

Where R and L are respectively the phase resistance and inductance and Ω_{max} is the maximum angular velocity of the motor.

The phase current maintains time at maximum speed of vehicle (corresponds to 120 electric degrees) is given by the following formula [9]:

$$t_p = \frac{1}{3}\times\frac{2\times\pi}{p\times\Omega_{max}} \quad (8)$$

The converter's continuous voltage takes the following form:

$$U_{dc} = \frac{2\times R\times I_p}{1 - \exp\left(-\frac{2\times\pi\times r}{3\times p\times\Omega_{max}\times\frac{L}{R}}\right)} + K_e\times\Omega_{max} \quad (9)$$

The phase inductance of the all configurations is expressed as follows [9]:

$$L = \frac{3\times\mu_0}{N_t}\times\left(\frac{A_t}{g + t_m} + \frac{A\times h_s}{2\times A_s}\right)\times N_s^2 \quad (10)$$

Where μ_0 is the air permeability, A_t is slot area, t_m is the magnet thickness, h_s is slot height, g is the air-gap thickness and A_s is the slot width.

The converter continuous voltage increases by increasing of the vehicle speed which validates the fact of calculating its value at maximum speed. Two important factors involving the increase of the converter continuous voltage:

- The increase of the motor electric constant.
- The reduction of the undulation factor.

Consequently, a compromise between the reduction of the converter continuous voltage directly related to the space reserved for the battery and the reduction of undulation factor is to be found.

2.3. Design Methodology

We choose the analytic modelling of the motor, because it's compatible to the optimisations approaches [10], [11], [12], [13]. The analytic sizing step is inverted starting from needs (torque, motor velocity) towards geometry dimensions. While, in classical analytic sizing, the computation starts from geometry towards needs. Indeed the limit of the plan torque/velocity of the motor is given. The electrical motor has to be able to function in this plan without constraints. Here, the problem resolution is completely reversed. For example, the operating temperature is fixed and then computed. It should be also recognised that the inversion of the problem facilitates its resolution. In the same way, the designer fixes the induction. The worksheet includes 200 items of computed elements. The difficulty is to classify these elements and to distinguish results from data.

The worksheet computes the geometrical dimensions of rotor and stator as well as windings, temperature, inductance, leakages and efficiency for different operating points and control modes.

A sizing program is developed with equations detailed below. The program inputs are:

1. Electric vehicle specifications.
2. Materials properties.
3. Configuration, i.e. magnet number and teeth number.
4. Inner and outer diameter of the motor.
5. Notebook data.
6. Current density in coils δ.
7. Rotor yoke B_{ry}, stator yoke B_{sy}, flux density in the air-gap B_g, reducer ratio and number of spire per phase N_s.

When inputs 3. and 4. are set, magnet shapes, teeth and slots are fixed. Then, the area of one tooth A_t and the average length of a spire L_{sp} are calculated from geometric equations.

This model is validated by finite elements method. Indeed, the motor is drawn according to its geometrical magnitudes extracted from analytical model with the software Maxwell-2d, and is simulated in dynamic and static in order to compare the results obtained with those found by the analytical method.

The coupling of this model to a model evaluating the power train losses and motor mass, poses an optimization problem with several variables and constraints. This latter is solved by the genetic algorithms (GAs) method [10], [11], [12], [13].

3. Dimensioning Torque

The back electromotive force (E.m.f) stage level is given by the following expression [10], [11], [12], [13]:

$$E= n\times N_s\times A\times B\times\Omega \times B_g \tag{11}$$

The instantaneous electromagnetic power $P_e(t)$ is expressed by the following relation

$$P_e(t)=\sum_{i=1}^{m} e_i(t)\times i_i(t) \tag{12}$$

Where $e_i(t)$ and $i_i(t)$ are respectively the back electromotive force and the current of the phase i.

Two phases are fed simultaneously and the currents of phases have the same wave-form as the electromotive force with a maximum value of motor phase current I. Consequently, for a constant speed, the electromagnetic power developed by the motor takes the following form:

$$P_e=2\times E\times I \tag{13}$$

The electromagnetic torque developed by the motor is expressed by:

$$T_m =2\times\frac{E\times I}{\Omega} \tag{14}$$

The electromagnetic torque developed by the motor results [9]:

$$T_m=2\times n\times N_s\times A\times B\times B_g\times I \tag{15}$$

The electromagnetic torque which the motor must develop so that the vehicle can move with a speed v is deduced from the dynamics fundamental relation related to the electric vehicle dynamic:

$$T_m=\frac{P_f}{\Omega}+T_d+\left(T_b+T_{vb}+T_{fr}\right)+\frac{T_r+T_a+T_c}{r_d}+\left(\frac{J}{R_w}+\frac{M_v\times R_w}{r_d}\right)\times\frac{dv}{dt} \tag{16}$$

Where r_d is the reduction ratio, M_v is the vehicle mass, R_w is the vehicle wheel radius, J is the motor moment of inertia and v is the vehicle velocity.

The different torques are expressed by the following equations [14], [15], [16]:

$$T_b=s\times\frac{v}{|v|} \tag{17}$$

$$T_{vb}=\chi\times v \tag{18}$$

$$T_{fr}=k\times v\times|v| \tag{19}$$

$$T_r=R_w\times f_r\times M_v\times g \tag{20}$$

$$T_a=R_w\times\frac{(M_{va}\times C_x\times A_f)}{2}\times V^2 \tag{21}$$

$$T_c=M_v\times g\times\sin(\lambda) \tag{22}$$

Where s is the dry friction coefficient, χ is the viscous friction coefficient, k is the fluid friction coefficient, λ is the angle that the road makes with the horizontal, M_{va} is the density of the air, C_x is the aerodynamic drag coefficient, r_p is a coefficient taking account of the mechanical losses in the motor and the transmission system, and A_f is the vehicle frontal area. The phase current becomes:

$$I=\frac{T_m}{2\times n\times N_s\times A\times B\times B_g} \quad (23)$$

The dimensioning current is expressed as follows:

$$I_{dim}=\frac{T_{dim}}{2\times n\times N_s\times A\times B\times B_g} \quad (24)$$

T_{dim} is the dimensioning torque. This torque is found by the genetic algorithm method on a standardized circulation mission in order to not exceed the limiting temperatures and to minimize the motor mass.

Several methods were proposed to define dimensioning sizes of the motor-reducer torque, based on simplified statistical tools [9]. A first method is based on the determination of the effective torque for a circulation mission in order to take into account the thermal aspect [9]. A second more elaborate approach consists in defining zones of strong occurrences and to take the sizes resulting from these zones like dimensioning sizes. Finally, a last simpler method consists in dividing the torque-speed plan into 4 zones, and to take the gravity center of each zone then to consider the gravity center of these four points balanced by the number of each zone points as dimensioning point. These methods have the advantage of quickly providing useful sizes for dimensioning and simulation, nevertheless they do not take into account the problem of electric vehicle consumption minimization. For our approach, the dimensioning torque will be iteratively calculated by the genetic algorithms method in order to satisfy a global optimization of autonomy while respecting the dimensional thermal stresses relating to our application specified by the schedule of conditions. To guide the algorithm to converge towards a powerful solution and in order to limit the space of research, the motor dimensioning torque must satisfy the following condition extracted inequality:

$$(1-\varepsilon)\times R_w\times\left(\frac{\frac{J}{R_w^2}+\frac{M_v}{r_d}}{t_d}\times V_b+\frac{M_v\times g\times\sin(\lambda)}{r_d}\right)\le T_{dim}(1+\varepsilon)\times R_w\times\left(\frac{\frac{J}{R_w^2}+\frac{M_v}{r_d}}{t_d}\times V_b+\frac{M_v\times g\times\sin(\lambda)}{r_d}\right) \quad (25)$$

The adjustment coefficient of the torque ε generally does not exceed 0.25 and will be adjusted by simulations of the propulsion system on normalised circulation missions.

4. Motor Sizing

The air-gap flux density is calculated for a maximal recovery position, or the magnet is in front of a main tooth. At this position the air-gap flux density is maximal. The distribution of the field lines to the level of a pole is illustrated by the figure 2:

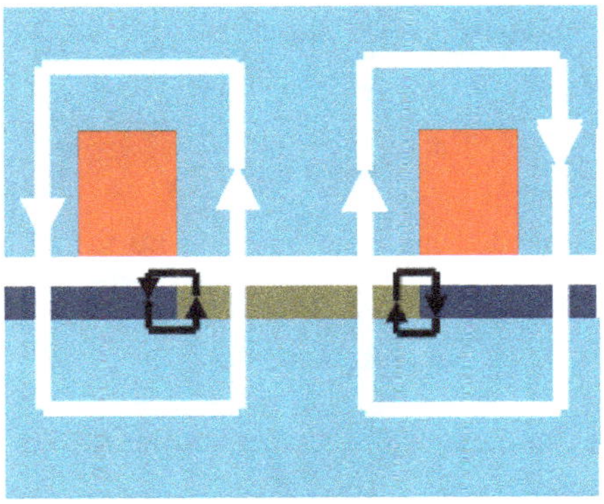

Figure 2. *Flux lines distribution at maximal recovery position.*

The flux decomposes itself in main flux assuring the traction of the rotor by interaction with the stator flux and in leakages flux between magnets.

As applying the Ampere theorem to the level of a stator pole, we can deduct the flux density due to the powering of a stator coil.

$$\oint_{\text{flux lines}}\vec{H}\times\vec{dl}=\frac{N_s}{2}\times I_{max}=2\times(H_{ri}\times t_m+H_{ri}\times g) \quad (26)$$

Where I_{max} is the maximal current feeding the motor, H is the magnetic field, H_{ri} is the air-gap magnetic field, t_m is the magnet thickness and μ_0 is the air permeability.

$$B_{ri}=\mu_0\times H_{ri} \quad (27)$$

Where B_{ri} is the flux density in the air-gap due to the powering of stator coil.

$$Bri=\frac{\mu_0}{4}\times\frac{N_s\times I_{max}}{t_m+g} \quad (28)$$

While applying the Ampere theorem, we can deduct the magnet thickness imposing a fixed flux density in the different zones of the motor while disregarding the flux density due to the powering of the stator coils, since the flux must cross two times the air-gap thickness and magnet with permeability very close to the air permeability.

$$\oint_{\text{flux lines}}\vec{H}\times\vec{dl}=0=2\times(H_m\times t_m+H_g\times g) \quad (29)$$

The air-gap flux density is linear according to the magnetic field for this working regime:

$$B_g=\mu_0\times H_g \quad (30)$$

While applying the flux conservation theorem to the level of the air-gap, we deduct the value of the air-gap flux density in function of the magnet flux density and the coefficient of the leakages flux.

$$B_m\times S_m\times K_{fu}=B_g\times S_m \quad (31)$$

The magnet flux density becomes:

$$B_m = \frac{B_g}{K_{fu}} \quad (32)$$

The magnet flux density is approached by the following linear equation:

$$B_m = \mu_0 \times \mu_m \times H_m + B_r \quad (33)$$

Where μ_m is the magnet's relative permeability, B_r is the remanence.

From the equation (29), (30), (31) and (32), we deduct the magnet thickness fixing the air-gap flux density equal to B_g,:

$$t_m = \mu_m \times \frac{B_g}{B_r - \frac{B_g}{K_{fu}}} \times g \quad (34)$$

Where $K_{fu} < 1$ is the magnet's leakage coefficient and g is the air-gap thickness. To avoid demagnetization, the phase currents must be lower then the demagnetization current I_d [8]:

$$I_d = \left(\frac{B_r - B_{min}}{\mu_m} \times t_m - B_{min} \times K_{fu} \times g \right) \times \frac{p}{2 \times \mu_0 \times N_s} \quad (35)$$

Where B_{min} is the minimum flux density allowed in the magnets and μ_0 is the air permeability. The rotor yoke thickness t_{ry} and stator yoke thickness t_{sy} derive from the flux conservation [9]:

$$t_{ry} = \frac{B_g}{B_{ry}} \times \frac{Min(A_t, A_m)}{2 \times A} \times \frac{1}{K_{fu}} \quad (36)$$

$$t_{sy} = \frac{B_g}{B_{sy}} \times \frac{Min(A_t, A_m)}{2 \times A} \quad (37)$$

Where A_t is the tooth area, A_m is the area of one magnet, B_{ry} and B_{sy} are respectively the flux densities in rotor and stator yokes. For the axial flux and trapezoidal wave-form motor configurations the slot height is [9]:

$$h_s = \frac{3.2.N_s}{2 N_t} \frac{I_{dim}}{\delta} \frac{1}{K_f} \frac{1}{A_s} \quad (38)$$

Where N_t is the number of principal teeth, δ is the current density in slots, K_f is the slot filling factor, A_s is the slot width and I_{dim} is the dimensioning current:

$$I_{dim} = \frac{T_{dim}}{K_e} \quad (39)$$

The slot width is expressed as follows:

$$A_s = B \times SIN\left(\frac{1}{2} \times \left(\frac{2 \times \pi}{N_t} - \alpha \times \beta \times \frac{\pi}{p} \times (1 - r_{did}) \right) \right) \quad (40)$$

Where r_{did} is the ratio between the angular width of the inserted tooth and that of the principal tooth. This ratio is optimised by finite elements simulations in order to reduce the flux leakages and to improve the back E.m.f. wave-form.

Optimisation problem

The optimization problem consists on the determination of the motor-converter sizes minimizing its mass and the electric vehicle consumption, while respecting the technological constraints of the application.

The motor weight is expressed as follows:

$$W_m = W_{sy} + W_t + W_c + W_{ry} + W_m \quad (41)$$

For the axial flux configurations the weight of stator yoke W_{sy}, tooth W_t, copper W_c, rotor yoke W_{ry}, and magnets W_m are expressed as follows:

$$W_{sy} = n \times d \frac{\pi}{4} \times (D_e^2 - D_i^2) \times t_{sy} \quad (42)$$

$$W_t = n \times d \times N_t \times A_t \times h_s \quad (43)$$

$$W_c = 3 \times n \times N_s \times L_{sp} \times \frac{I_{dim}}{\delta} \times d_c \quad (44)$$

$$W_{ry} = \pi \times \left(\left(\frac{D_e}{2} \right)^2 - \left(\frac{D_i}{2} \right)^2 \right) \times t_{ry} \times d \quad (45)$$

$$W_m = 2 \times n \times p \times A_m \times t_m \times d_m \quad (46)$$

Where d is the density of the metal sheet, d_c is the density of copper, d_m is the magnet density, A_a is the magnet angular width, A_d is the angular width of principal teeth and A_e is the slot angular width.

For the trapezoidal wave-form configurations, the copper losses are expressed by the following relation:

$$P_c = 2 \times R \times I^2 \quad (47)$$

The phase resistance is given by the following expression:

$$R = r_{cu}(T_b) \times \frac{N_s \times L_{sp}}{S_c} \quad (48)$$

Where r_{cu} is the copper receptivity, L_{sp} is the average length of spire, T_b is the copper temperature and S_c is the active section of one conductor:

$$S_c = \frac{I_{dim}}{\delta} \quad (49)$$

The iron losses are expressed by the following relation [14], [15], [16], [17], [18], [19]:

$$P_{fer} = C \times f^{1.5} \times (n \times W_t \times B_g^2 + n \times W_{sy} \times B_{sy}^2) \quad (50)$$

Where c is the core loss, f is the motor powering frequency, W_t is the teeth weight, W_{sy} is the stator yoke weight, B_g is the ai-rgap flux density and B_{sy} is the flux density in stator yoke. The mechanical losses are expressed by the following relation [11]:

$$P_m = (T_b + T_{vb} + T_{fr}) \times \Omega \quad (51)$$

Where Ω is the angular speed of the electric motor.

The losses in the static converter are nearly hopeless, they are not held in account in the model of power train losses calculation.

Since always two phases are feeding, the losses in the battery are expressed by the following expression:

$$P_b = 2 \times R_{batt} \times I_b^2 \quad (52)$$

Where R_{batt} is the internal resistance of the battery.

The power train losses is expressed consequently by the next equation:

$$P_{ptl} = P_c + P_{fer} + P_m \quad (53)$$

The model of the power train losses is coupled to the electric vehicle power train model. This structure of this model is illustrated by the figure 3.

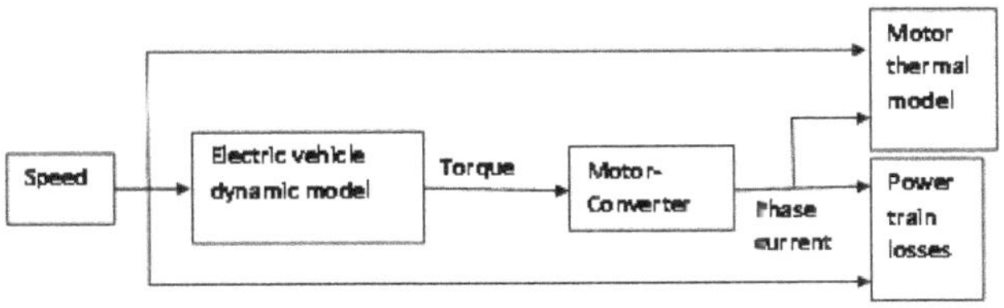

Figure 3. *Electric vehicle power train model.*

5. Genetic Algorithms Optimization of the Motor Mass and the Power Train Losses

The function to optimize is expressed by the following expression:

$$F_o = W_m + a P_{ptl} \quad (54)$$

Where "a" is a coefficient fixing the influence degree of P_{ptl} at the global objective function compared to W_m. Indeed, "a" brings closer the value of ($a P_{ptl}$) to the value of W_m.

The optimisation problem consists in optimising the F_o with respect to the problem constraints. In fact, Genetic Algorithms (GAs) are used to find optimal values of the switched frequency f_{sw}, radius of the wheel in meters R_w, the gear ratio r_d, The adjustment coefficient of the torque ε, the internal diameter D_i, the external diameter D_e, the flux density in the air-gap B_g, the current density in the coils δ, the flux density in the rotor yoke B_{ry}, the flux density in the stator yoke B_{sy} and the number of phase spires N_s [14], [15], [16], [17], [18], [19].

The beach of variation of each parameter $x_i \in (f_{sw}, R_w, r_d, \varepsilon, D_i, D_e, B_g, \delta, B_{ry}, B_{sy}, N_s)$ must respect the following constraint: $x_{imin} \leq x_i \leq x_{imax}$. The values of the lower limit x_{imin} and the upper limit x_{imax} are established following technological, physical and expert considerations, for example:

- The internal and external diameter of the motor is delimited by the space reserved for the motor.

$$\begin{cases} \text{maximise}(F_o) \\ \text{with :} \\ 500 \leq f_{sw} \leq 5000 \text{ (Hz)} \\ 0.25 \leq R_w \, 0.35 \text{ (m)} \\ 1 \leq r_d \leq 8 \\ 0 \leq \varepsilon \leq 0.25 \\ 0.025 \leq D_i \leq 0.25 \text{ (m)} \\ 0.26 \leq D_e \leq 0.5 \text{ (m)} \\ 0.1 \leq B_g \leq 1.04 \text{ (T)} \\ 5 \leq \delta \leq 7 \text{ (A/mm}^2\text{)} \\ 0.1 \leq B_{ry} \leq 1.06 \text{ (T)} \\ 0.1 \leq B_{sy} \leq 1.6 \text{ (T)} \\ 10 \leq N_s \leq 1000 \\ r \leq 10\,\% \\ Udc \leq 100 \text{(V)} \\ I \leq I_d \text{(A)} \\ T_a \leq 50 \text{ (°C)} \\ T_b \leq 90 \text{ (°C)} \end{cases} \quad (55)$$

Where T_a and T_b are respectively the magnets and the coils temperatures.

The F_o model is coupled to a program of optimization by the method of the genetic algorithm. The progress of the program of optimization of the F_o with constraints is described by this organization diagram (figure 4) [14], [15], [16], [17], [18]:

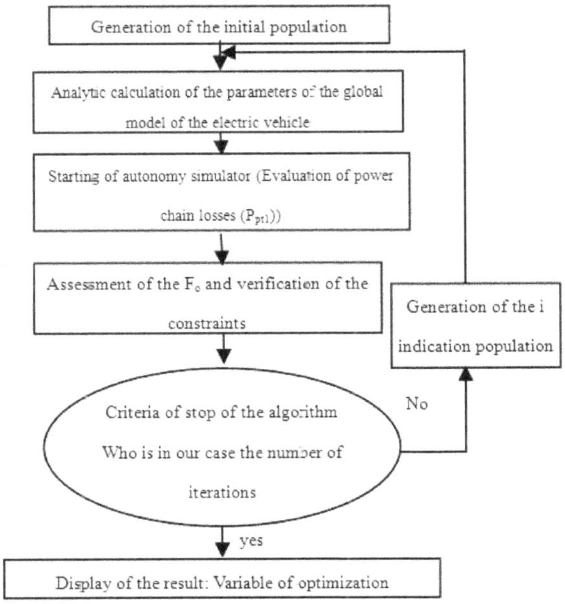

Figure 4. *Progress of the optimization program.*

6. Conclusion

An analytical model dimensioning the whole motor converter is developed. This model is coupled to an optimization program in order to find the design and control parameters of the whole motor-converter minimizing the power train energy losses and the electric vehicle cost. This study encourages the manufacture procedure of electric vehicles in big series [9], [14], [15], [16], [17], [18], [19].

References

[1] Chaithongsuk, S., Nahid-Mobarakeh, B., Caron, J., Takorabet, N., & Meibody-Tabar, F. : Optimal design of permanent magnet motors to improve field-weakening performances in variable speed drives. Industrial Electronics, IEEE Transactions on, vol 59 no 6, p. 2484-2494, 2012.

[2] Rahman, M. A., Osheiba, A. M., Kurihara, K., Jabbar, M. A., Ping, H. W., Wang, K., & Zubayer, H. M.: Advances on single-phase line-start high efficiency interior permanent magnet motors. Industrial Electronics, IEEE Transactions on, vol 59 no 3, p. 1333-1345, 2012.

[3] C.C Hwang, J.J. Chang : Design and analysis of a high power density and high efficiency permanent magnet DC motor, Journal of Magnetism and Magnetic Materials, Volume 209, Number 1, February 2000, pp. 234-236(3)-Publisher: Elsevier.

[4] MI. Chunting CHRIS: Analytical design of permanent-magnet traction-drive motors" Magnetics, IEEE Transactions on Volume 42, Issue 7, July 2006 Page(s):1861 - 1866 Digital Object Dentifier 10.1109/TMAG.2006.874511.

[5] S.TOUNSI, R.NÉJI, F.SELLAMI: Conception d'un actionneur à aimants permanents pour véhicules électriques, Revue Internationale de Génie Électrique volume 9/6 2006 - pp.693-718.

[6] Sid Ali. RANDI: Conception systématique de chaînes de traction synchrones pour véhicule électrique à large gamme de vitesse. Thèse de Doctorat 2003, Institut National Polytechnique de Toulouse, UMRCNRS N° 5828.

[7] C. PERTUZA: Contribution à la définition de moteurs à aimants permanents pour un véhicule électrique routier. Thèse de docteur de l'Institut National Polytechnique de Toulouse, Février 1996.

[8] S. TounsI, R. NEJI and F. SELLAmI: Mathematical model of the electric vehicle autonomy. ICEM2006 (16th International Conference on Electrical Machines), 2-5 September 2006 Chania-Greece, CD: PTM4-1.

[9] R. NEJI, S. TOUNSI, F. SELLAMI: Contribution to the definition of a permanent magnet motor with reduced production cost for the electrical vehicle propulsion. Journal European Transactions on Electrical Power (ETEP), Volume 16, issue 4, 2006, pp. 437-460.

[10] P. BASTIANI: Stratégies de commande minimisant les pertes d'un ensemble convertisseur machine alternative : application à la traction électrique. Thèse INSA 01 ISAL 0007, 2001.

[11] G. Henriot: Traité théorique et pratique des engrenages : théorie et technologie 1. tome 1 Edition Dunod 1952.

[12] D-H. Cho, J-K. Kim, H-K. Jung and C-G. Lee: Optimal design of permanent-magnet motor using autotuning Niching Genetic Algorithm, IEEE Transactions on Magnetics, Vol. 39, No. 3, May 2003.

[13] Islam, M. S., Islam, R., & Sebastian, T.: Experimental verification of design techniques of permanent-magnet synchronous motors for low-torque-ripple applications. Industry Applications, IEEE Transactions on, vol 47 no 1, p. 88-95, 2011.

[14] Parasiliti, F., Villani, M., Lucidi, S., & Rinaldi, F. : Finite-element-based multiobjective design optimization procedure of interior permanent magnet synchronous motors for wide constant-power region operation. Industrial Electronics, IEEE Transactions on, vol 59 no 6, p. 2503-2514, 2012.

[15] Mahmoudi, A., Kahourzade, S., Rahim, N. A., & Ping, H. W.: Improvement to performance of solid-rotor-ringed line-start axial-flux permanent-magnet motor. Progress In Electromagnetics Research, 124, p. 383-404, 2012.

[16] Duan, Y., & Ionel, D. M.: A review of recent developments in electrical machine design optimization methods with a permanent-magnet synchronous motor benchmark study. Industry Applications, IEEE Transactions on, vol 49 no 3, p. 1268-1275, 2013.

[17] Liu, G., Yang, J., Zhao, W., Ji, J., Chen, Q., & Gong, W.: Design and analysis of a new fault-tolerant permanent-magnet vernier machine for electric vehicles. Magnetics, IEEE Transactions on, vol 48 no 11, p. 4176-4179, 2012.

[18] Lee, S., Kim, K., Cho, S., Jang, J., Lee, T., & Hong, J.: Optimal design of interior permanent magnet synchronous motor considering the manufacturing tolerances using Taguchi robust design. Electric Power Applications, IET, vol 8 no 1, 23-28, 2014.

[19] TOUNSI, R. NEJI and F. SELLAMI : Electric vehicle control maximizing the autonomy: 3rd International Conference on Systems, Signal & Devices (SSD'05), SSD-PES 102, 21-24 March 2005, Sousse, Tunisia.

Methodology for electrothermal characterization of permanent magnet motor and its equivalent to coiled rotor

Souhir Tounsi

School of Electronics and Telecommunications of Sfax, Sfax University, Sfax, Tunisia

Email address:

souhir.tounsi@isecs.rnu.tn

Abstract: In this paper, we present a methodology for electrothermal characterization of two configurations: permanent magnet motor and its equivalent to wound rotor. This modeling approach is in the aim to evaluating the different temperatures in different active parts of the two configurations, to choose the type and characteristics of the cooling system to use. A comparative study between the two solutions is presented.

Keywords: Electrothermal, Modeling, Nodal Method, Simulation, Coiled Motor, Permanents Magnets Motor

1. Introduction

The main objective of the electrothermal modeling of thermal stresses is to respect the good working of the electric motor. In fact, exceeding the melting temperature of the coils insulation set at 100 ° C leads to the deterioration of the coils and subsequently damage the motor. In addition, knowledge of different temperatures in the active areas of the engine allows firstly to take into account the change of the B-H magnets characteristic (critical temperature is not to exceed 200 ° C for Sm-Co magnet), and secondly the change in resistance phases of electric motor, strongly influencing the electrical, magnetic and mechanical behavior of engine. This knowledge temperatures determines the nature and power of the cooling system to integrate in the two engine selected configurations. The fisrt configuration is permanent magnets synchrnous motor (PMSM) and the second is a coiled rotor synchronous motor (CRSM). In this context, we study the two cases of motors to choose the most efficient and consistent with a cooling system the least powerful and consequentely the least expensive.

We set a goal to not exceed the limits of the following temperatures:

200 ° C for permanent magnets (Sm-Co).

100 ° C for the insulator coil.

This survey must succeed to the choice of the structure of the motor compatible to the least expensive cooling system.

2. Choice of Modeling Approach

Conventional analytical methods, describes heat transfer with an acceptable complexity. However these methods require precise knowledge of many coefficients (thermal conductivity, heat transfer coefficient, emissivity) that's often difficult to obtain.

Finite elements methods require an important memory resources and computation time. They are therefore not compatible with our approach should allow to model a magnetic component in its environment ie part of a chain of power. In addition, the model developed should make account of all the phenomena involved in a magnetic component (electric phenomena, magnetic and thermal).

The nodal method provides better accuracy compromised results and simulation time. It is therefore compatible to optimization approaches such as the performance of electric vehicle. It seems however best suited to our concerns and lends itself well to an experimental approach. The component to be modeled is divided into insulated areas interconnected by a thermal resistance, the center of a zone is called node. A thermal capacity and a heat source are associated with each zone. A system of differential equations is obtained by writing the heat balance at the various nodes. A first approximation is to consider thermal resistance as constant (for better accuracy thermal resistances can be modeled using analytical relations).

For the components of our synchronous motor, the

equivalent circuit is limited to a few resistors and capacitors whose values can be obtained by calculation. This method meets our specifications [1].

3. Nodal Thermal Model of PMSM and CRSM Structures

The system to be modeled is a synchronous electric motor with axial flux and permanent magnet, or with a coiled rotor (Figure 1). For the two configurations the stator is with same structure and mechanical, electrical and magnetic characteristics are equivalent.

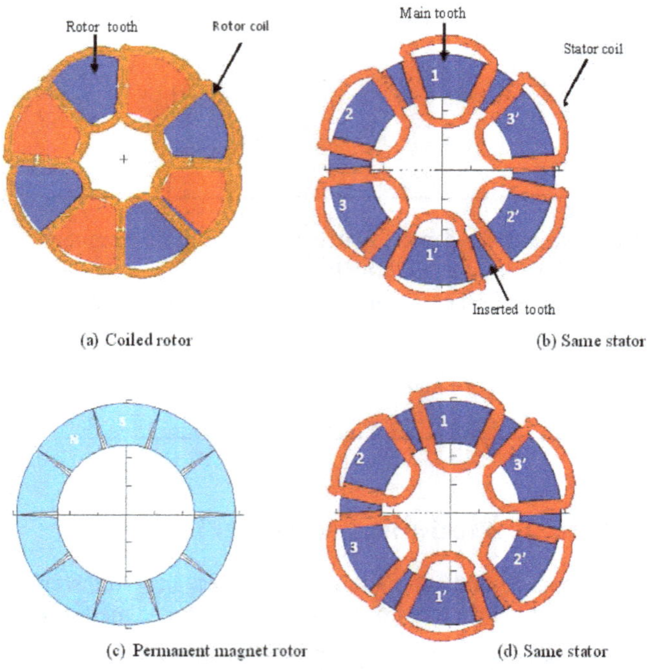

Figure 1. *PMSM and CRSM equivalents structures.*

The 3 D structure of PMSM is illustrated by figure 2.

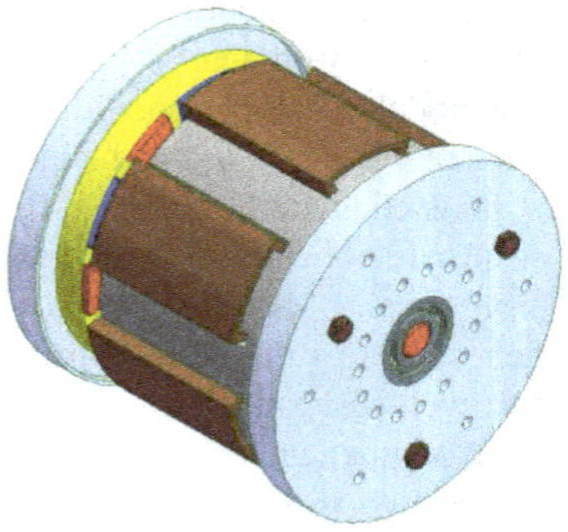

Figure 2. *3 D structure of PMSM*

3.1. Assumptions Used for Modeling

We recall firstly our assumptions about thermal modeling of magnetic components to justify the principles of selected measures that lead to the measurement of average temperatures. Temperatures are assumed uniform in the material and in the different coils.

We consider an ambient temperature of 27 ° C and we assume a maximum permissible temperature rise of 100 ° C for the stator winding and the insulation. This warming-up is determined by the temperature holding of the conductors and the slots bottom insulators. at the same way, the maximum warming-up of the magnets (Sm-Co) must be taken in account and we admit a warming-up of the order of 200 °C. In the context of our model, we consider only perfect contacts, and their conduction resistances are low and broken into account in the thermal model of the PMSM and CRSM.

The heat exchange is assumed in the axial direction, because the length of the two structures in this direction is much lower than that in the radial direction, in addition heat exchange sections perpendicular to the axis of the motor is much greater than those radials [2].

3.2. Heat Transfer in the PMSM and CRSM Structures

Heat transfer in the PMSM is illustrated by the following Figure 3:

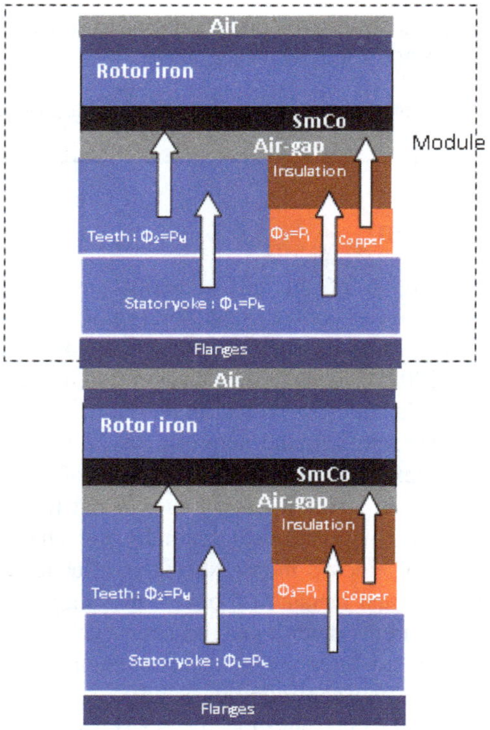

Figure 3. *The heat transfer in the PMSM.*

The heat transfer in the CRSM is illustrated by the figure 4 [3]:

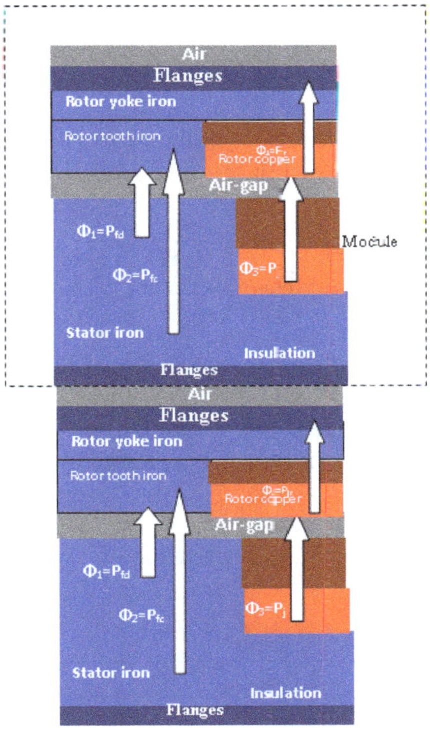

Figure 4. *The heat transfer in the CRSM.*

4. Transient Thermal Model

We selected a model using thermal-electrical analogy. In order to simplify the model, we have the following assumptions:

- Uniform heat generation.
- Uniformity of physical properties across the element.
- Uniformity of the exchanges on each of the faces.

The machine is divided into simple volume elements exchanging heat between them by conduction or convection. Copper losses include slots losses and end winding losses. The transien thermal model of the PMSM structure can then be represented by an analog electrical network, as described in figure 5 [1], [2], [3], [4] and [5].

The transient thermal model of the CRSM structure is represented by an analog electrical network, as described in figure 6.

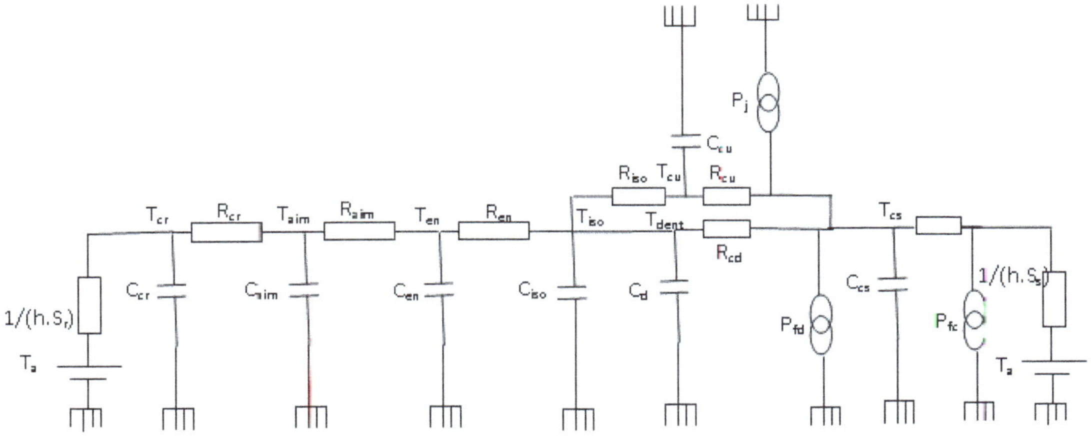

Figure 5. *Transient thermal model of the PMSM actuator.*

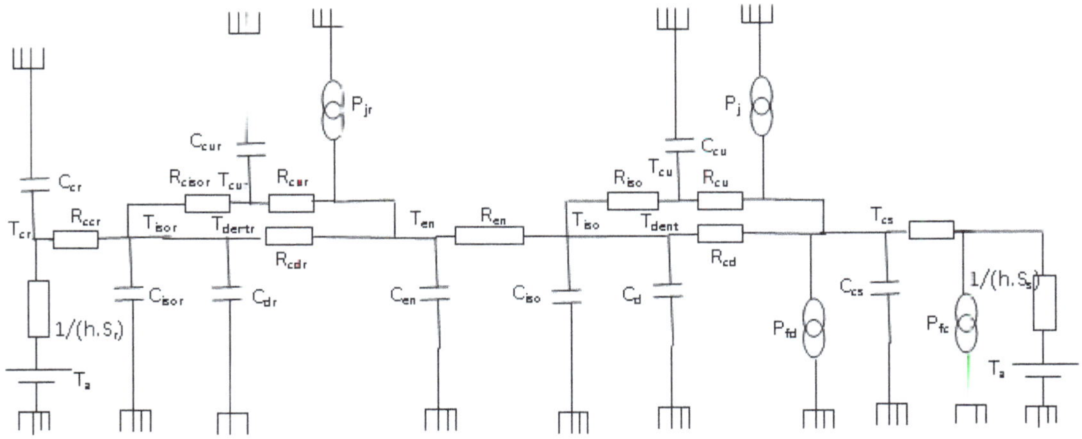

Figure 6. *Transient thermal model of the CRSM actuator.*

For the models presented in figure 5 and 6, we define two isothermal zones constituted by the magnetic material on the one hand and by the winding on the other hand.

Previously defined areas are warm seat due to copper losses and iron losses.

P_j, P_{fc} and P_{fd} corresponding respectively to the total losses in the coils, in the stator yoke and in the stator teeth. The variables Ti correspond to the temperatures at different points of the machine. The terms of thermal resistances are deduced from the resolution of the heat equation border areas.

4.1. Calculation of Conduction Resistances

For the sake of simplification, the conduction along the transverse axis of the stator is not taken although it may be essential, especially in the windings. Heat transfer in a stator element therefore allows one preferred direction, the axial direction. This is reflected by the following heat equation [1].

$$\frac{\overrightarrow{\Phi_T}}{S_{et}} = -\lambda \times \overrightarrow{\text{grad}T} = -\lambda \times \frac{T_1 - T_2}{x_1 - x_2}\vec{x} \tag{1}$$

Equation 2 can be derived from a general formula of thermal conduction resistance to an axial flux distribution:

$$R = \frac{E}{\lambda \times S_{et}} \tag{2}$$

Where λ is the thermal conductivity, S_{et} is the heat exchange section and Φ_T is the total heat flux exchanged and E is the thickness of the heat exchange.

The conduction resistances can be deduced from the geometrical equations of PMSM and CRSM.

For PMSM structures, the conduction resistances of the materials constituting the rotor are expressed by the following relationships:

- The conduction resistance of the rotor yoke is expressed by the following relationship:

$$R_{cr} = \frac{H_{cr}}{\lambda_{fer} \times \left(\pi \times \frac{D_e^2 - D_i^2}{4}\right)} \tag{3}$$

Where λ_{fer} is the thermal conductivity of the iron, H_{cr} is the rotor yoke thikness, D_e and D_i are respectively the exetrnel and internal diameter of the motor.

- The conduction resistance of the magnet is expressed by the following relationship:

$$R_{caim} = \frac{H_a}{\lambda_a \times 2 \times p \times S_a} \tag{4}$$

Where λ_a is thermal conductivity of the magnets, p is the nomber of pole pairs, S_a is the magnet section and H_a is magnet thikness.

- The conduction resistance of the air-gap is expressed by the following relationship:

$$R_{cen} = \frac{e}{\lambda_{air} \times \left(\pi \times \frac{D_e^2 - D_i^2}{4}\right)} \tag{5}$$

Where λ_{air} is the thermal conductivity of air and e is the air-gap thikness.

For CRSM structures, the conduction resistances of the materials constituting the rotor are expressed by the following relationships:

- The conduction resistance of the copper rotor is expressed by the following relationship:

$$R_{ccr} = \frac{\dfrac{N_{sr} \times I_e}{\delta \times L_p \times \frac{1}{6} \times \frac{(D_e + D_i)}{4}}}{\lambda_{fer} \times \left(L_{er} \times (E_{cs} + L_{er} \times 2) \times 2 + 2 \times \left(\left(\frac{D_e + D_i}{4}\right) \times L_p \times \frac{2}{3}\right) \times L_{er}\right) \times 2 \times p} \tag{6}$$

Where L_{er} is the width of a rotor slot:

$$L_{er} = L_p \times \frac{1}{6} \times \left(\frac{D_e + D_i}{4}\right) \tag{7}$$

and E_{cs} is equal to:

$$E_{cs} = \left(\frac{D_e - D_i}{2}\right) \tag{8}$$

L_p is the polar pitch :

$$L_p = \frac{\pi}{p} \tag{9}$$

The conduction resistance of the rotor teeth is expressed by the following relationship:

$$R_{cdr} = \frac{\dfrac{N_{sr} \times I_e}{\delta \times L_p \times \frac{1}{6} \times \frac{(D_e + D_i)}{4} \times K_f}}{\lambda_{fer} \times L_p \times \frac{2}{3} \times \left(\frac{D_e + D_i}{4}\right) \times E_{cs} \times 2 \times p} \tag{10}$$

- The conduction resistance of the rotor insulation is expressed by the following relationship:

$$R_{cisor} = \frac{\dfrac{(1 - K_f) \times N_{sr} \times I_e}{\delta \times L_p \times \frac{1}{6} \times \frac{(D_e + D_i)}{4} \times K_f}}{\lambda_{iso} \times \left(L_{er} \times (E_{cs} + L_{er} \times 2) \times 2 + 2 \times \left(\left(\frac{D_e + D_i}{4}\right) \times L_p \times \frac{2}{3}\right) \times L_{er}\right) \times 2 \times p} \tag{11}$$

Where λ_{iso} is the thermal conductivity of insulation and K_f is load factor of slots.

- The conduction resistance of the rotor yoke is expressed by the following relationship:

$$R_{ccr} = \frac{H_{cr}}{\lambda_{fer} \times \pi \times \left(\frac{D_e^2 - D_i^2}{4} \right)} \quad (12)$$

We recall that for both structures, the stator structure is the same. The conduction resistances respectively of the coils (R_{cu}), insulation (R_{iso}) of main teeth (R_{cd}) and stator yoke (R_{cs}) for the two structures are equal.

- The conduction resistance of the coils is expressed by the following relationship:

$$R_{cu} = \frac{K_f \times H_d}{\lambda_c \times N_d \times \left(2 \times L_{enc} \times \frac{D_e - D_i}{2} + \frac{D_i}{2} \times A_{dent1} \times L_{enc} + \frac{D_e}{2} \times A_{dent2} \times L_{enc} \right)} \quad (13)$$

Where λ_c is the thermal conductivity of copper and H_d is the stator tooth high.

- The conduction resistance of the insulator is expressed by the following relationship:

$$R_{iso} = \frac{(1 - K_f) \times H_d}{\lambda_{iso} \times N_d \times \left(2 \times L_{enc} \times \frac{D_e - D_i}{2} + \frac{D_i}{2} \times A_{dent1} \times L_{enc} + \frac{D_e}{2} \times A_{dent2} \times L_{enc} \right)} \quad (14)$$

Where λ_{iso} is the thermal conductivity of the insulation, A_{dent1} the lower main tooth angle, A_{dent2} is higher angle of the main tooth, N_d is the tooth number.

The slot width L_{enc} is expressed as follows:

$$L_{enc} = B \sin\left(\frac{1}{2} \left(\frac{2\pi}{N_d} - \alpha \beta \frac{\pi}{p} (1 - r_{did}) \right) \right) \quad (15)$$

Where r_{did} is the ratio between the angular width of the inserted tooth and that of the principal tooth. This ratio is optimised by finite elements simulations in order to reduce the flux leakages and to improve the back electromotiv force wave-form.

Where B is equal to :

$$B = \frac{D_e + D_i}{2} \quad (16)$$

The PMSM and CRSM configuration is caracterized by variation law of the pole pairs number (p) according to an integer number n varying from one to infinity, the ratio (r) of the number of principal teeth (N_d) by the number of pole pairs, the ratio (v) between the angular width between two principal teeth and that of a principal tooth, the ratio (α) between the angular width of a principal tooth and that of a magnet and the ratio (β) between the angular width of a magnet and the polar step.

The conduction resistance of the main teeth is expressed by the following relationship:

$$R_{cd} = \frac{H_d}{\lambda_{fer} \times (N_d \times S_d)} \quad (17)$$

Where H_d is the teeth high, S_d is the main tooth section.

- The conduction resistance of the stator yoke is expressed by the following relationship:

$$R_{cs} = \frac{H_{cs}}{\lambda_{fer} \times \left(\pi \times \frac{D_e^2 - D_i^2}{4} \right)} \quad (18)$$

Wher H_{cs} is the stator yoke thikness.

- The conduction resistance of the air between two modules is expressed as follows:

$$R_{Air} = \frac{E_{air}}{\lambda_{air} \times \frac{\pi \times D_e^2}{4}} \quad (19)$$

Where E_{air} is the thickness of air between two modules.

- The conduction resistance of the flanges is expressed as follows:

$$R_{fls} = R_{flr} = \frac{E_{fl}}{\lambda_{fer} \times \frac{\pi \times D_e^2}{4}} \quad (20)$$

Where E_{fl} is the thickness of the flanges.

4.2. Convection Resistances Computation

The convective heat transfer is the preferred mode of transfer within the fluids. Then it is generally much more important than conduction. We must distinguish between natural convection and forced convection.

Density differences related to differences in temperature cause movements of the fluid which is heated in contact with hot body and thus carries the heat to colder areas: the natural convection. The yoke and the outer flanges of the machine in the absence of external fan, undergo this transfer mode. In the internal parts, not brewed areas are rare due to the rotor. The heat exchange coefficient between the housing and the ambient air, can be between 20 and 50 W. K^{-1}.m^{-2} for machine natural ventilation and can exceed 80 W. K^{-1}.m^{-2} for machines with forced ventilation.

The only network element that refers to a transfer by convection is R_{ext} , which represents the overall thermal resistance between the surface of the casing, the flanges and the ambient air.

$$R_{ex} = \frac{1}{h.S_s} = \frac{1}{h.S_r} \quad (21)$$

Where h is the heat transfer coefficient, S_s is the area of heat exchange with the stator and S_r is the area of heat exchange with the rotor.

The calculation of external surface of the actuator requires some remarks. Should we consider only the side surfaces (flanges) of the machine or also include the outer surface of the cylinder? Model assumptions, namely a consideration of phenomena only in the axial axis, incite choose the first solution. Wherein the surfaces of heat exchange by convection are expressed by the following relationship:

$$S_s = S_r = \pi \times \frac{D_e^2}{4} \tag{22}$$

4.3. Calculation of Thermal Capacity

We will study the thermal phenomena of a transient point of view, it is therefore necessary to involve the heat capacities of the materials constituting the components of the machine.

The terms of thermal capacity is calculated from the following relationship between the mass of materials and their massive heat capacity using the following equation [6] and [7]:

$$C = \rho \times V \times c = M \times c \tag{23}$$

With ρ is the density of the material, V is the volume of the material, c is the mass heat capacity of the material and M is the mass of the material.

4.4. Heat Flux

The heat flux Φ_1 corresponds to iron losses in the stator yoke (P_{fc}). This flux propagates from the center of gravity of the stator yoke. It is expressed by the following equation [8]:

$$\Phi_1 = P_{fc} = q \times f^{1.5} \times M_{cs} \times B_{cs}^2 \tag{24}$$

Where q is the quality factor of the metal sheets, M_{cs} is the mass of the stator yoke and B_{cs} is the magnetic induction in the stator yoke.

The heat flux Φ_2 corresponds to the iron losses in the stator yoke (P_{fd}). This flux propagates from the center of gravity of the stator teeth. It is expressed by the following equation [8]:

$$\Phi_2 = P_{fd} = q \times f^{1.5} \times M_{ds} \times B_d^2 \tag{25}$$

Where M_{ds} is the mass of the stator teeth and B_d is the magnetic induction in the stator teeth.

The heat flux Φ_3 corresponds to the copper losses in the stator (P_j). This flux propagates from the center of gravity of the stator copper. It is expressed by the following equation:

$$\Phi_3 = P_j = \frac{3}{2} \times R \times I^2 \tag{26}$$

Where R is the resistance of the stator winding, it is expressed by the following relationship:

$$R = \frac{r_{cu} \times N_s \times L_{sp}}{\dfrac{C_{dim}}{\sqrt{2} \times \delta \times K_e}} \tag{27}$$

Where r_{cu} is the resistivity of copper, L_{sp} is the average length of one turn, N_s is the number of phase spires, C_{dim} is the dimensionnig torque of the motor, δ is the current density in the copper and K_e is the back electromotive force constant.

$$L_{sp} = (A_{dent1} + A_{enc1}) \times \left(\frac{D_i}{2} - \frac{L_{enc}}{2}\right) + (A_{dent2} + A_{enc2}) \times \left(\frac{D_e}{2} - \frac{L_{enc}}{2}\right) + 2 \times \left(\left(\frac{D_e - D_i}{2}\right) + L_{enc}\right) \tag{28}$$

For CRSM structure the flux Φ_4 corresponds to rotor copper losses (P_{jr}). This flux propagates from the center of gravity of the rotor copper, it is expressed by the following equation:

$$\Phi_4 = R_e \times I_e^2 \tag{29}$$

Where R_e is the resistance of the rotor winding, is expressed by the following relationship:

$$R_e = \frac{2 \times p \times r_{cu} \times N_{sb} \times L_{sp}}{\dfrac{I_e}{\delta}} \tag{30}$$

where N_{Sb} is the number of conductors of the rotor winding, I_e is the excitation current and L_{msr} is the average length of a rotor coil:

$$L_{msr} = 2 \times \left(\left(\frac{D_e - D_i}{2}\right) + L_{er}\right) + \left(\left(\frac{D_e + D_i}{4}\right) \times L_p \times \frac{2}{3} + 2 \times L_{er}\right) \tag{31}$$

5. Results of Simulations

The loss model for both PMSM and CRSM structures for an electromagnetic switch power chain is located under the environment of Matlab / Simulink. Heat fluxes are calculated based on the inverse modeling approach of the power chain (figure 7).

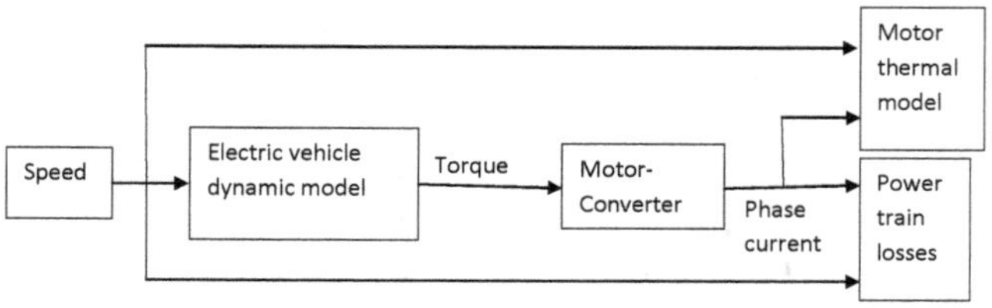

Figure 7. *Electric vehicle power train model.*

Simulation results are obtained for the standard travels (1027 s) during repeated over a period of 200 000 s (Figure 8).

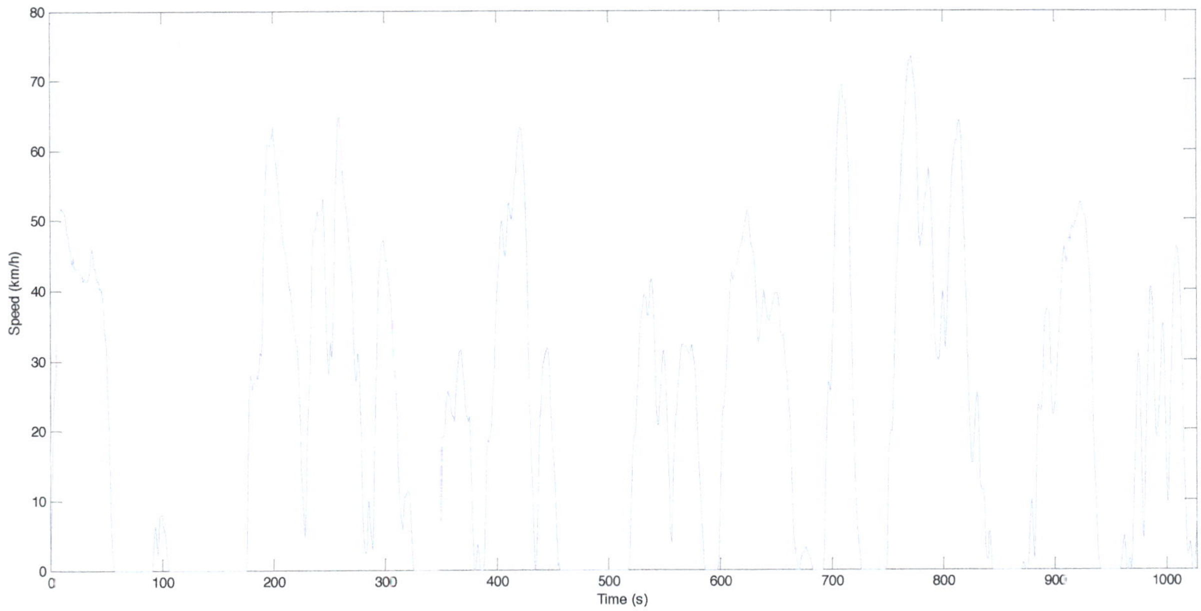

Figure 8. *Standardized travels.*

5.1. PMSM Structure

The simulation of the thermal model of the PMSM structure, with natural convection with air (convection coefficient equal to 30 W/(m².K)), shows the evolution of temperatures in different active parts of the motor (figure 9).

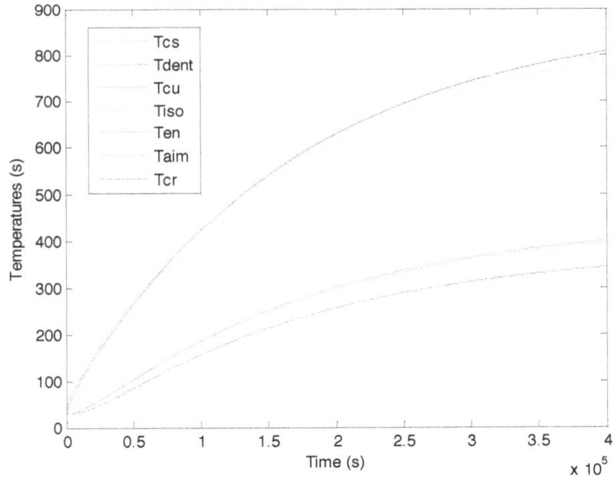

Figure 9. *Evolution of temperatures in the different active parts of the motor (h = 30 W/(m².K)).*

Where Tfl is the average temperature of the flanges, Tcr is the average temperature of the rotor yoke, Taim is the average temperature of magnets, Ten is the average temperature of the air gap, Tiso is the average temperature of the insulation, Tcu is the average temperature of the copper, Tdent is the average temperature of the stator teeth and Tc is the average temperature of the stator yoke .

This figure shows that there's an exceeding of 700 ° C for the resin, which proves the need for a cooling system. Several simulations are undertaken for several values of coefficient of forced convection in water, have led to the fixing of this coefficient to 5000 W/(m².K).

Evolutions of temperature in the different active parts of the engine for operation with a cooling system with forced convection in water (h = 5000 W/(m².K)) is illustrated in figure 10.

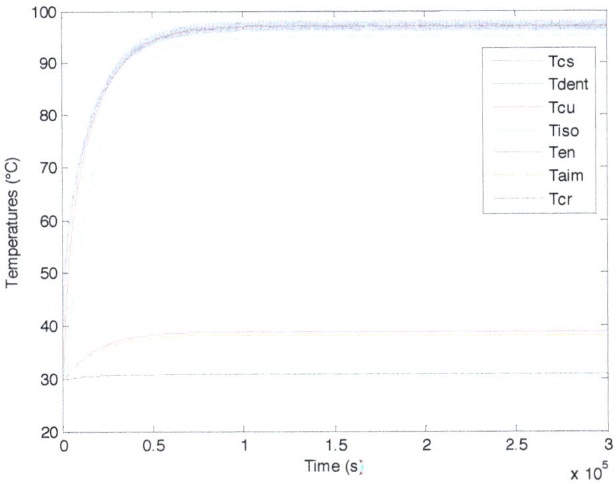

Figure 10. *Evolution of temperatures in the different active parts of the motor (h = 5000 W/(m².K)).*

This figure shows that the temperature of the insulation is reduced to 90 ° C, acceptable value.

5.2. CRMS Structure

The simulation of the thermal model of the CRSM structure, with natural convection with air (convection coefficient equal to 30 W/(m².K)), shows the evolution of temperatures in different active parts of the motor (Figure 11).

Where Tfl is the average temperature of the flanges, Tcr is the average temperature of the rotor yoke, Tdentr is the average temperature of the rotor teeth , Tcur is the average temperature of the rotor copper , Ten is the average temperature of the air gap , Tiso is average temperature of the insulation, Tcu is the average temperature of the stator copper, Tdent is the average temperature of the stator teeth and Tcs is the average temperature of the stator yoke .

This figure shows that there's an exceeding of 1400 ° C for the resin, which proves the need for a cooling system. Simulations are launched for same value of convection coefficient for a forced cooling system thein PMSM structure (5000 W/(m².K)) leds to the evolutions of temperature in the different active parts of the engine figure 12.

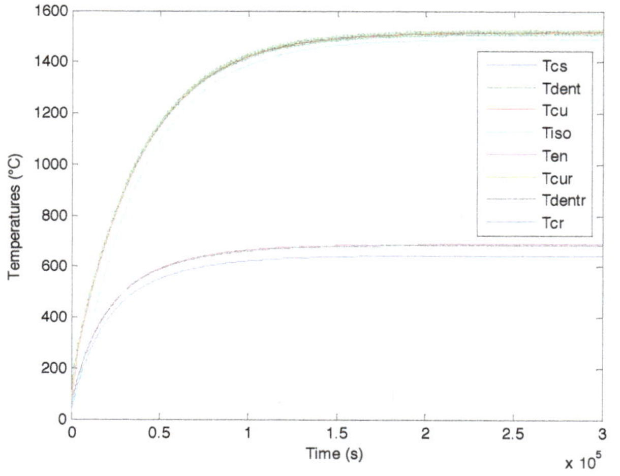

Figure 11. *Evolution of temperatures in the different active parts of the motor (h = 30 W/(m².K)).*

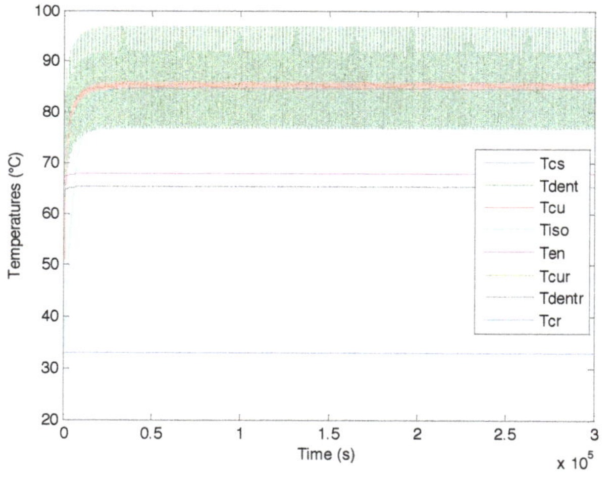

Figure 12. *Evolution of temperatures in the different active parts of the motor (h = 5000 W/(m².K)).*

This figure shows that the temperatures of copper and insulation are reduced to 84 ° C, acceptable value for a good functionning of engine.

5.3. Comparison between the Two Structures Engine

Thermal analysis relied on the nodal thermal models of both PMSM and CRSM structures shows that for the same functionning mode and for the same stator, the components of a CRSM structure heats more than the components of a PMSM structure. This property is justified by the fact that the CRSM structure has additional rotor copper losses. Accordingly, the PMSM structure system requires the least powerful cooling system and the less expensive. This structure is chosen for further study.

6. Conclusion

In this paper we present a methodology for electrothermal modeling of two engine configurations, one with permanent magnet and the other with wound rotor to select the configuration compatible with the less powerful and the least expensive cooling system. This approach is based on modeling with nodal method because it provides a fast and acceptable precision for our application, it is integrable to optimization approaches of the design parameters of electric cars. This study has led to the choice of the configuration with permanent magnet since the simulation results show that the cooling system to be integrated to this configuration is the less powerful and less expensive by the following.

List of Symbols

C_{flr}	Rotor flanges capacity
C_{cr}	Rotor yoke capacity
C_{aim}	Magnet capacity
C_{en}	Air-gap capacity
C_{iso}	Insulation capacity
C_{cu}	Copper capacity
C_d	Teeth capacity
C_{cs}	Stator yoke capacity
C_{Air}	Air capacity
C_{fls}	Stator flanges capacity
P_j	Copper losses of stator
P_{fd}	Iron losses of stator teeth
P_{fc}	Iron losses of stator yoke
T_{flr}	Average temperature of rotor flanges
T_{cr}	Average temperature of rotor yoke
T_{aim}	Average temperature of magnet
T_{en}	Average temperature of air-gap
T_{iso}	Average temperature of stator Insulation
T_{cu}	Average temperature of stator copper
T_{dent}	Average temperature of stator teeth
T_{cs}	Average temperature of stator yoke
T_{Air}	Average temperature of air
T_{fls}	Average temperature of stator flanges
T_a	Ambient temperature
R_{flr}	Thermal conduction resistance of rotor flanges
R_{cr}	Thermal conduction resistance of rotor yoke
R_{aim}	Thermal conduction resistance of magnet
R_{en}	Thermal conduction resistance of air-gap
R_{iso}	Thermal conduction resistance of of Insulation
R_{cu}	Thermal conduction resistance of stator copper
R_{cd}	Thermal conduction resistance of teeth
R_{cs}	Thermal conduction resistance of stator yoke
R_{Air}	Thermal conduction resistance of air
R_{fls}	Thermal conduction resistance of stator flanges

C_{cr}	Capacity of rotor copper
C_{dr}	Capacity of rotor teeth
C_{ccr}	Capacity of rotor yoke
C_{isor}	Capacity of rotor insulation
T_{dentr}	Average temperature of rotor teeth
T_{isor}	Average temperature of rotor insulation
T_{cur}	Average temperature of rotor copper
R_{ccr}	Thermal conduction resistance of rotor yoke
R_{cdr}	Thermal conduction resistance of rotor teeth
R_{cisor}	Thermal conduction resistance of rotor insulation
R_{cur}	Thermal conduction resistance of rotor copper
P_{jr}	Rotor copper losses
h	Coefficient of thermal convection

References

[1] S. TOUNSI et R. NEJI: «Design of an Axial Flux Brushless DC Motor with Concentrated Winding for Electric Vehicles», Journal of Electrical Engineering (JEE), Volume 10, 2010 - Edition: 2, pp. 134-146.

[2] A. AMMOUS, B. ALLARD, H. MOREL: «Transient temperature mesurements and modeling of IGBT's under short circuit», IEEE transaction electronic devices, vol. 13, n° 1, 1998, p. 12-25.

[3] S.TOUNSI, R.NÉJI, F.SELLAMI : « Conception d'un actionneur à aimants permanents pour véhicules électriques », Revue Internationale de Génie Électrique volume 9/6 2006 - pp.693-718.

[4] Tounsi S., Neji R., Sellami F., « Modélisation des pertes dans la chaîne de traction du véhicule électrique », CTGE 2004, 19-21 Février, Tunis, Tunisie, p. 291-297.

[5] M.A.FAKHFAKH, M. HADJ KASEM, S. TOUNSI et R. NEJI: «Thermal Analysis of Permanent Magnet Synchronous Motor for Electric Vehicle», Journal of Asian Electric Vehicles, volume 6, Number 2, December 2008, pp. 1145-1151.

[6] S. MEZANI, « Modélisation électromagnétique et thermique des moteurs à induction, en tenant compte des harmoniques d'espace ». Thèse de doctorat, Institut National polytechnique de LORRAINE, 2004.

[7] Q.Pan et A. Razek, « Phénomènes magnéto-thermiques dans les machines asynchrones à cage. Analyse par éléments finis », Revue Générale de Thermique, n°348, pp. 720-726, décembre 1990.

[8] R. NEJI, S. TOUNSI et F. SELLAMI: «Contribution to the definition of a permanent magnet motor with reduced production cost for the electrical vehicle propulsion», European Transactions on Electrical Power (ETEP), 2006, 16: pp. 437-460.

A Comparative Analysis of Cogging Torque Reduction in BLDC Motor Using Bifurcation and Slot Opening Variation

G. Suresh Babu[1], T. Murali Krishna[1], B. Vikram Reddy[2]

[1]Department of Electrical & Electronics Engineering, Chaitanya Bharathi Institute of Technology, Gandipet, Hyderabad, India
[2]Department of Electrical & Electronics Engineering, Chaitanya Bharathi Institute of Technology, Hyderabad, India

Email address:
gsb67@cbit.ac.in (G. S. Babu), tmurali5@gmail.com (T. M. Krishna)

Abstract: The utility of PM-BLDC machines is extending its tentacles in industrial arena. The key features of BLDC machines include high starting torque density and extending speed range, though the cogging torque is a threat for its performance. Various techniques have been devised to minimize cogging torque, out of which two approaches Bifurcation and Slot Opening methods have been focused in this paper. Usage of SPEED Software in comparing the reduction of cogging torque for the above two techniques is the highlight of this paper.

Keywords: Cogging Torque, BLDC Motor, Bifurcation

1. Introduction

Permanent-Magnet BLDC machines are propelled in industrial applications and the lack of sliding contacts makes them reliable and their characteristics make them suitable for sensor less drive applications [2], [3].

Nevertheless, they are affected by a few drawbacks such as high costs of PM materials and cogging torque which lowers torque quality and affects smooth running of the machine, producing vibrations, ripples in output torque and mechanical noise. In these machines, Internal Permanent Magnet machines show better performance in flux weakening operation [4] and achieve higher flux density due to the small air gap that allows imposing a magnetizing current effectively but have higher torque ripple and many design issues compared to Surface Permanent Magnet machines [5], [2].

Load torque comprises cogging torque ripple and the load torque. The cogging torque reduction methods can be obtained from analytical expression (derived by the energy method and the Fourier series analysis) or by FEA simulations. This paper presents the analysis and comparison of different low cost cogging torque reduction methods which can be practically applied to IPM machines. The peak-to-peak torque ripple of the cogging torque, peak-to-peak rated torque profile, mean value of rated torque and efficiency will be used as index values to compare and evaluate the different methods [6] by FEA simulation on a known 4-pole machine, used as a test bed to identify the most effective ones for a given starting geometry.

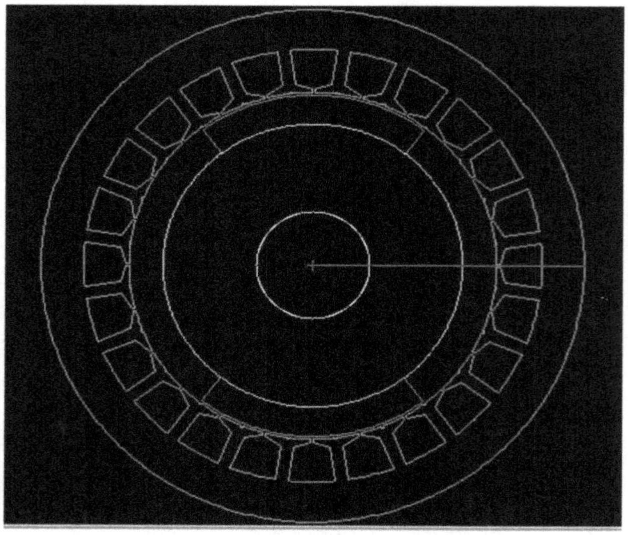

Fig. 1. Motor Simulation in SPEED considered for the Project.

2. Cogging Torque

Cogging torque is caused by the alignment of stator and

rotor at low speeds and produced by the force of attraction between the stator teeth and PM rotor. This is torque which is present even without excitation. A descriptive equation according to the definition is given as,

$$T_{cogg} = -\left(\frac{\partial W}{\partial \theta}\right).$$

Cogging torque produces zero net work, but it acts as a disturbance superimposed on the electromagnetic torque generated during machine operation and the cogging torque period is linked with the number of slots and poles.

3. Cogging Torque Reduction Methods

This paper presents cogging torque reduction methods for the commercial and military grade machines with less cost and effective means i.e., Bifurcation with reduced Tooth Width (TW) and Slot Opening (SO) variations

A. Reduced Stator Tooth with Bifurcation

This is one of the methods in reducing cogging torque in BLDC motors i.e. the reduction of stator tooth with bifurcation [1].

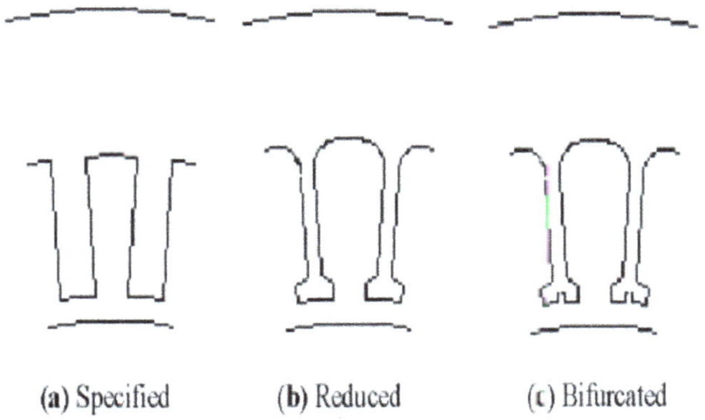

Fig. 2. Reduced Stator TW and Bifurcation modification to Original Motor.

In fig (a), the original stator slots have been showed. In this case, the reluctance variation with respect to the rotor rotation is large and the cogging torque produced is also very large. The magnitude of cogging torque depends on the variation of reluctance w.r.t. rotor rotation.

In fig (b), the slots width is reduced. In this case the net amount of iron to which the PM of rotor is reduced and the force of attraction between the stator teeth and rotor PM decreases. This reduces the cogging torque. The net amount of cogging torque reduced in this method is low.

In fig (c), the stator tooth is bifurcated. This helps in maintaining the change in airgap reluctance constant w.r.t. the rotor rotation. Hence the cogging torque is reduced to the greater amount of nearly 40% when compared to normal design. This bifurcation results in reducing the efficiency of 2%.

The resultant graphs for the three cases are plotted.

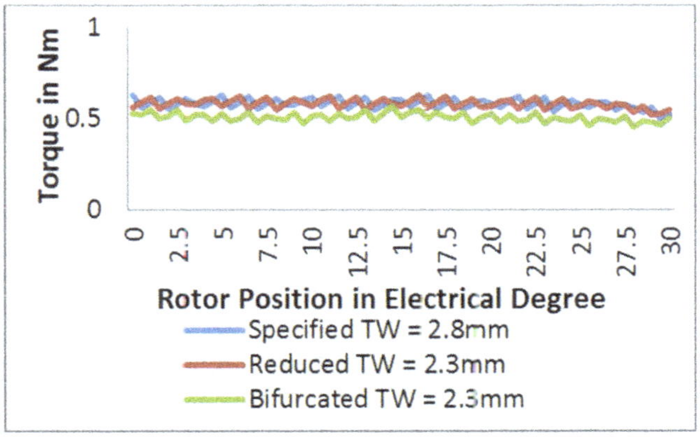

Fig. 3. Graph showing different Torque for Reduced Stator TW and Bifurcation modification to Original Motor.

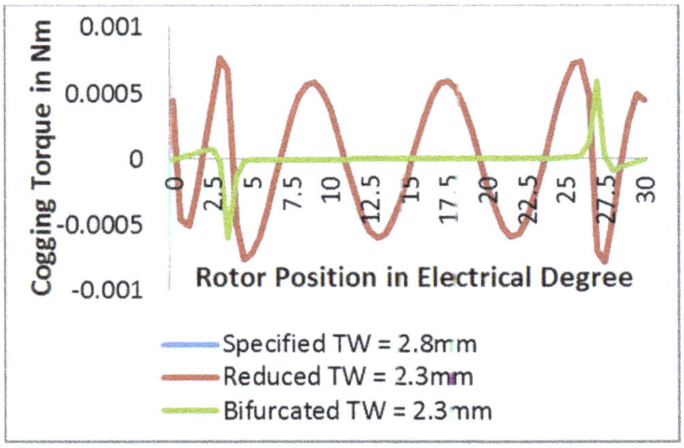

Fig. 4. Graph showing different Cogging Torque for Reduced Stator TW and Bifurcation modification to Original Motor.

B. Slot Opening Variation

Slot opening has an effective impact on the cogging torque. This is the direct and noticeable one. Since the cogging torque is generated by the interaction of the stator teeth and the rotor magnetic field, the slot opening width has a significant impact on this phenomenon. Moreover, also the back-EMF harmonic content depends on the air gap flux density distribution. Therefore, the slot opening width has to be chosen very carefully in order to optimize the machine design. Generally the cogging torque and the harmonic content of the back-EMF decrease as the slot opening become smaller. Since the slot opening influences the winding manufacturing and costs, the choice of the optimal width is a trade-off between cost and performance.

The slot opening shouldn't be minimum, this causes problems for cooling and the rated loading cannot be imposed. And also it shouldn't be more, because this causes increasing the cogging torque. Simulation is done for various slot opening cases to do trade off in reducing maximum cogging torque.

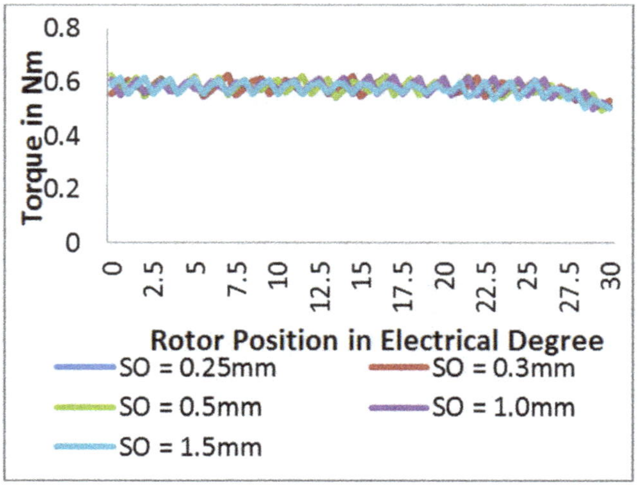

Fig. 5. Graph showing different Torque for Slot Opening variation.

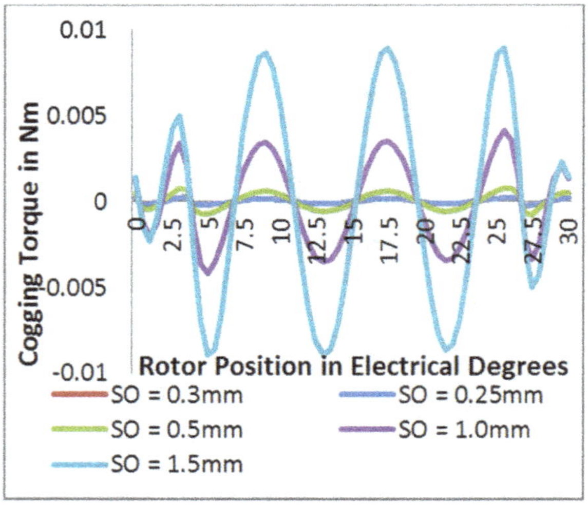

Fig. 6. Graph showing different Cogging Torque for Slot Opening variation.

4. Results Comparision

Table I. Tabular representation of Results comparison.

METHOD		T_{rated} [Nm]	T_{rated} [%]	T_{cogg} [Nm]	T_{cogg} [%]	Eff	Eff [%]
Slot Opening	0.25mm	0.516	1.9	1.8e-4	-76.83	89.19	1.4
	0.3mm	0.515	1.7	2.55e-4	-67.18	89.09	1.3
	0.5mm	0.506	0	7.7e-4	0	87.89	0
	1mm	0.429	-15.2	4.17e-3	436.67	75.05	-14.6
	1.5mm	0.27	-46.64	8.9e-3	1045.4	47.71	-45.72
Reduced TW & Bifurcation	2.8mm	0.506	0	7.78e-4	0.12	87.89	0
	2.3mm	0.499	-1.3	7.7e-4	0	86.49	-1.6
	2.3mm (Bif)	0.455	-10.07	5.91e-4	-23.93	90.33	2.7

5. Conclusion

Two methods are compared on a common reference machine (24-slot 4-pole IPM machine) by extensive FEM simulations. The results summarized in Table I show that cogging torque reduction techniques developed can be easily applied to IPM machines. For best results, during optimization it is advisable not to focus only on cogging torque reduction but to monitor the side effects as well.

From this paper, Slot Opening variation method produces on efficient solution in reducing cogging torque and is less cost when compared to the Bifurcation method.

But Slot Opening method suffers from a tradeoff between cost and total rating usage. If Slot Opening is minimum, the motor cannot be used to its full potential and only 70% of it can be utilized. But there is no trade off to Bifurcation Method but Tooth Width should be sufficient to hold the weight of conductor. This method increases efficiency.

6. Motor Specifications

V_{dc}	48V
I_{rated}	5A
Connection	Star
Speed	4140 rpm
Number of phases	3
Outer radius	48mm
Inner radius	32.5mm
Stator yoke	7.5mm
No. of slots	24
Rotor yoke	16.5mm
Shaft radius	10mm
Air gap	0.5mm
Magnet length	5.5mm
No. of poles	4
Magnet pole arc	150°

Acknowledgements

The authors express deep sense of gratitude to the management of CBIT and authorities of RCI, Hyderabad for having encouraged their sincere attempts in achieving greater heights.

References

[1] R. Somanatham, P. V. N. Prasad, and A. D. Rajkumar, "Reduction of Cogging Torque in PMBLDC Motor with Reduced Stator Tooth Width and Bifurcated Surface Area using Finite Element Analysis," IEEE 0-7803-9772-X, June 2006.

[2] N. Bianchi and T. M. Jahns, "Design, analysis, and control of interior PM synchronous machines," in IEEE IAS Annu. Meeting, Seattle, Oct. 12, 2004.

[3] D. Novotny and T. Lipo, Vector Control and Dynamics of AC Drives. Oxford, U.K.: Oxford Science Publications, 2000.

[4] T. M. Jahns, "Flux-weakening regime operation of an interior permanent-magnet synchronous motor drive," IEEE Trans. Ind. Appl., vol. IA-23, no. 4, pp. 681–689, Jul./Aug. 1987.

[5] G. Pellegrino, A. Vagati, P. Guglielmi, and B. Boazzo, "Performance comparison between surface-mounted and interior pm motor drives for electric vehicle application," IEEE Trans. Ind. Electron., vol. 59, no. 2, pp. 803–811, Feb. 2012.

[6] G.-H. Kang, Y.-D. Son, G.-T. Kim, and J. Hur, "A novel cogging torque reduction method for interior-type permanent-magnet motor," IEEE Trans. Ind. Appl., vol. 45, no. 1, pp. 161–167, Jan.–Feb. 2009.

[7] N. Bianchi and S. Bolognani, "Design techniques for reducing the cogging torque in surface-mounted PM motors," IEEE Trans. Ind. Appl., vol. 38, no. 5, pp. 1259–1265, Sep./Oct. 2002.

Permissions

All chapters in this book were first published in EPES, by Science Publishing Group; hereby published with permission under the Creative Commons Attribution License or equivalent. Every chapter published in this book has been scrutinized by our experts. Their significance has been extensively debated. The topics covered herein carry significant findings which will fuel the growth of the discipline. They may even be implemented as practical applications or may be referred to as a beginning point for another development.

The contributors of this book come from diverse backgrounds, making this book a truly international effort. This book will bring forth new frontiers with its revolutionizing research information and detailed analysis of the nascent developments around the world.

We would like to thank all the contributing authors for lending their expertise to make the book truly unique. They have played a crucial role in the development of this book. Without their invaluable contributions this book wouldn't have been possible. They have made vital efforts to compile up to date information on the varied aspects of this subject to make this book a valuable addition to the collection of many professionals and students.

This book was conceptualized with the vision of imparting up-to-date information and advanced data in this field. To ensure the same, a matchless editorial board was set up. Every individual on the board went through rigorous rounds of assessment to prove their worth. After which they invested a large part of their time researching and compiling the most relevant data for our readers.

The editorial board has been involved in producing this book since its inception. They have spent rigorous hours researching and exploring the diverse topics which have resulted in the successful publishing of this book. They have passed on their knowledge of decades through this book. To expedite this challenging task, the publisher supported the team at every step. A small team of assistant editors was also appointed to further simplify the editing procedure and attain best results for the readers.

Apart from the editorial board, the designing team has also invested a significant amount of their time in understanding the subject and creating the most relevant covers. They scrutinized every image to scout for the most suitable representation of the subject and create an appropriate cover for the book.

The publishing team has been an ardent support to the editorial, designing and production team. Their endless efforts to recruit the best for this project, has resulted in the accomplishment of this book. They are a veteran in the field of academics and their pool of knowledge is as vast as their experience in printing. Their expertise and guidance has proved useful at every step. Their uncompromising quality standards have made this book an exceptional effort. Their encouragement from time to time has been an inspiration for everyone.

The publisher and the editorial board hope that this book will prove to be a valuable piece of knowledge for researchers, students, practitioners and scholars across the globe.

List of Contributors

Mashauri Adam Kusekwa
Electrical Engineering department, Dar es Salaam Institute of Technology (DIT), Dar es Salaam-Tanzania

Akpama Eko James
Dept. of Elect/Elect/ Engineering Cross River University of Technology, Calabar/Nigeria

Linus Anih
Dept. of Electrical Engineering Nniversity of Nigeria, Nsukka, Enugu/Nigeria

Ogbonnaya Okoro
Dept. of Elect/Elect/ Engineering, Micheal Okpara University Agriculture, Umudike/Nigeria

Amjad Khan, Mohammed Zakir Bellary and Mohammad Ziaullah
Department of Electronics and communication Engineering, P.A College of Engineering, Mangalore, Karnataka, India

Abdul Razak Kaladgi
Department of Mechanical Engineering, P.A College of Engineering, Mangalore, Karnataka, India

S. Boopathi
Embedded System Technologies, Knowledge Institute of Technology, Tamilnadu, India

M. Jagadeeshraja and L. Manivannan
Department of Electrical and Electronics Engineering, Knowledge Institute of Technology, Tamilnadu, India

M. Dhanasu
Steel Authority of India, Tamilnadu, India

Danish Chaudhary and Aziz Ahmed
Dept. of Electrical & Electronics Engg, Alfalah University, Dhauj, Faridabad, India

Anwar Shahzad Siddiqui
Dept. of Electrical Engineering, Jamia Milia Islamia, New Delhi, India

Parsa Sedaghatmanesh
Electrical Power Engineering, Islamic Azad University of Saveh, Markazi, Iran

Mohammad Taghipour
Industrial Engineering, Science & Research Branch of Islamic Azad University, Tehran, Iran

Jun Dong and Rong Li
School of Economics and Management, North China Electric Power University, Beijing, China

Titu Bhowmick and Dharmasa
Department of Electrical and Computer Engineering, Caledonian College of Engineering, AL Hail, Oman

Amjad Khan, Mohammed Zakir Bellary and Mohammad Ziaullah
Department of Electronics and communication Engineering, P.A College of Engineering, Mangalore, Karnataka, India

Abdul Razak Kaladgi
Department of Mechanical Engineering, P.A College of Engineering, Mangalore, Karnataka, India

Ahmed Hussain Elmetwaly
Dept. of Electrical Engineering, El Shorouk Academy, Cairo, Egypt

Abdelazeem Abdallah Abdelsalam
Dept.of Electrical Engineering, Suez Canal University, Ismailia, Egypt

Azza Ahmed Eldessouky and Abdelhay Ahmed Sallam
Dept. of Electrical Engineering, Port-Said University, Port-Said, Egypt

Devendra Manikrao Holey and Vinod Kumar Chandrakar
Department of Electrical Engineering, G. H. Raisoni College of Engineering, Nagpur, India

Amaize Aigboviosa Peter
Department of Electrical and Information Engineering, College of Engineering, Covenant University, Ota, Nigeria

Ignatius Kema Okakwu and Abel Ehimen Airoboman
Department of Electrical/Electronics Engineering, University of Benin, Benin City, Nigeria

Emmanuel Seun Oluwasogo
Department of Electrical and Computer Engineering, Kwara State University, Malete, Nigeria

Akintunde Samson Alayande
Department of Electrical Engineering, Faculty of Engineering and the Built Environment, Tshwane University of Technology, Pretoria, South Africa

Tae-Sik Kong, Hee-Dong Kim, Tae-Sung Park, Kyeong-Yeol Kim and Ho-Yol Kim
Korea Electric Power Corporation (KEPCO) Research Institute, Daejeon, South Korea

Atiqur Rahman and Miftah Al Karim
Department of Electrical and Electronic Engineering, American International University-Bangladesh (AIUB), Dhaka, Bangladesh

Valentyna Novosad
The Scientific Company "MAE", Kyiv, Ukraine

Kitheka Joel Mwithui and David Murage
Department of Electrical and Electronic Engineering, Jomo Kenyatta University of Agriculture and Technology, Nairobi, Kenya

Michael Juma Saulo
Department of Electrical and Electronic Engineering, Technical University of Mombasa, Mombasa, Kenya

Mihail Hristov Antchev
Department of Power electronics, Technical university-Sofia, Sofia, Bulgaria

Moez Hadj Kacem and Rafik Neji
Electrical Engineering Department, National School of Engineers of Sfax , Sfax University, Sfax, Tunisia

Souhir Tounsi
Industrial Informatic Department, National School of Electronics and Telecommunications of Sfax, Sfax University, Sfax, Tunisia

Souhir Tounsi
National School of Electronics and Telecommunications of Sfax, Sfax University, SETIT Research Unit, Sfax, Tunisia

Baek Ju Sung
Korea Institute of Machinery&Materials, Daejeon, Korea

Aicha Khlissa, Houcine Marouani and Souhir Tounsi
School of Electronics and Telecommunications of Sfax, Sfax University, Sfax, Tunisia

Mihail Antchev
Power Electronics Department, Faculty of Electronic Engineering and Technologies, Technical University-Sofia, Sofia, Bulgaria

Angelina Tomova-Mitovska
Schneider Electric Slovakia, Bratislava, Slovakia

Wiem Nhidi
National School of Engineers of Gabes (ENIG), Sfax University, SETIT Research Unit, Gabes, Tunisia

Souhir Tounsi
Ational School of Electronics and Telecommunications of Sfax, Sfax University, SETIT Research Unit, Sfax, Tunisia

Mohamed Salim Bouhlel
Institut Superieur de Biotechnologie de Sfax (ISBS), Sfax University, SETIT Research Unit, Sfax, Tunisia

Zahari Ivanov
Department of Electrical Supply, Electrical Equipment and Electrical Transport, Faculty of Electrical Engineering, Technical University - Sofia, Sofia, Bulgaria

Hristo Antchev
Power Electronics Department, Faculty of Electronic Engineering and Technologies, Technical University - Sofia, Sofia, Bulgaria

Souhir Tounsi
National School of Electronics and Telecommunications of Sfax-(SETIT): Research Unit, Sfax University, Sfax, Tunisia

J. Beiza, H. Mohebalizadeh and A. Kh. Hamidi
Department of Electrical Engineering, Shabestar Branch, Islamic Azad University, Shabestar, Iran

Kostadin Milanov
Department of Electrical Apparatus, Faculty of Electrical Engineering, Technical University - Sofia, Sofia, Bulgaria

Department of Electrical Apparatus, Faculty of Electrical Engineering, Technical University - Sofia, Sofia, Bulgaria

Mintcho Mintchev
Department of Electrical Apparatus, Faculty of Electrical Engineering, Technical University - Sofia, Sofia, Bulgaria

Hristo Antchev
Power Electronics Department, Faculty of Electronic Engineering and Technologies, Technical University - Sofia, Sofia, Bulgaria

Mariem Ben Amor
National School of Engineers of Sfax (ENIS), Sfax University, SETIT Research Unit, Sfax, Tunisia

Souhir Tounsi
National School of Electronics and Telecommunications of Sfax, Sfax University, SETIT Research Unit, Sfax, Tunisia

Mohamed Salim Bouhlel
Higher Institute of Biotechnology of Sfax (ISBS), Sfax University, SETIT Research Unit, Sfax, Tunisia

Souhir Tounsi
School of Electronics and Telecommunications of Sfax, Sfax University, Sfax, Tunisia

G. Suresh Babu and T. Murali Krishna
Department of Electrical & Electronics Engineering, Chaitanya Bharathi Institute of Technology, Gandipet, Hyderabad, India

B. Vikram Reddy
Department of Electrical & Electronics Engineering, Chaitanya Bharathi Institute of Technology, Hyderabad, India

Aicha Khlissa, Houcine Marouani and Souhir Tounsi
School of Electronics and Telecommunications of Sfax, Sfax university (B.P. 1163, 3018 Sfax -Tunisie

Index

www.ingramcontent.com/pod-product-compliance
Lightning Source LLC
Chambersburg PA
CBHW080522200326
41458CB00012B/4302
9781632385376